Study Guide, Volume 2

for

Tipler and Mosca's
# Physics for Scientists and Engineers
## *Fifth Edition*

TODD RUSKELL
*Colorado School of Mines*

GENE MOSCA
*U.S. Naval Academy*

W. H. Freeman and Company
*New York*

Printed in the United States of America

ISBN: 0-7167-8331-2

First printing 2003

W. H. Freeman and Company
41 Madison Avenue
New York, NY 10010
Houndmills, Basingstoke RG21 6XS, England

# CONTENTS

# To the Student

This Study Guide was written to help you master Chapters 21 through 41 of Paul Tipler and Eugene Mosca's *Physics for Scientists and Engineers,* Fifth Edition. Each chapter of the Study Guide contains the following sections:

**I.   Key Ideas.** A brief overview of the important concepts presented in the chapter.

**II.   Physical Quantities and Key Equations.** A list of the constants, units, and basic equations introduced in the chapter.

**III. Potential Pitfalls.** Warnings about mistakes that are commonly made.

**IV.   True or False Questions and Responses.** Statements to test whether or not you understand essential definitions and relations. All the false statements are followed by explanations of why they are false. In addition, many of the true statements are followed by explanations of why they are true.

**V.   Questions and Answers.** Questions that require mostly qualitative reasoning. A complete answer is provided for each question so that you can compare it with your own.

**VI. Problems, Solutions, and Answers.** With few exceptions, the problems come in pairs; the first of the pair is followed by a detailed solution to help you develop a model, which you can then implement in the second problem. One or more hints and a step-by-step guide accompany most problems. These problems will help you build your understanding of the physical concepts and your ability to apply what you have learned to physical situations.

## What Is the Best Way to Study Physics?

Of course there isn't a single answer to that. It is clear, however, that you should begin early in the course to develop the methods that work best for you. The important thing is to find the system that is most comfortable and effective for you, and then stick to it.

In this course you will be introduced to numerous concepts. It is important that you take the time to be sure you understand each of them. You will have mastered a concept when you fully understand its relationships with other concepts. Some concepts will *seem* to contradict other concepts or even your observations of the physical world. Many of the True or False statements and the Questions in this Study Guide are intended to test your understanding of concepts. If you find that your understanding of an idea is incomplete, don't give up; pursue it until it becomes clear. We recommend that you keep a list of the things that you come across in your studies that you do not understand. Then, when you come to understand an idea, remove it from your list. After you complete your study of each chapter, bring your list to your most important resource, your physics instructor, and ask for assistance. If you go to your instructor with a few well-defined questions, you will very likely be able to remove any remaining items from your list.

Like the example problems presented in the textbook, the problem solutions presented in this Study Guide *start with basic concepts,* not with formulas. We encourage you to follow this practice. Physics is a collection of interrelated basic concepts, not a seemingly infinite list of disconnected, highly specific formulas. Don't try to memorize long lists of specific formulas, and then use these formulas as the starting point for solving problems. Instead, focus on the concepts first and be sure that you understand the ideas before you apply the formulas.

Probably the most rewarding (but challenging) aspect of studying physics is learning how to apply the fundamental concepts to specific problems. At some point you are likely to think, "I understand the theory, but I just can't do the problems." If you can't do the problems, however, you probably don't understand the theory. Until the physical concepts and the mathematical equations become your tools to apply at will to specific physical situations, you haven't really learned them. There are two major aspects involved in learning to solve problems: drill and skill. By *drill* we mean going through a lot of problems that involve the direct application of a particular concept until you start to feel familiar with the way it applies to physical situations. Each chapter of the Tipler textbook contains about 35 single-concept problems for you to use as drill. *Do a lot of these!*—at least as many as you need in order to feel comfortable handling them.

By *skill* we mean the ability both to recognize which concepts are involved in more advanced, multi-concept problems, and to apply those concepts to particular situations. The text has several intermediate-level and advanced-level problems that go beyond the direct application of a single concept. As you develop this skill you will master the material and become empowered. As you find that you can deal with more complex problems—even some of the advanced-level ones—you will gain confidence and enjoy applying your new skills. The examples in the textbook and the problems in this Study Guide are designed to provide you with a pathway from the single-concept to the intermediate-level and advanced-level problems.

A typical physics problem describes a physical situation-such as a child swinging on a swing-and asks related questions. For example: If the speed of the child is 5.0 m/s at the bottom of her arc, what is the maximum height the child will reach? Solving such problems requires you to apply the concepts of physics to the physical situation, to generate mathematical relations, and to solve for the desired quantities. The problems presented here and in your textbook are exemplars; that is, they are examples that deserve imitation. When you master the methodology presented in the worked-out examples, you should be able to solve problems about a wide variety of physical situations.

To be successful in solving physics problems, study the techniques used in the worked-out example problems. A good way to test your understanding of a specific solution is to take a sheet of paper, and-without looking at the worked-out solution-reproduce it. If you get stuck and need to refer to the presented solution, do so. But then take a fresh sheet of paper, start from the beginning, and reproduce the entire solution. This may seem tedious at first, but it does pay off.

This is not to suggest that you reproduce solutions by rote memorization, but that you reproduce them by drawing on your understanding of the relationships involved. By reproducing a solution in its entirety, you will verify for yourself that you have mastered a particular example problem. As you repeat this process with other examples, you will build your very own personal base of physics knowledge, a base of knowledge relating occurrences in the world around you—the physical universe—and the concepts of physics. The more complete the knowledge base that you build, the more success you will have in physics.

All the problems in the Study Guide are accompanied by step-by-step suggestions on how to solve them. As previously mentioned, the problems come in pairs, with a detailed solution for the odd-numbered problem. We suggest that you start with an odd-numbered problem and study the problem steps and the worked-out solution. Be sure to note how the solution implements each step of the problem-solving process. Then try the second, "interactive" problem in the pair. When attacking a problem, read the problem statement several times to be sure that you can picture the problem being presented. Then make an illustration of this situation. Now you are ready to solve the problem.

You should budget time to study physics on a regular, preferably daily, basis. Plan your study schedule with your course schedule in mind. One benefit of this approach is that when you study on a regular basis, more information is likely to be transferred to your long-term memory than when you are obliged to cram. Another benefit of studying on a regular basis is that you will get much more from lectures. Because you will have already studied some of the material presented, the lectures will seem more relevant to you. In fact, you should try to familiarize yourself with each chapter before it is covered in class. An effective way to do this is first to read the Key Concepts of that Study Guide chapter. Then thumb through the textbook chapter, reading the headings and examining the illustrations. By orienting yourself to a topic *before* it is covered in class, you will have created a receptive environment for encoding and storing in your memory the material you will be learning.

Another way to enhance your learning is to explain something to a fellow student. It is well known that the best way to learn something is to teach it. That is because in attempting to articulate a concept or procedure, you must first arrange the relevant ideas in a logical sequence. In addition, a dialogue with another person may help you to consider things from a different perspective. After you have studied a section of a chapter, discuss the material with another student and see if you can explain what you have learned.

We wish you success in your studies and encourage you to contact us at truskell@Mines.edu if you find errors, or if you have comments or suggestions.

## Acknowledgments from Todd Ruskell

I want to thank Eugene Mosca, the primary author of the Study Guide that accompanied the fourth edition of the textbook. Much of the material in this edition comes either directly or indirectly from his efforts. I also want to thank Paul Tipler and Eugene Mosca and for writing a textbook that has been a delight to work with. I would like to thank the publishing staff, especially Brian Donnellan, for his patience with me throughout this project. I am deeply indebted to the reviewers of this Study Guide, especially Anthony Buffa. All of their efforts have made this a better Study Guide than it could otherwise have been. Finally, I wish to thank my wife Susan for her patience and support.

July 2003

Todd Ruskell
Colorado School of Mines

# Chapter 21

# The Electric Field I: Discrete Charge Distributions

## I. Key Ideas

***Section 21-1. Electric Charge.*** Electric charge is a fundamental property of matter. Some of its manifestations—for instance, that rubbing some substances together causes them to attract small objects—have been known for thousands of years. Electric charge occurs only as positive or negative. Charge is also quantized, that is, amounts of charge are observed only as integer multiples of the **fundamental unit of charge** $e$, which is equal to the magnitude of the charge of an electron. Charge is also conserved; that is, the total amount of charge in an isolated environment does not change. The **law of conservation of charge** is a fundamental law of nature. Under certain conditions, charged particles such as electrons are created, but the creation of a particle with charge $-e$ is always accompanied by the simultaneous creation at the same location of a second particle with charge $+e$. The SI unit of charge is the **coulomb (C).** The fundamental unit of charge is related to the coulomb by

$$e = 1.602 \times 10^{-19} \text{ C} \qquad \text{Fundamental unit of charge}$$

***Section 21-2. Conductors and Insulators.*** In many materials, such as copper and other metals, some of the electrons are free to move about the entire material. Such materials are called **conductors.** In other materials, such as wood or glass, all the electrons are bound to nearby atoms. These materials are called **insulators.** The number of free electrons in a metal depends on the metal, but typically there is about one per atom. A net charge can be given to a conductor by adding or removing free electrons.

When a positively charged glass rod is brought near a metal, the free electrons in the metal are overall attracted by the net positive charge of the rod and migrate toward it. If, while the positively charged rod remains nearby, the conductor is momentarily touched by a second conductor, free electrons will flow from the second conductor to the first conductor. This results in the first conductor acquiring a net negative charge and the second conductor acquiring a net positive charge without any charge being transferred to or from the glass rod. This indirect method of charging a metal object is called **electrostatic induction** or **charging by induction.** A convenient large conducting object is the earth itself. When a metal object is brought into contact with the earth it is said to be **grounded.**

*Section 21-3. Coulomb's Law.* The force between point electric charges is given by **Coulomb's law,** which states

> The force exerted by one point charge on another varies inversely as the square of the distance separating the charges and is proportional to the product of the charges. The force is repulsive if the charges have the same sign and attractive if they have opposite signs.

In mathematical form this law is expressed as

$$\vec{F}_{1,2} = \frac{kq_1q_2}{r_{1,2}^2}\hat{r}_{1,2} \qquad\qquad \text{Coulomb's law}$$

where $\vec{F}_{1,2}$ is the force exerted by $q_1$ on $q_2$, $\hat{r}_{1,2} = \vec{r}_{1,2}/r_{1,2}$ is a unit vector pointing from $q_1$ to $q_2$, and $k$ is the **Coulomb constant,** which has the value

$$k = 8.99 \times 10^9 \ \text{N} \cdot \text{m}^2/\text{C}^2 \qquad\qquad \text{Coulomb constant}$$

The *magnitude* of the electric force exerted by $q_1$ on $q_2$ a distance $r$ away is

$$F = \frac{k|q_1q_2|}{r_{1,2}^2}$$

The electric force between point charges is analogous to the gravitational force between point masses, but the electrostatic force can be either attractive or repulsive.

If there are more than two point charges, the force exerted on one due to all the others is just the vector sum of the individual forces due to each, and the force exerted by each is unaffected by the presence of the other charges. This is called the **principle of superposition of electric forces.**

*Section 21-4. The Electric Field.* It's convenient to describe the forces exerted by electric charges in terms of an electric field. The field describes a "condition in space" at a point such that, if a hypothetical **test charge** $q_0$ were placed at that point, the ratio of the force $\vec{F}$ on it to its charge is the value of the electric field $\vec{E}$ at the point:

$$\vec{E} = \frac{\vec{F}}{q_0} \quad (q_0 \text{ small}) \qquad\qquad \text{Definition of electric field}$$

where $q_0$ is sufficiently small so as not to disturb the charge distribution producing the electric field $\vec{E}$.

The electric field $\vec{E}$ at a point $P$ due to a point charge $q_i$ is

$$\vec{E}_{i,P} = \frac{kq_i}{r_{i,P}^2}\hat{r}_{i,P} \qquad\qquad \text{Coulomb's law for the electric field due to a point charge}$$

where $\vec{r}_{i,P}$ is the vector from the charge location (the source point) to $P$ (the **field point**), and $\hat{r}_{i,P}$ is a unit vector in the same direction. The electric field at a field point is the vector sum of the electric fields at that point due to all the point charges present:

$$\vec{E}_P = \sum_i \vec{E}_{i,P} = \sum_i \frac{kq_i}{r_{i,P}^2} \hat{r}_{i,P}$$        Electric field due to several point charges

Thus, the total field is a vector quantity.

A system of two point charges that are equal in magnitude, opposite in sign, and separated by a small distance $L$ is called an **electric dipole.** An electric dipole is characterized by its **electric dipole moment $\vec{p}$,** which is defined as

$$\vec{p} = q\vec{L}$$        Definition of electric dipole moment

where $q$ is the magnitude of one of the charges and $\vec{L}$ is the vector from the location of the negative charge to the location of the positive charge.

***Section 21-5. Electric Field Lines.*** An individual vector quantity, such as the velocity of a projectile, is best illustrated by a straight arrow pointing in the direction of the vector with a length that is proportional to the magnitude of the vector. However, a vector field, such as the velocity of the water in a flowing stream, is best illustrated by a representative number of directed, curved lines that bend so as to remain tangent to the direction of the vector field. In this representation the number of lines per unit area passing through a small surface oriented at right angles to the field direction is proportional to the local magnitude of the field vector. Arrowheads are drawn on the field line to show which way the vector points.

When the vector field is an electric field, the lines are called **electric field lines.** The rules for illustrating electric fields via electric field lines are

**1.** Electric field lines begin on positive charges (or at infinity) and end on negative charges (or at infinity). In regions containing no charge the lines are continuous (they don't stop or start).

**2.** Near a point charge the lines are distributed uniformly in all directions about the charge.

**3.** The number of lines leaving a positive charge or terminating on a negative charge is proportional to the magnitude of the charge.

**4.** The number of lines per unit area passing through a small surface oriented at right angles to the field direction is proportional to the magnitude of the field at that location.

**5.** At large distances from a system of charges, the field lines are equally spaced and radial, as if they came from a single point charge equal to the net charge of the system.

**6.** No two field lines can cross.

Figure 21-1*a* is an illustration of the electric field of a positive point charge using arrows to represent the field, and Figure 21-1*b* is an illustration of the same electric field using field lines instead of arrows.

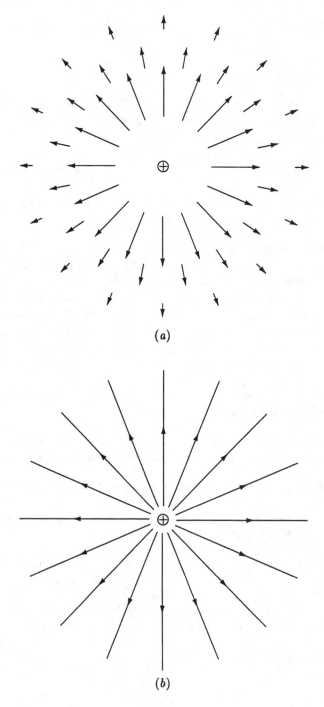

(*a*)

(*b*)

Figure 21-1

***Section 21-6. Motion of Point Charges in Electric Fields.*** When a particle with a charge $q$ is placed in an electric field $\vec{E}$, it experiences a force $q\vec{E}$. If this is the only force acting on the particle, then, in accordance with Newton's 2nd Law $\left( \vec{F} = m\vec{a} \right)$, the acceleration of the particle is

$$\vec{a} = \frac{q}{m}\vec{E}$$

where $m$ is the mass of the particle. (It is fairly common for electrons to acquire speeds of 10% of the speed of light or more. For speeds this high, Newton's laws are inadequate, and the more generally valid laws of the theory of relativity must be used.)

***Section 21-7. Electric Dipoles in Electric Fields.*** Atoms and molecules, even though they are electrically neutral, are affected by electric fields. An atom consists of a small positive nucleus surrounded by one or more electrons. We can think of the atom as a small, massive, positively charged nucleus surrounded by a negatively charged electron cloud. If the electron cloud is spherically symmetric, its center of charge is at the center of the atom, coinciding with the center of the nucleus. Such an atom does not have an electric dipole moment and is said to be **nonpolar.** When a nonpolar atom is placed in an electric field, the force exerted by the field on the negative electron cloud is oppositely directed to the force exerted by the field on the positively charged nucleus. These forces cause the centers of charge of the atom's negative and positive charges to move in opposite directions until the attractive forces the charges exert on each other balance the forces exerted by the external electric field. When this happens the atom is like an electric dipole. The dipole moment of a nonpolar atom or molecule in an external electric field is called an **induced dipole moment.**

In some molecules, water for example, the center of positive charge does not coincide with the center of negative charge even in the absence of an external electric field. These **polar molecules** have a permanent electric dipole moment.

When an electric dipole $\vec{p}$ is placed in a uniform electric field $\vec{E}$, the force on the positive charge and the force on the negative charge are oppositely directed but do not always act along the same line. As shown in Figure 21-2, these two forces then tend to align the dipole with the electric field direction.

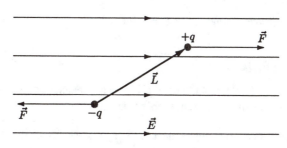

Figure 21-2

The torque vector $\vec{\tau}$ exerted by the electric field on the dipole is given by

$$\vec{\tau} = \vec{p} \times \vec{E}$$                                    Torque on an electric dipole

The dipole has a minimum amount of potential energy when it is aligned with the electric field. The formula for the potential energy $U$ of a dipole in a uniform electric field is

$$U = -\vec{p} \cdot \vec{E}$$    Potential energy of an electric dipole

## II.  Physical Quantities and Key Equations

**Physical Quantities**

*Fundamental charge*                     $e = 1.602 \times 10^{-19}$ C

*Coulomb constant*                       $k = 8.99 \times 10^{9}$ N·m²/C²

**Key Equations**

*Coulomb's law*                          $\vec{F}_{1,2} = \dfrac{kq_1 q_2}{r_{1,2}^2} \hat{r}_{1,2}$

*Definition of Electric field*           $\vec{E} = \dfrac{\vec{F}}{q_0}$   ($q_0$ small)

*Electric field due to a point charge*   $\vec{E}_P = \dfrac{kq_i}{r_{i,P}^2} \hat{r}_{i,P}$

*Electric field due to several point charges*   $\vec{E}_P = \sum_i \vec{E}_{i,P} = \sum_i \dfrac{kq_i}{r_{i,P}^2} \hat{r}_{i,P}$

*Definition of electric dipole moment*   $\vec{p} = q\vec{L}$

*Torque on an electric dipole*           $\vec{\tau} = \vec{p} \times \vec{E}$

*Potential energy of an electric dipole*   $U = -\vec{p} \cdot \vec{E}$

## III.  Potential Pitfalls

We sometimes speak of the fundamental unit of charge $e$ as "the charge of an electron," but $e$ is a positive quantity. The charge of an electron, properly, is $-e$.

When using Coulomb's law to compute electric force, it is easy to get the signs wrong. To keep the signs straight, just remember that charges of the same sign repel and charges of opposite sign attract.

When the electric fields of two or more point charges superpose, the resultant electric field is the vector sum of the individual fields.

You have to be forever careful about signs. The force on a positive charge is in the direction of the electric field $\vec{E}$, but the force on a negative charge is directed opposite to $\vec{E}$.

An electric dipole consists of two charges that are equal in magnitude and of opposite sign.

The equation $\vec{E} = \vec{F} / q_0$ defines the electric field, while the equation $\vec{E} = \sum \left[ \left( kq_i / r_{i,0}^2 \right) \hat{r}_{i,0} \right]$ may be used to compute the field due to a distribution of point charges.

## IV. True or False Questions and Responses

**True or False**

____ 1. Conservation of charge refers to the fact that electric charge can be found only in integral multiples of the fundamental charge $e$.

____ 2. Coulomb's experiments with a torsion balance demonstrated that the electric force varies inversely as the square of the distance between point charges.

____ 3. Positively charged particles repel each other, whereas negatively charged particles attract each other.

____ 4. The force exerted by an electric field $\vec{E}$ on a test charge $q_0$ is independent of the magnitude of $q_0$.

____ 5. As written in Key Equations, Coulomb's law is automatically consistent with Newton's 3rd Law.

____ 6. Gravitational forces are ignored in problems on the atomic scale because gravitational forces and electric forces cannot act simultaneously on the same particle.

____ 7. If an otherwise free point charge is in a region in which there is an electric field $\vec{E}$, its acceleration is necessarily along the direction of the field.

____ 8. Electric field lines can cross only if several charges are present.

____ 9. In a system of charges that has zero net charge overall, there are no electric field lines.

____ 10. Electric field lines start on negatively charged particles.

____ 11. The electric field due to an electric dipole is always parallel to the direction of the electric dipole moment $\vec{p}$.

____ 12. Because an electric dipole has a zero net charge, a uniform electric field exerts no net force on it.

____ 13. The torque exerted on an electric dipole by an electric field tends to align the dipole moment with the field.

_____ 14. The potential energy of an electric dipole in a uniform electric field $\vec{E}$ is a minimum when the angle between the electric dipole moment $\vec{p}$ and $\vec{E}$ is 90°.

_____ 15. Far from a dipole, its electric field decreases with the square of the distance from it.

**Responses to True or False**

1. False. Conservation of charge means that the net charge of a system does not change as long as charge neither enters nor leaves the system by crossing the system's boundary.

2. True.

3. False. Positively charged particles repel each other and negatively charged particles repel each other. However, a negatively charged particle attracts a positively charged particle and vice versa.

4. False. The magnitude of the force $\vec{F}$ exerted by an electric field on a test charge $q_0$ is proportional to $|q_0|$.

5. True. $\vec{F}_{1,2} = -\vec{F}_{2,1}$, as $\hat{r}_{1,2} = -\hat{r}_{2,1}$.

6. False. The gravitational forces are ignored because they are utterly negligible compared with the electric forces involved.

7. True—at least if by "along" is meant in either the same or in the opposite direction. If no other force acts, $\vec{a} = \vec{F}/m = (q/m)\vec{E}$, so if $q$ is positive, $\vec{a}$ and $\vec{E}$ are in the same direction, and if $q$ is negative, $\vec{a}$ and $\vec{E}$ are oppositely directed.

8. False. Electric field lines do not cross. If two electric field lines did cross, then at the point of intersection the electric field would point in two directions. Such a situation is an absurdity.

9. False. Consider, for example, the electric field of an electric dipole.

10. False. They *start* on positively charged particles and *end* on negatively charged particles.

11. False. The electric field of a dipole is the superposition of the electric fields of a negatively charged particle and a positively charged particle. It is illustrated in Figure 21-21 on page 667 of the text.

12. True. In a uniform electric field the two forces acting on the dipole are equal in magnitude and opposite in direction. However, a uniform electric field may exert a torque on the dipole.

13. True.

14. False. The potential energy is a minimum when the dipole moment and the electric field are in the same direction. This is a position of stable equilibrium.

15. False. It decreases with the cube of the distance from it.

# V.  Questions and Answers

**Questions**

1.  After combing your hair with a plastic comb you find that when you bring it near a small bit of paper, the bit of paper moves toward the comb. Then, shortly after the paper touches the comb, it moves away from the comb. Explain these observations.

2.  After combing your hair with a plastic comb you find that when you bring the comb near an empty aluminum soft-drink can that is lying on its side on a nonconducting table top, the can rolls toward the comb. After being touched by the comb the can is still attracted by the comb. Explain.

3.  At point $P$ in Figure 21-3, $\vec{E}$ is found to be zero. What can you say about the signs and magnitudes of the charges?

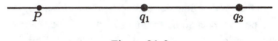

Figure 21-3

4.  A positively charged glass rod attracts a lighter object suspended by a thread. Does it follow that the object is negatively charged? If, instead, the rod repels it, does it follow that the suspended object is positively charged?

5.  Some days it can be frustrating to attempt to demonstrate electrostatic phenomena for a class. An experiment that works beautifully one day may fail the next day if the weather has changed. Air conditioning helps a lot. Why is this?

6.  Consider the electric field lines due to the presence of the two charges shown in Figure 21-4. (*a*) On which charge do field lines begin? On which do they end? (*b*) Suppose you draw a total of eight lines terminating on the $-2\,\text{pC}$ charge. How many lines should you draw beginning on the $+3\,\text{pC}$ charge?

Figure 21-4

Must all the lines drawn beginning on one charge terminate on the other charge?

7.  The two charges shown in Figure 21-4 are separated by about 500 nm. Consider a 1-m-radius spherical surface centered at the location of the $+3\,\text{pC}$ charge. If a total of 200 field lines terminate on the $-2\,\text{pC}$ charge, about how many field lines pass through the spherical surface and how are they distributed on this surface?

8.  What are the advantages of thinking of the force on a charge at a point $P$ as being exerted by an electric field at $P$, rather than by other charges at other locations? Is the convenience of the field as a calculational device worth inventing a new physical quantity? Or is there more to the field concept than this?

**Answers**

1.  When the plastic comb is run through your hair, electrons are transferred from your hair to it, so the comb acquires a net negative charge. The electric field of the comb polarizes (that is, induces an electric dipole moment in) the bit of paper along the field direction. This results in a positive charge on the edge of the paper nearest the comb and a negative charge of equal magnitude on the edge farthest from the comb. Because the electric field of the comb is nonuniform, it attracts the nearby positive charge more strongly than it repels the more distant negative charge. When the bit of paper actually touches the comb, some negative charge is transferred from the comb to it, giving it a net negative charge. Because like charges repel, the bit of paper is now repelled by the comb.

2.  When the plastic comb is run through your hair, electrons are transferred from your hair to it, so the comb acquires a net negative charge. As shown in Figure 21-5a, this charge repels the free electrons in the aluminum, which results in some of the free electrons moving to the part of the can away from the comb. If the net charge on the can is zero, this leaves the part of the can next to the comb positively charged (with an electron deficit) and the part of the can away from the comb negatively charged (with an electron excess). This results in a net attractive force exerted by the comb on the can because the magnitude of the attractive force exerted on the nearby positive charge exceeds the magnitude of the repulsive force exerted on the more distant negative charge. When the comb touches the can, some electrons are transferred from the comb to the can, giving the can a net negative charge. When the negatively charged comb is again brought near the can, the force is again attractive, even though the net charge on the can is negative as shown in Figure 21-5b. There is still an induced positive charge on the can near the comb, and the attractive force on this positive charge exceeds the repulsive force on the larger, but considerably more distant, negative charge.

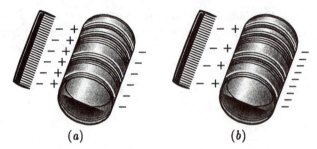

(a)                    (b)

Figure 21-5

3.  The two electric fields at $P$ are equal in magnitude and oppositely directed. The electric field due to a positively charged particle is directed away from the particle, whereas the electric field due to a negatively charged particle is directed toward the particle. Thus, the charges must be of opposite sign. The magnitude of $q_1$ must be less than the magnitude of $q_2$ because the distance between $P$ and $q_1$ is less than that between $P$ and $q_2$.

4.  This is similar to the situation described in Question 1. When the suspended object is attracted to the glass rod, it is not necessarily charged at all. However, when the object is repelled, it must have the same charge as the glass rod.

5.  The main reason is humidity. The higher the humidity, the greater the rate at which static charges leak off charged objects.

6.  (*a*) Field lines begin on the positive charge and end on the negative charge. (*b*) You should draw four field lines per picocoulomb, for a total of 12 lines. Not all the lines beginning on one charge must end on the other. Four of the lines beginning on the $+3\,\text{pC}$ charge go off to infinity.

7.  The radius of the sphere is large compared to the distance between the two charges. Thus, on the surface of the sphere the electric field due to the presence of the two charges will be almost identical with the electric field that would exist if, instead of the two charges, a single charge of $+1\,\text{pC}$ were located at the center of the sphere. There would be 100 lines passing through the spherical surface, and they would be (almost) uniformly distributed.

8.  In electrostatics, the field is just a computational device and using it is merely a matter of convenience. However, in electrodynamics the field is necessary if energy and momentum are to be conserved.

## VI.  Problems, Solutions, and Answers

**Example #1.**  A charge of $+3.00\,\mu\text{C}$ is located at the origin and a second charge of $+2.00\,\mu\text{C}$ is located on the *xy* plane at the point ($x = 30.0\,\text{cm}$, $y = 20.0\,\text{cm}$). Determine the electric force exerted by the $3\,\mu\text{C}$ charge on the $2\,\mu\text{C}$ charge.

**Picture the Problem.**  Coulomb's force law will be used to calculate the force.

| | |
|---|---|
| 1. Write out Coulomb's force law as a calculational guide. | $\vec{F}_{1,2} = \dfrac{kq_1 q_2}{r_{1,2}^2}\,\hat{r}_{1,2}$ |
| 2. Draw a sketch of the two charges and include the *x* and *y* axes. Label the charge at the origin $q_1$, and the other charge $q_2$. Draw the vector $\vec{r}_{1,2}$, which goes from $q_1$ to $q_2$. | 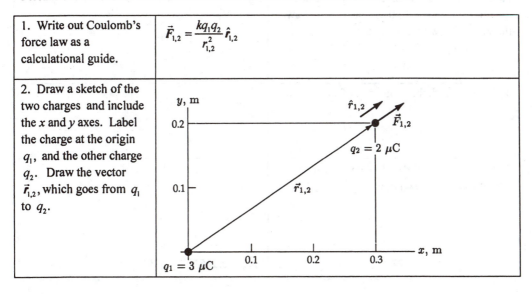 |

| 3. Evaluate both the magnitude of $\vec{r}_{1,2}$ and $\hat{r}_{1,2}$. | $\vec{r}_{1,2} = \vec{r}_2 - \vec{r}_1 = 0.3\,\mathrm{m}\,\hat{i} + 0.2\,\mathrm{m}\,\hat{j}$ <br><br> $r_{1,2} = \sqrt{(0.3\,\mathrm{m})^2 + (0.2\,\mathrm{m})^2}$ <br><br> $\hat{r}_{1,2} = \dfrac{\vec{r}_{1,2}}{r_{1,2}} = \dfrac{0.3\,\mathrm{m}\,\hat{i} + 0.2\,\mathrm{m}\,\hat{j}}{\sqrt{(0.3\,\mathrm{m})^2 + (0.2\,\mathrm{m})^2}}$ |
|---|---|
| 4. Substitute all values into the Coulomb force law, and calculate the force. | $\vec{F}_{1,2} = \dfrac{(8.99\times10^9\ \mathrm{N\cdot m^2/C^2})(3\times10^{-6}\,\mathrm{C})(2\times10^{-6}\,\mathrm{C})}{\left[(0.3\,\mathrm{m})^2 + (0.2\,\mathrm{m})^2\right]^{3/2}}\left(0.3\,\mathrm{m}\,\hat{i} + 0.2\,\mathrm{m}\,\hat{j}\right)$ <br><br> $= 0.345\,\mathrm{N}\,\hat{i} + 0.230\,\mathrm{N}\,\hat{j}$ |

**Example #2—Interactive.**  A charge of $-3.00\,\mu\mathrm{C}$ is located at the origin and a second charge of $+2.00\,\mu\mathrm{C}$ is located on the $xy$ plane at the point ($x = 30.0\,\mathrm{cm}$, $y = 20.0\,\mathrm{cm}$). Determine the electric force exerted by the $+2\,\mu\mathrm{C}$ charge on the $-3\,\mu\mathrm{C}$ charge.

**Picture the Problem.**  Coulomb's force law will be used to calculate the force. **Try it yourself.** Work the problem on your own, in the spaces provided, to get the final answer.

| 1. Write out Coulomb's force law as a calculational guide. |  |
|---|---|
| 2. Draw a sketch of the two charges and include the $x$ and $y$ axes.  Label the charge at the origin $q_2$, and the other charge $q_1$.  Draw the vector $\vec{r}_{1,2}$, which goes from $q_1$ to $q_2$. |  |
| 3. Evaluate both the magnitude of $\vec{r}_{1,2}$ and $\hat{r}_{1,2}$. |  |
| 4. Substitute all values into the Coulomb force law, and calculate the force. | $\vec{F}_{1,2} = 0.345\,\mathrm{N}\,\hat{i} + 0.230\,\mathrm{N}\,\hat{j}$ |

**Example #3.**  You are at a party watching a sporting event on television. The television signal is transmitted to your receiving dish by a 100-kg communications satellite in a geosynchronous orbit some 5.63 earth radii ($3.59\times10^7\,\mathrm{m}$) overhead. A spherical party balloon with a 25.0-cm diameter contains helium at room temperature (20.0°C) and at a pressure of 1.30 atm. If one electron could be stripped from one out of every ten helium atoms in the balloon and transferred to the satellite, with what force would the balloon attract the satellite? What effect would this force have on the motion of the satellite?

**Picture the Problem.** This is a Coulomb's force law problem, but before you can calculate the force, you first have to determine the charge on the balloon and satellite, by determining how many helium atoms are in the balloon.

| | |
|---|---|
| 1. Write out Coulomb's force law as a calculational guide. We know the force will be attractive, so we just need the magnitude of the force. | $$F_E = \frac{k|q_1 q_2|}{r_{1,2}^2}$$ |
| 2. Use the ideal-gas law to find an expression for the number of helium atoms in the balloon. Use this to determine the charge that accompanies the stripped electrons. | $$PV = nRT$$ $$N\,\text{atoms} = (n\,\text{mol})N_A$$ $$PV = NRT/N_A$$ $$N = PVN_A/(RT)$$ $$= \frac{(1.3\,\text{atm})\left[\frac{4}{3}\pi(0.125\,\text{m})^3\right]\left(6.02\times10^{23}\,\text{mol}^{-1}\right)}{\left[0.0821\,\text{L}\cdot\text{atm}/(\text{mol}\cdot\text{K})\right](293\,\text{k})}\frac{10^3\,\text{L}}{1\,\text{m}^3}$$ $$= 2.66\times10^{23}$$ |
| 3. Substitute the charge of the balloon and satellite into Coulomb's force law to calculate the attractive force of the balloon on the satellite. | $$F_E = \frac{k(Ne/10)(Ne/10)}{(5.63R_E)^2}$$ $$= \frac{(8.99\times10^9\,\text{N}\cdot\text{m}^2/\text{C}^2)\left[(2.66\times10^{23})(1.60\times10^{-19}\,\text{C})/10\right]^2}{\left[5.36(6.37\times10^6\,\text{m})\right]^2}$$ $$= 140\,\text{N}$$ |
| 4. Compare this force to the gravitational force of the earth on the satellite to determine if the electrostatic force is a significant effect. Since the satellite is 6.63 earth radii from the center of the earth, the force on it is simply $1/6.63^2$ the weight of the satellite on the surface of the earth. | $$F_g = mg\frac{1}{6.63^2} = \frac{(100\,\text{kg})(9.81\,\text{N/kg})}{6.63^2} = 22.3\,\text{N}$$ |

**Comments.** The electrostatic force on the satellite would be approximately 6.3 times the force of gravity, which would significantly alter the orbit of the satellite. However, it is not technologically feasible to isolate this large a charge on a balloon, a satellite, or any other similar object, as we shall see in Chapter 22.

**Example #4—Interactive.** At a party you are talking with a friend who is about an arm's length from you. Make an order-of-magnitude estimate of the forces of attraction between you and your friend if 1% of your electrons could be transferred to your friend. Compare this force with the weight of an object with mass $M_E$ (the mass of the earth).

**Picture the Problem.** This is a Coulombs' force law problem. To determine the number of electrons in your body you must first make a reasonable estimate of your mass. Then assume you are made entire of water. **Try it yourself.** Work the problem on your own, in the spaces provided, to get the final answer.

| | |
|---|---|
| 1. Estimate your mass. | |
| 2. Assume both you and your friend are made 100% of water. Estimate the number of electrons in this much water. Water has a molecular mass of 18 g/mol and there are ten electrons for each water molecule. | |
| 3. Determine the charge associated with this many electrons. After the charge is transferred, you will have as much positive charge as your friend has negative charge. | |
| 4. Estimate the length of your arm, and calculate the electrostatic force between you and your friend. | |
| 5. Compare the value obtained in step 4 with the gravitational force on an object with mass equal to the mass of the earth located on the surface of the earth. | |

**Example #5.** An electron, released in a region where the electric field is uniform, is observed to have an acceleration of $3.00 \times 10^{14}$ m/s$^2$ in the positive $x$ direction. Determine the electric field producing this acceleration. Assuming the electron is released from rest, determine the time required for it to reach a speed of 11,200 m/s, the escape speed from the earth's surface.

**Picture the Problem.** You will need to use Newton's 2nd law, kinematics, and the relationship between force and electric field to solve this problem.

| 1. Write out Newton's 2nd law for this particle. We know the force is an electrostatic force due to the electric field. Solve for the electric field. You will need to look up the mass of an electron. | $\vec{F} = m\vec{a}$ <br><br> $q\vec{E} = m\vec{a}$ <br><br> $\vec{E} = \dfrac{m\vec{a}}{-e} = \dfrac{\left(9.11 \times 10^{-31}\,\text{kg}\right)\left(3 \times 10^{14}\,\text{m/s}\,\hat{\imath}\right)}{-1.60 \times 10^{-19}\,\text{C}} = -1710\,\text{N/C}\,\hat{\imath}$ |
|---|---|
| 2. Use kinematics to solve for the time required to reach the escape speed from the earth's surface. | $v_x = v_{x,0} + a_x t$ <br><br> $t = \dfrac{v_x - v_{x,0}}{a_x} = \dfrac{11{,}200\,\text{m/s} - 0}{3 \times 10^{14}\,\text{m/s}^2} = 3.73 \times 10^{-11}\,\text{s}$ |

**Example #6—Interactive.** When a test charge of $+5.00\,\mu\text{C}$ is placed at a certain point $P$, the force that acts on it is 0.0800 N, directed northeast. (*a*) What is the electric field at $P$? (*b*) If the $+5.00\,\mu\text{C}$ test charge is replaced by a $-2.00\,\mu\text{C}$ charge, what force would act on it?

**Picture the Problem.** For this problem, all you need is the relationship between the electrostatic force and the electric field. **Try it yourself.** Work the problem on your own, in the spaces provided, to get the final answer.

| 1. Using the vector relationship between force and electric, find the electric field at point $P$. | |
|---|---|
| | $\vec{E} = 16{,}000\,\text{N/C, northeast}$ |
| 2. Using the electric field from step 1, calculate the force on the $-2.00\,\mu\text{C}$ charge. | |
| | $\vec{F} = 0.0320\,\text{N, southwest}$ |

**Example #7.** Three point charges are placed on the $xy$ plane: a $+50.0$-nC charge at the origin, a $-60.0$-nC charge on the $x$ axis at $x = 10.0\,\text{cm}$, and a $+150$-nC charge at the point (10.0 cm, 8.00 cm). (*a*) Find the total electric force on the $+150.0$-nC charge. (*b*) What is the electric field at the location of the $+150$-nC charge due to the presence of the $+50.0$- and $-60.0$-nC charges?

**Picture the Problem.** For this problem, we will use the superposition of electric forces and electric fields.

| 1. The total force on the $+150.0$-nC charge is the sum of the forces from the other two charges. Let $q_1$ be the charge at the origin, $q_2$ be the $-60.0$-nC charge, and $q_3$ be the $+150$-nC charge. Use the superposition principle to write an expression for the total force on $q_3$. | $\vec{F}_3 = \vec{F}_{1,3} + \vec{F}_{2,3}$ <br><br> $= \dfrac{kq_1 q_3}{r_{1,3}^2}\hat{r}_{1,3} + \dfrac{kq_2 q_3}{r_{2,3}^2}\hat{r}_{2,3}$ |
|---|---|

| | |
|---|---|
| 2. Draw an illustration of the three charges, making sure to properly draw the $\vec{r}$ vectors between the charges. | 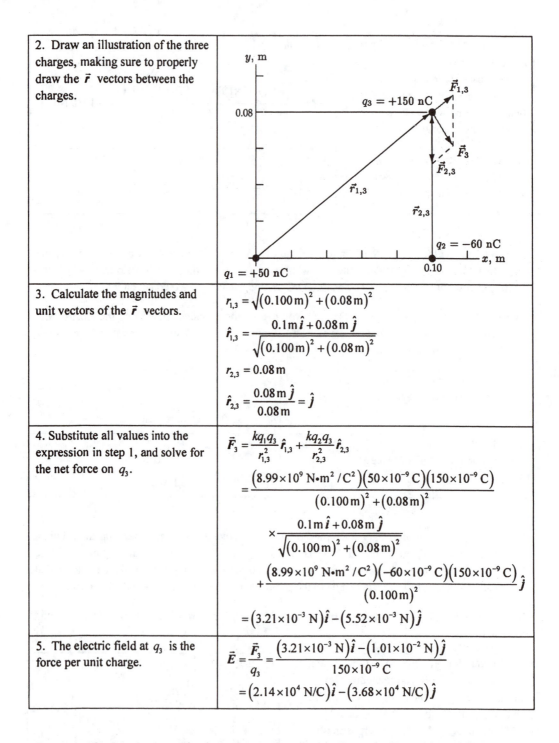 |
| 3. Calculate the magnitudes and unit vectors of the $\vec{r}$ vectors. | $r_{1,3} = \sqrt{(0.100\,\text{m})^2 + (0.08\,\text{m})^2}$ <br><br> $\hat{r}_{1,3} = \dfrac{0.1\,\text{m}\,\hat{i} + 0.08\,\text{m}\,\hat{j}}{\sqrt{(0.100\,\text{m})^2 + (0.08\,\text{m})^2}}$ <br><br> $r_{2,3} = 0.08\,\text{m}$ <br><br> $\hat{r}_{2,3} = \dfrac{0.08\,\text{m}\,\hat{j}}{0.08\,\text{m}} = \hat{j}$ |
| 4. Substitute all values into the expression in step 1, and solve for the net force on $q_3$. | $\vec{F}_3 = \dfrac{kq_1q_3}{r_{1,3}^2}\hat{r}_{1,3} + \dfrac{kq_2q_3}{r_{2,3}^2}\hat{r}_{2,3}$ <br><br> $= \dfrac{(8.99\times10^9\ \text{N·m}^2/\text{C}^2)(50\times10^{-9}\ \text{C})(150\times10^{-9}\ \text{C})}{(0.100\,\text{m})^2 + (0.08\,\text{m})^2}$ <br><br> $\times \dfrac{0.1\,\text{m}\,\hat{i} + 0.08\,\text{m}\,\hat{j}}{\sqrt{(0.100\,\text{m})^2 + (0.08\,\text{m})^2}}$ <br><br> $+ \dfrac{(8.99\times10^9\ \text{N·m}^2/\text{C}^2)(-60\times10^{-9}\ \text{C})(150\times10^{-9}\ \text{C})}{(0.100\,\text{m})^2}\hat{j}$ <br><br> $= (3.21\times10^{-3}\ \text{N})\hat{i} - (5.52\times10^{-3}\ \text{N})\hat{j}$ |
| 5. The electric field at $q_3$ is the force per unit charge. | $\vec{E} = \dfrac{\vec{F}_3}{q_3} = \dfrac{(3.21\times10^{-3}\ \text{N})\hat{i} - (1.01\times10^{-2}\ \text{N})\hat{j}}{150\times10^{-9}\ \text{C}}$ <br><br> $= (2.14\times10^4\ \text{N/C})\hat{i} - (3.68\times10^4\ \text{N/C})\hat{j}$ |

**Example #8—Interactive.** The three charges described in Example #7 remain at their locations and a charge of $-30.0\,\text{nC}$ is placed on the $y$ axis at $y = 8.00$ cm. (a) Find the total electric force on the $+150.0$-nC charge. (b) What is the electric field at the location of the $+150.0$-nC charge due to the presence of the $+50.0$-, $-60.0$-, and $-30.0$-nC charges?

**Picture the Problem.** The new net force will be the sum of the force calculated in Example #7, and the force as a result of the new −30.0-nC charge. **Try it yourself.** Work the problem on your own, in the spaces provided, to get the final answer.

| | |
|---|---|
| 1. Calculate the force on the +150.0-nC charge due to the added −30-nC charge. | |
| 2. Add the force calculated in step 1 to the force calculated in Example #7. | $\vec{F}_3 = \left(-8.36 \times 10^{-4}\,\text{N}\right)\hat{i} - \left(5.52 \times 10^{-3}\,\text{N}\right)\hat{j}$ |
| 3. The electric field is the force per unit charge. | $\vec{E} = \left(-5.57 \times 10^{3}\,\text{N/C}\right)\hat{i} - \left(3.68 \times 10^{4}\,\text{N/C}\right)\hat{j}$ |

**Example #9.** An electron is traveling to the right along the $x$ axis with kinetic energy $K$, which is along the axis of a cathode ray tube as shown in Figure 21-6. There is an electric field $\vec{E} = (2.00 \times 10^{4}\,\text{N/C})\hat{j}$ between the deflection plates, which are 6.00 cm long and are separated by 2.00 cm. Determine the minimum initial kinetic energy the electron can have and still avoid colliding with one of the plates.

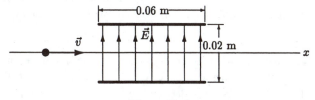

Figure 21-6

**Picture the Problem.** Determine the electron's acceleration. You will need to use two-dimensional kinematics relationships to find the initial kinetic energy.

| | |
|---|---|
| 1. Determine the acceleration of the electron from its mass and the electrostatic force on it. If the positive $y$ axis is directed upward, then the acceleration of the electron will be in the $-y$ direction. | $\vec{a} = \dfrac{\vec{F}}{m} = \dfrac{-e\vec{E}}{m}$ |

| | |
|---|---|
| 2. Write a kinematic expression for the $y$ position of the electron as a function of time. The electron is initially on the $x$ axis, and the initial $y$ component of the velocity is zero. | $\Delta y = v_{0,y}t + \frac{1}{2}a_y t^2$ $y = -\frac{eE}{2m}t^2$ |
| 3. The $x$ component of the force is zero, so the electron moves with constant speed in the positive $x$ direction. The initial velocity is entirely in the $x$ direction. | $\Delta x = v_{0,x}t + \frac{1}{2}a_x t^2$ $x = v_0 t$ |
| 4. Solve the expression in step 3 for time, and substitute into the expression from step 2. | $t = \frac{x}{v_0}$ $y = -\frac{eE}{2m}\left(\frac{x}{v_0}\right)^2$ |
| 5. Rearrange the expression in step 4 to solve for the initial kinetic energy of the electron. | $K = \frac{1}{2}mv_0^2 = -\frac{eEx^2}{4y}$ |
| 6. Solve for the minimum kinetic energy by substituting values of $x = 6\,\text{cm}$ and $y = -1\,\text{cm}$ for the situation of the electron just missing the plates. | $K_{min} = -\frac{\left(1.60\times10^{-19}\,\text{C}\right)\left(2\times10^4\,\text{N/C}\right)\left(0.06\,\text{m}\right)^2}{4\left(-0.01\,\text{m}\right)}$ $= 2.88\times10^{-16}\,\text{J}$ |

**Example #10—Interactive.** The electric field is zero everywhere except in the region $0 \leq x \leq 3$ cm, where there is a uniform electric field of 100 N/C in the $+y$ direction. An electron is moving along the negative $x$ axis with a velocity $\vec{v} = (1.00\times10^6\,\text{m/s})\hat{i}$. When the electron passes through the region $0 \leq x \leq 3$ cm the electric field exerts a force on it. (a) When the $x$ coordinate of the electron's position is 3.00 cm, what is its velocity and what is the $y$ coordinate of its position? (b) When the $x$ coordinate of its position equals 10.0 cm, what is its velocity and what is the $y$ coordinate of its position?

**Picture the Problem.** Find the acceleration of the electron while it is in the region of the electric field. Use two-dimensional kinematics to find the velocity of the electron just at the point it leaves the region with an electric field. Once the electron leaves the electric field, there will be no force acting on it, so it will not experience any acceleration, just constant-velocity two-dimensional motion. **Try it yourself.** Work the problem on your own, in the spaces provided, to get the final answer.

| | |
|---|---|
| 1. Draw a sketch to help you visualize the problem. | |

| | |
|---|---|
| 2. Find an expression for the acceleration of the electron while under the influence of the electric field. | |
| 3. From the equations for motion in the $x$ direction, find an expression for the time it will take the electron to travel from $x = 0$ to $x = 3\,\text{cm}$. | |
| 4. From the equations for motion in the $y$ direction, determine the $y$ component of the velocity when the electron is at $x = 3\,\text{cm}$. The $x$ component of the velocity is the same as before. Why? | $\vec{v}(x = 3\,\text{cm}) = (1.00 \times 10^6 \text{ m/s})\hat{i} - (5.27 \times 10^5 \text{ m/s})\hat{j}$ |
| 5. From the equations for motion in the $y$ direction, determine the $y$ position of the electron when $x = 3\,\text{cm}$. | $y = -0.00790\,\text{m}$ |
| 6. Determine the accleration of the electron after it leaves the electric field region, and use this to calculate the velocity of the electron when it is at $x = 10\,\text{cm}$. | $\vec{v}(x = 10\,\text{cm}) = (1.00 \times 10^6 \text{ m/s})\hat{i} - (5.27 \times 10^5 \text{ m/s})\hat{j}$ |
| 7. Determine the time required for the electron to move from $x = 3$ to $x = 10\,\text{cm}$ and use this time to find the $y$ position of the electron when it is at $x = 10\,\text{cm}$. | $y = -0.0448\,\text{m}$ |

**Example #11.** A charge $q_1 = +2q$ is at the origin and a charge $q_2 = -q$ is on the $x$ axis at $x = a$. Find expressions for the total electric field $\vec{E}$ on the $x$ axis in each of the regions (a) $x < 0$, (b) $0 < x < a$, and (c) $a < x$. (d) Determine all points on the $x$ axis where the electric field is zero. (e) Make a plot of $E_x$ versus $x$ for all points on the $x$ axis, $-\infty < x < \infty$.

**Picture the Problem.** For each of the regions, find an expression for the total $x$ component of the electric field due to each charge. Determine if the electric field is zero at any location, and do the plot.

| | |
|---|---|
| 1.  Sketch the situation for part (*a*). | |
| 2.  For part (*a*), the $\hat{r}$ vector for each charge will be in the $-\hat{i}$ direction.  Remember that the $x$ coordinate will always be negative for this case. | $$\vec{E} = \vec{E}_1 + \vec{E}_2 = \frac{kq_1}{r_1^2}\hat{r}_1 + \frac{kq_2}{r_2^2}\hat{r}_2$$ $$= \frac{2kq}{(-x)^2}\left(-\hat{i}\right) + \frac{-kq}{(-x+a)^2}\left(-\hat{i}\right)$$ $$= \left(-\frac{2kq}{x^2} + \frac{kq}{(a-x)^2}\right)\hat{i}$$ |
| 3.  Sketch the situation for part (*b*). | |
| 4.  For part (*b*) the $x$ coordinate is always a positive number. | $$\vec{E} = \vec{E}_1 + \vec{E}_2 = \frac{kq_1}{r_1^2}\hat{r}_1 + \frac{kq_2}{r_2^2}\hat{r}_2$$ $$= \frac{2kq}{x^2}\left(\hat{i}\right) + \frac{-kq}{(a-x)^2}\left(-\hat{i}\right)$$ $$= \left(\frac{2kq}{x^2} + \frac{kq}{(a-x)^2}\right)\hat{i}$$ |
| 5.  Sketch the situation for part (*c*). | |
| 6.  In part (*c*) the $x$ coordinate will always be a positive number.  The $\hat{r}$ vectors will both also be in the $+\hat{i}$ direction. | $$\vec{E} = \vec{E}_1 + \vec{E}_2 = \frac{kq_1}{r_1^2}\hat{r}_1 + \frac{kq_2}{r_2^2}\hat{r}_2$$ $$= \frac{2kq}{x^2}\left(\hat{i}\right) + \frac{-kq}{(x-a)^2}\left(\hat{i}\right)$$ $$= \left(\frac{2kq}{x^2} - \frac{kq}{(x-a)^2}\right)\hat{i}$$ |

7. Part (*d*). For values of *x* less than zero, there are no places where the electric field is exactly zero. The electric field is everywhere dominated by the charge of $2q$, although the electric field does approach zero as $x \to \infty$.

For $0 < x < a$ the electric field is everywhere positive.

There is a zero of the electric field for $a < x$. The solution $x = a\left(2 - \sqrt{2}\right) = 0.586a$ is not within the region of interest, so we have to throw that one out.

$$E_x = 0 = \frac{2kq}{x^2} - \frac{kq}{(x-a)^2}$$

$$\frac{2kq}{x^2} = \frac{kq}{(x-a)^2}$$

$$x^2 = 2(x-a)^2 = 2x^2 - 4ax + 2a^2$$

$$0 = x^2 - 4ax + 2a^2$$

$$x = \frac{4a \pm \sqrt{16a^2 - 8a^2}}{2} = 2a \pm a\sqrt{2}$$

$$= a\left(2 \pm \sqrt{2}\right)$$

$$x = a\left(2 + \sqrt{2}\right)$$

8. A sketch of the *x* component of the electric field is shown at the right.

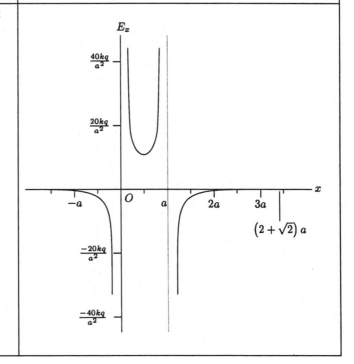

**Example #12—Interactive.** A charge $q_1 = -q$ is on the *y* axis at $y = a$, and a charge $q_2 = +q$ is on the *y* axis at $y = -a$. Make a plot of the *y* component $E_y$ of the electric field versus *x* for all points on the *x* axis, $-\infty < x < \infty$.

**Picture the Problem.**  Add the $y$ components of the electric field from each point charge together to find $E_y(x)$ and then plot that function.  **Try it yourself.**  Work the problem on your own, in the spaces provided, to get the final answer.

| | |
|---|---|
| 1. Find an expression for the $y$ component of the electric field as a function of $x$. | $$E_y = \frac{2kqa}{\left(x^2 + a^2\right)^{3/2}}$$ |
| 2. Plot the function you found in step 1. | 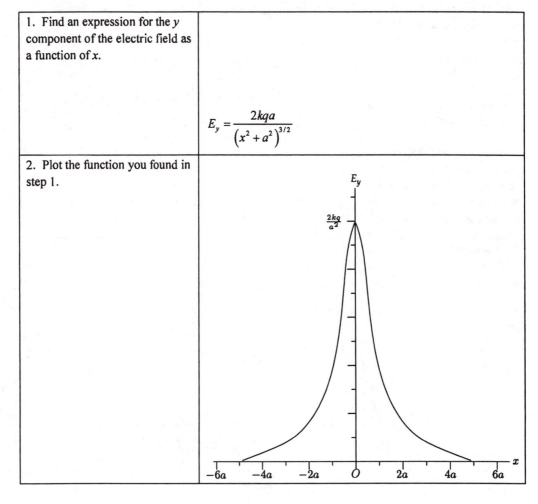 |

# Chapter **22**

# The Electric Field II: Continuous Charge Distributions

## I.  Key Ideas

It is accepted knowledge that charge is discretely distributed in space. However, situations often occur in which it is expedient to treat charges as if their spatial distribution were continuous. This approximation is analogous to the manner in which mass is often considered to be continuously distributed, even though it is accepted knowledge that most of the mass of an atom is located in the atomic nucleus, which is extremely small compared to the atom.

We define the **volume charge density** $\rho$ as the charge per unit volume, measured in coulombs per cubic meter:

$$\rho = \frac{dQ}{dV} \qquad\qquad \text{Volume charge density}$$

If the charge is distributed in a thin layer on the surface of an object, we define the **surface charge density** $\sigma$ as the charge per unit area:

$$\sigma = \frac{dQ}{dA} \qquad\qquad \text{Surface charge density}$$

If the charge is distributed along a line in space, we define the **linear charge density** $\lambda$ as the charge per unit length:

$$\lambda = \frac{dQ}{dL} \qquad\qquad \text{Linear charge density}$$

***Section 22-1. Calculating $\vec{E}$ from Coulomb's Law.*** The electric field $\vec{E}$ at point $P$ due to an element $dq$ of a charge distribution is given by

$$d\vec{E} = \frac{k\,dq}{r^2}\hat{r} \qquad\qquad \text{Electric field of an element of charge}$$

where $\vec{r}$ is a vector from the source point where the charge $dq$ is located to the field point $P$ at which the field is to be evaluated, and $\hat{r} = \vec{r}/r$ is a unit vector in the direction of $\vec{r}$. In accordance with the principle of superposition, the total electric field $\vec{E}$ at $P$ is found by integrating this expression over the entire charge distribution, which occupies volume $V$:

$$\vec{E} = \int_V \frac{k\,dq}{r^2}\hat{r} \qquad\qquad \text{Electric field of continuous distribution of charge}$$

where $dq = \rho\,dV$. If the charge is distributed on a surface, we use $dq = \sigma\,dA$ and integrate over the surface. If the charge is along a line, we use $dq = \lambda\,dL$ and integrate over the line. In the text, expressions for the electric fields due to several important charge distributions are determined by evaluating this integral. The resulting expressions are

$$E_y = \frac{2k\lambda}{y}\sin\theta_0$$

$$= \frac{2k\lambda}{y}\frac{\frac{1}{2}L}{\sqrt{\left(\frac{1}{2}L\right)^2 + y^2}} \qquad \vec{E} \text{ on the perpendicular bisector of a uniformly charged finite line}$$

$$E_y = \frac{2k\lambda}{y} \qquad\qquad \vec{E} \text{ at a distance } y \text{ from a uniformly charged infinite line}$$

$$E_x = \frac{kQx}{\left(x^2 + a^2\right)^{3/2}} \qquad\qquad \vec{E} \text{ on the axis of a uniformly charged ring}$$

$$E_x = 2\pi k\sigma\left(1 - \frac{|x|}{\sqrt{x^2 + R^2}}\right) \qquad\qquad \vec{E} \text{ on the axis of a uniformly charged disk}$$

$$E_x = 2\pi k\sigma \qquad\qquad \vec{E} \text{ for a uniformly charged infinite plane}$$

***Section 22-2 Gauss's Law.*** The electric flux $d\phi$ through an element of surface area $dA$ is the component $E_n$ of the electric field normal to the area multiplied by the area. That is, $d\phi = E_n\,dA$. The component $E_n$ of the electric field normal to the area can be expressed $E_n = \vec{E}\cdot\hat{n}$, where $\hat{n}$ is a unit vector that is normal to the element of area. Thus the flux of the electric field through an element of area can be expressed $d\phi = \vec{E}\cdot\hat{n}\,dA$. The **electric flux** $\phi$ through a surface $S$ is

$$\phi = \int_S \vec{E}\cdot\hat{n}\,dA \qquad\qquad \text{Electric flux}$$

The number of electric field lines passing through a surface is proportional to the electric flux through the surface. Thus, the electric flux through a surface can be determined either by direct integration or by counting field lines.

The electric field strength is represented by the number of electric field lines per unit area for an area that is perpendicular to the field lines. Determining the net flux out of a closed surface is mathematically equivalent to counting the number of field lines leaving the surface. (The count is the number leaving less the number entering the surface.) Field lines originate from a positive charge in proportion to the magnitude of the charge. Also, field lines terminate at a negative charge in proportion to the magnitude of the charge. It follows that the net flux out of a closed surface is proportional to the net charge inside the surface.

For example, if the surface is a sphere centered on an isolated charge $q$, the field intensity on the surface decreases as the square of the radius of the sphere, and the surface area of the sphere increases as the square of the radius. Thus, the product of the field intensity and the area is constant, independent of the radius of the sphere. The net charge enclosed is also independent of the radius of the sphere. This result is called **Gauss's law**:

$$\phi_{net} = \oint_S \vec{E} \cdot \hat{n} \, dA = 4\pi k Q_{inside} = \frac{Q_{inside}}{\varepsilon_0} \qquad \text{Gauss's law}$$

where $Q_{inside}$ is the net charge inside the surface $S$ and $\varepsilon_0 = 1/(4\pi k)$ is called the **permittivity of free space**. The integral sign with a circle indicates that the integral is to be taken over an entire *closed* surface. By convention $\hat{n}$ is the unit normal directed *out* of the surface.

*Section 22-3. Calculating $\vec{E}$ from Gauss's Law.* Gauss's law is useful in situations where a closed surface (also called a **gaussian surface**) can be found over which the flux integral can be evaluated. Symmetry along with other considerations is often used in the evaluation of this integral. For a given charge distribution, the challenge is to conceive of a closed surface through which the flux can be evaluated. Gaussian surfaces are typically segmented, with the surface of some segments parallel to the electric field and the surface of other segments perpendicular to the field. For the area segments parallel to the field, $\hat{n}$ is perpendicular to the field. Thus $\vec{E} \cdot \hat{n} = E_n = 0$, and the flux through such segments is zero. For the area segments perpendicular to the field, $\hat{n}$ is parallel to the field, and $\vec{E} \cdot \hat{n} = E_n = \pm E$. These segments are selected so that $E$, the magnitude of the field, is constant everywhere on the segment. The flux through a segment is then $E_n A_{seg}$, where $A_{seg}$ is the area of the segment.

There are three types of symmetry for which Gauss's law can be applied in a straightforward manner:

**1.** *Spherical symmetry.* A charge distribution has point symmetry if the distribution depends only upon the distance from a point. Consider a uniformly charged spherical shell that is centered at point $P$. This charge distribution has point symmetry. For charge distributions with point (that is, spherical) symmetry, $\vec{E}$ must be radial, so the gaussian surface of choice is a spherical surface that is everywhere perpendicular to $\vec{E}$. Because $E_n = E_r = \pm E$ is constant for all points at the same distance from the center, $\phi_{net} = E_r 4\pi r^2$ for a spherical surface of radius $r$ centered on $P$.

**2.** *Cylindrical symmetry.* A charge distribution has line symmetry if the distribution depends only upon the distance from a straight line that extends to infinity in both directions. Suppose the region of space within a distance $a$ of the $x$ axis were filled with a uniform volume charge density. This charge distribution has line symmetry about the $x$ axis. For charge distributions with line symmetry, $\vec{E}$ must be directed at right angles with the symmetry line, like the bristles of a bottle brush. The gaussian surface of choice is a can-shaped surface with the axis of the can coincident with the line. On the ends of the can, $\hat{n}$ is perpendicular to the field $\vec{E}$, so $E_n = \vec{E} \cdot \hat{n} = 0$ and the flux out of the ends is zero. On the curved side of the can, $\vec{E}$ is parallel to $\hat{n}$ so

$E_n = \vec{E} \cdot \hat{n} = E_r = \pm E$. Because $E_r$ is constant everywhere on this segment, $\phi_{net} = E_r 2\pi r \ell$, where $r$ and $\ell$ are the radius and length of the can.

**3.** *Planar symmetry.* A charge distribution has planar symmetry if the distribution depends only upon the distance from an infinite plane. For charge distributions with planar symmetry, $\vec{E}$ must be directed either toward or away from the symmetry plane. Suppose the region of space within a distance $a$ of the $yz$ plane is filled with a volume charge density $\rho = Cx^2$, $-a < x < a$. This charge density has planar symmetry about the $yz$ plane because it depends only upon the distance from the plane. For charge distributions with planar symmetry the gaussian surface of choice is a can-shaped surface with the ends of the can parallel with—and equidistant from—the plane. On the curved side of the can, $\hat{n}$ is perpendicular to the field $\vec{E}$. Because $\vec{E} \cdot \hat{n} = E_n = \pm E$ is the same on both ends of the can, and because $E$ is the same everywhere on each end, $\phi_{net} = 2E_n A$, where $A$ is the area of one end of the can.

In the text, Gauss's law is used to determine the electric field due to a variety of symmetric charge distributions. The resulting expressions for the electric fields are

$$E_r = 0 \quad r < R$$
$$E_r = \frac{1}{4\pi\varepsilon_0}\frac{Q}{r^2} \quad r > R \qquad \vec{E} \text{ for a uniformly charged thin spherical shell}$$

$$E_r = \frac{1}{4\pi\varepsilon_0}\frac{Q}{R^3}r \quad r \leq R$$
$$E_r = \frac{1}{4\pi\varepsilon_0}\frac{Q}{r^2} \quad r \geq R \qquad \vec{E} \text{ for a uniformly charged solid sphere}$$

$$E_r = 0 \quad r < R$$
$$E_r = \frac{\sigma R}{\varepsilon_0 r} = \frac{1}{2\pi\varepsilon_0}\frac{\lambda}{r} \quad r > R \quad \vec{E} \text{ for a uniformly charged thin, infinitely long cylindrical shell}$$

$$E_r = \frac{\rho}{2\varepsilon_0}r = \frac{\lambda}{2\pi\varepsilon_0 R^2}r \quad r \leq R$$
$$E_r = \frac{\rho R^2}{2\varepsilon_0 r} = \frac{1}{2\pi\varepsilon_0}\frac{\lambda}{r} \quad r \geq R \qquad \vec{E} \text{ for a uniformly charged infinitely long cylinder}$$

$$E_n = \frac{\sigma}{2\varepsilon_0} \qquad \vec{E} \text{ for a uniformly charged infinite plane}$$

**Section 22-4. Discontinuity of $E_n$.** The normal component of the electric field changes by $\sigma/\varepsilon_0$ across any surface with a surface charge density $\sigma$. This can be derived from Gauss's law. Thus,

$$E_{n2} - E_{n1} = \frac{\sigma}{\varepsilon_0} \qquad \text{Discontinuity of } E_n$$

*Section 22-5. Charge and Field at Conductor Surfaces.* A material that has a large number of free charge carriers is called a conductor. In metals the free charge carriers are electrons that are not bound to the side of the atom but are free to move about the material. If a conductor is placed in an external electric field, the field exerts forces on the free charge carriers causing them to move about until the electric field is zero everywhere within the conductor. Then there are no longer forces on the charge carriers, so they no longer move about; that is, the distribution of charge carriers is static. The conductor is then said to be in **electrostatic equilibrium.** If a conductor is in electrostatic equilibrium, the electric field at the surface of the conductor must be perpendicular to the surface. If there were a component of the electric field tangent to the surface, it would exert a force on the free charge carriers at the surface, causing them to move and redistribute themselves until there is no longer a tangential component of the electric field. If the normal component of the electric field is very strong, it can pull the charge carriers off the surface. This is known as field emission. Normally we will assume that the fields are not this strong.

Consider a gaussian surface that is completely imbedded in the material of a conductor that is in electrostatic equilibrium. The electric field is zero everywhere within the conducting material, so the electric field is zero at all points on the gaussian surface. Therefore, the flux out of the surface is zero and, in accordance with Gauss's law, the net charge inside the surface is zero. This result is true for a gaussian surface of any size or shape imbedded anywhere within the conductor. Therefore, the charge density is zero everywhere within a conductor in static equilibrium. This means that any net electric charge must reside on the surface of any conductor in static equilibrium.

Because the electric field in the material of a conductor is zero everywhere, the normal component of the electric field outside the conductor, in accordance with our earlier result of the discontinuity of $E_n$, is $E_n = \sigma / \varepsilon_0$. That is,

$$E_n = \frac{\sigma}{\varepsilon_0}$$    Electric field just outside the surface of a conductor

## II.  Physical Quantities and Key Equations

**Physical Quantities**

*Permittivity of free space*    $$\varepsilon_0 = \frac{1}{4\pi k} = \frac{1}{4\pi \left(8.99 \times 10^9 \text{ N} \cdot \text{m}^2/\text{C}^2\right)}$$
$$= 8.85 \times 10^{-12} \text{ C}^2/\left(\text{N} \cdot \text{m}^2\right)$$

**Key Equations**

*Electric field of a continuous distribution of charge*    $$\vec{E} = \int_V \frac{k\, dq}{r^2} \hat{r}$$

*$\vec{E}$ on the perpendicular bisector of a uniformly charged finite line*

$$E_y = \frac{2k\lambda}{y}\sin\theta_0 = \frac{2k\lambda}{y}\frac{\frac{1}{2}L}{\sqrt{\left(\frac{1}{2}L\right)^2 + y^2}}$$

*$\vec{E}$ at a distance y from a uniformly charged infinite line*    $E_y = \dfrac{2k\lambda}{y}$

*$\vec{E}$ on the axis of a uniformly charged ring*    $E = \dfrac{kQx}{\left(x^2 + a^2\right)^{3/2}}$

*$\vec{E}$ on the axis of a uniformly charged disk*    $E = 2\pi k\sigma\left(1 - \dfrac{|x|}{\sqrt{x^2 + R^2}}\right)$

*$\vec{E}$ for a uniformly charged infinite plane*    $E = 2\pi k\sigma = \dfrac{\sigma}{2\varepsilon_0}$

*Electric flux*    $\phi = \displaystyle\int_S \vec{E}\cdot\hat{n}\,dA$

*Gauss's law*    $\phi_{net} = \displaystyle\oint_S \vec{E}\cdot\hat{n}\,dA = 4\pi kQ_{inside} = \dfrac{Q_{inside}}{\varepsilon_0}$

*$\vec{E}$ for a uniformly charged thin spherical shell*

$$E_r = 0 \qquad\qquad r < R$$
$$E_r = \frac{1}{4\pi\varepsilon_0}\frac{Q}{r^2} \qquad r > R$$

*$\vec{E}$ for a uniformly charged solid sphere*

$$E_r = \frac{1}{4\pi\varepsilon_0}\frac{Q}{R^3}r \qquad r \leq R$$
$$E_r = \frac{1}{4\pi\varepsilon_0}\frac{Q}{r^2} \qquad r \geq R$$

*$\vec{E}$ for a uniformly charged thin, infinitely long cylindrical shell*

$$E_r = 0 \qquad\qquad\qquad r < R$$
$$E_r = \frac{\sigma R}{\varepsilon_0 r} = \frac{1}{2\pi\varepsilon_0}\frac{\lambda}{r} \qquad r > R$$

*$\vec{E}$ for a uniformly charged infinitely long cylinder*

$$E_r = \frac{\rho}{2\varepsilon_0}r = \frac{\lambda}{2\pi\varepsilon_0 R^2}r \qquad r \leq R$$
$$E_r = \frac{\rho R^2}{2\varepsilon_0 r} = \frac{1}{2\pi\varepsilon_0}\frac{\lambda}{r} \qquad r \geq R$$

*Discontinuity of $E_n$*

$$E_{n2} - E_{n1} = \frac{\sigma}{\varepsilon_0}$$

*Electric field just outside the surface of a conductor*    $E_n = \dfrac{\sigma}{\varepsilon_0}$

*Two useful integrals*

$$\int \frac{dx}{\left(x^2 + a^2\right)^{3/2}} = \frac{x}{a^2\sqrt{x^2 + a^2}} + c$$

$$\int \frac{x\,dx}{\left(x^2 + a^2\right)^{3/2}} = \frac{-1}{\sqrt{x^2 + a^2}} + c$$

## III. Potential Pitfalls

Gauss's-law calculations of the electric field $\vec{E}$ are useful only when symmetry or some other consideration specifies the direction of the field. A gaussian surface should be configured such that $E_n$ is the same everywhere on some segments of the surface and zero on the other segments. Outside a conductor, symmetry is used to determine whether $E_n$ is zero or the same everywhere. Within the material of a conductor $E_n$ is zero everywhere.

When calculating the electric field by evaluating the integral

$$\vec{E} = \int_V \frac{k\,dq}{r^2}\hat{r}$$

you need to evaluate it one component at a time. That is:

$$E_x = \int_V \frac{k\,dq}{r^2}\hat{r}\cdot\hat{i}, \ldots$$

In electrostatics the formulas for the perimeter, surface area, and the volume of common figures such as the circular disk, the sphere, and the cylinder are frequently encountered. If you need to review these formulas, you can find them in *Appendix D of the Volume 1 textbook on page AP-15.*

## IV. True or False Questions and Responses

**True or False**

_____ 1.  Near an isolated infinitely long straight uniformly charged wire, the magnitude of the electric field decreases with the square of the distance from the wire.

_____ 2.  Near an isolated uniformly charged plane, the magnitude of the electric field decreases with the first power of the distance from the plane.

_____ 3.  Outside an isolated uniformly charged spherical shell, the magnitude of the electric field decreases with the square of the distance from the center of the shell.

_____ 4.  Inside the cavity of an isolated uniformly charged spherical shell, the magnitude of the electric field is everywhere zero.

_____ 5.  Inside an isolated, infinitely long, uniformly charged cylindrical shell, the magnitude of the electric field is everywhere zero.

_____ 6.  If the net electric flux out of a closed surface is zero, the electric field must be zero everywhere on the surface.

_____ 7.  If the net electric flux out of a closed surface is zero, the charge density must be zero everywhere inside the surface.

_____ 8.  The electric field is zero everywhere within the material of a conductor in electrostatic equilibrium.

_____ 9.  The tangential component of the electric field is zero at all points just outside the surface of a conductor in electrostatic equilibrium.

_____ 10.  The normal component of the electric field is the same at all points just outside the surface of a conductor in electrostatic equilibrium.

_____ 11.  The area of the surface of the water in a swimming pool of radius $r$ is $2\pi r^2$.

_____ 12.  The surface area of a sphere of radius $r$ is $4\pi r^3$.

_____ 13.  The volume of a sphere of radius $r$ is $4\pi r^3$.

_____ 14.  The surface area of a soup can of height $\ell$ and radius $r$ is $2\pi r^2 + 2\pi r\ell$.

_____ 15.  The volume inside a soup can of height $\ell$ and radius $r$ is $\pi r^2 \ell$.

**Responses to True or False**

1.  False. Near an infinitely long straight uniformly charged wire, the magnitude of the electric field decreases with the first power of the distance from the wire.

2.  False. Near a uniformly charged plane, the magnitude of the electric field does not vary with the distance from the plane.

3.  True.

4.  True.

5.  True.

6.  False. A zero net electric flux out of a surface can be accomplished with a positive flux out of some parts of the surface and a negative flux out of other parts. For example, see Figure 22-12 on page 691 of the text.

7. False. If the net flux out of the surface is zero, the total charge inside the surface must be zero. This can be accomplished with positive charges at some locations and negative charges at other locations inside the surface.

8. True.

9. True.

10. False. The normal component of the electric field is $\sigma/\varepsilon_0$. However, the surface charge density $\sigma$ is not necessarily the same at all points on the surface.

11. False. The area of a circle of radius $r$ is $\pi r^2$.

12. False. The surface area of a sphere of radius $r$ is $4\pi r^2$.

13. False. The volume of a sphere of radius $r$ is $4\pi r^3/3$.

14. True. The area of each end is $\pi r^2$ and, when straightened out, the side of the can is a rectangle of length $\ell$ and width $2\pi r$.

15. True. The volume is the area of the base times the height.

## V.  Questions and Answers

### Questions

1. Is the electric field $\vec{E}$ in Gauss's law only that part of the electric field due to the charge inside a surface, or is it the total electric field due to all charges both inside and outside the surface?

2. Inside a spherical charge distribution of constant volume charge density, why is it that as one moves out from the center, the electric field increases as $r$ rather than decreases as $1/r^2$?

3. Why is the expression $E = 2\pi k\sigma\left[1 - |x|/\left(x^2 + R^2\right)^{1/2}\right]$ for the electric field along the axis of a uniformly charged thin disk different than the expression $E = 2\pi k\sigma$ for the electric field of a uniformly charged infinite plane when Gauss's law *seemingly* gives the same result for these two cases?

4. The electric field just outside the surface of a conductor is *twice* that due to an infinite uniformly charged plane having the *same* surface charge density. Why aren't they the same? Why the factor of two?

### Answers

1. The electric field $\vec{E}$ in Gauss's law is the total electric field due to all charges both inside and outside the surface. However, the results are unaffected if $\vec{E}$ represents the electric field due only to the charges inside the surface. The net contribution to the electric flux out of a surface due to any charge located outside that surface is zero.

2.  The electric field a distance $r$ from the center of a uniformly charged sphere is $kQ/r^2$, where $Q$ is the charge inside a sphere of radius $r$. Thus, inside the sphere, as one moves out from the center $Q$ increases with the cube of the distance from the center of the sphere. Thus the electric field increases as $r$ rather than decreases as $1/r^2$.

3.  Gauss's law, correctly applied, does not give the same result for the two cases.

    For the uniformly charged infinite plane we consider a soup can-shaped gaussian surface with the ends of the can both parallel to and equidistant from the plane. From the symmetry we know that on the curved side of the can the normal component of the electric field is everywhere zero, so the flux through this side is zero. Also from symmetry, we know that on the ends of the can the normal component of the electric field is uniform and equal to its magnitude. Thus the total flux out of the can is $\phi = 2E_n A = 4\pi k\sigma A$, so $E_n = 2\pi k\sigma$.

    For the uniformly charged disk we again consider a soup can-shaped gaussian surface with the ends of the can both parallel to and equidistant from the disk. There is insufficient symmetry to allow us to conclude either that on the curved side of the can the normal component of the electric field is zero or that on the ends of the can the normal component of the electric field is uniform and equal to its magnitude.

4.  As shown in Figure 22-30 on page 703 of the text, just outside the surface of a conductor the electric field is the superposition of two electric fields, one due to the nearby surface charge and the other due to all other charges. These two fields cancel just inside the surface of the conductor so they must have equal magnitudes. Just outside the surface the two fields are in the same direction so the net electric field is twice as large as the electric field due to the nearby surface charge alone.

## VI.  Problems, Solutions, and Answers

**Example #1.**  Find the $y$ component of the electric field at point $P$ on the $y$ axis at $y = 3.00\,\text{m}$ due to the charge on a wire on the $x$ axis between $x = 1.00\,\text{m}$ and $x = 3.00\,\text{m}$ whose uniform charge per unit length is $4.00\,\mu\text{C}/\text{m}$.

**Picture the Problem.**  Sketch the situation to help with the visualization.  You will need to integrate the electric field produced at $P$ due to small, differential charge elements along the charged wire.

| | |
|---|---|
| 1. Sketch the situation. On the sketch, draw a differential element of charge $dq$ that occupies a space $dx$ located at a position $x$. You should also draw $\vec{r}$ for the charge element, which points from the charge element toward the point where the electric field is being calculated. Include a sketch of the electric field vector at point $P$, as well as the $y$ component of that field. | 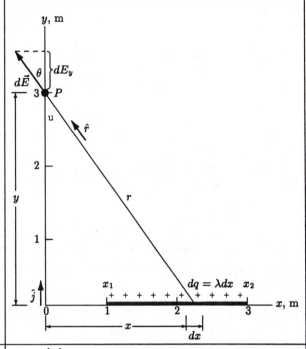 |
| 2. Write out the integral form of Coulomb's law as a guide in solving the rest of the problem. We are not interested in the entire vector electric field, but just the $y$ component, so find a general expression for that. | $\vec{E} = \int \dfrac{k\,dq}{r^2}\hat{r}$<br><br>$E_y = \int \dfrac{k\,dq}{r^2}\hat{r}\cdot\hat{j} = \int \dfrac{k\,dq}{r^2}\cos\theta = \int \dfrac{k\,dq}{r^2}\dfrac{y}{r}$ |
| 3. Determine an expression for the charge element $dq$. | $dq = \lambda\,dx = \left(4\,\mu\text{C}/\text{m}\right)dx$ |
| 4. Determine an expression for $r$. | $r = \left|\vec{r}\right| = \left|-x\hat{i} + y\hat{j}\right| = \sqrt{x^2 + y^2}$ |
| 5. Substitute all values into the expression found in step 2, and integrate to find the $y$ component of the electric field. The limits of integration are given by the location of the charged wire. | $E_y = \int_{x_1}^{x_2} \dfrac{k\lambda y\,dx}{r^3} = k\lambda y \int_{x_1}^{x_2} \dfrac{dx}{\left(x^2 + y^2\right)^{3/2}}$<br><br>$= ky\lambda \left.\dfrac{x}{y^2\sqrt{x^2+y^2}}\right|_{x_1}^{x_2} = \dfrac{k\lambda}{y}\left(\dfrac{x_2}{\sqrt{x_2^2+y^2}} - \dfrac{x_1}{\sqrt{x_1^2+y^2}}\right)$<br><br>$= \dfrac{\left(8.99\times10^9 \text{ N}\cdot\text{m}^2/\text{C}^2\right)\left(4\times10^{-6} \text{ C/m}\right)}{3\,\text{m}}$<br><br>$\times\left(\dfrac{3\,\text{m}}{\sqrt{\left(3\,\text{m}\right)^2 + \left(3\,\text{m}\right)^2}} - \dfrac{1\,\text{m}}{\sqrt{\left(1\,\text{m}\right)^2 + \left(3\,\text{m}\right)^2}}\right)$<br><br>$= 4690\,\text{N/C}$ |

**Example #2—Interactive.**  Find the $x$ component of the electric field at point $P$ of the previous problem.

**Picture the Problem.**  Sketch the situation to help with the visualization.  You will need to integrate the electric field produced at $P$ due to small, differential charge elements along the charged wire.  **Try it yourself.**  Work the problem on your own, in the spaces provided, to get the final answer.

| | |
|---|---|
| 1.  Sketch the situation.  On the sketch, draw a differential element of charge $dq$ that occupies a space $dx$ located at a position $x$.  You should also draw $\vec{r}$ for the charge element, which points from the charge element toward the point where the electric field is being calculated.  Include a sketch of the electric field vector at point $P$, as well as the $x$ component of that field. | |
| 2.  Write out the integral form of Coulomb's law as a guide in solving the rest of the problem.  We are not interested in the entire vector electric field, but just the $x$ component, so find a general expression for that. | |
| 3.  Determine an expression for the charge element $dq$. | |
| 4.  Determine an expression for $r$. | |
| 5.  Substitute all values into the expression found in step 2, and integrate to find the $x$ component of the electric field.  The limits of integration are given by the location of the charged wire. | $E_x = -2900 \, \text{N/C}$ |

**Example #3.**  Find the $x$ component of the electric field at point $P$ on the $x$ axis at $x = 40.0\,\text{cm}$ due to the charge on a semicircular segment of wire in the upper half of the $yz$ plane and centered at the origin, as shown in Figure 22-1. The segment has a uniform charge per unit length of 4.00 nC/m, and the radius $a$ of the semicircle is 30.0 cm.

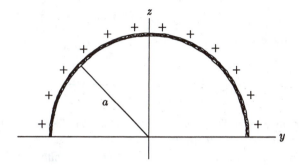

Figure 22-1

**Picture the Problem.** Once again make a sketch of the situation. This one will have to be three-dimensional. The electric field will be the integral of the electric field created by several differential charge elements along the semicircular wire.

| | |
|---|---|
| 1. Sketch the situation. On the sketch, draw a differential element of charge $dq$ that occupies a length $dL$ located at an angular position $\phi$. You should also draw $\vec{r}$ for the charge element, which points from the charge element toward the point where the electric field is being calculated. Include a sketch of the electric field vector at point $P$, as well as the $x$ component of that field. | 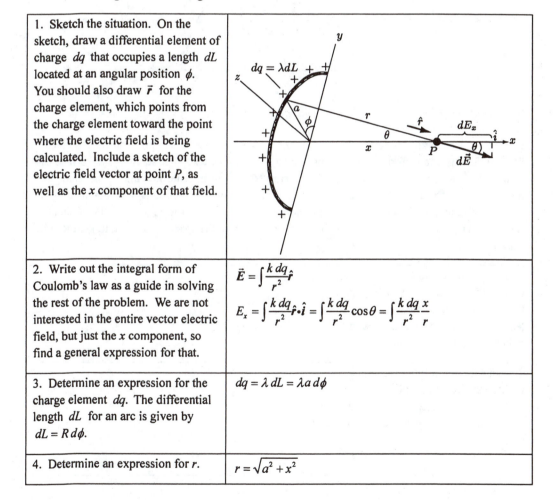 |
| 2. Write out the integral form of Coulomb's law as a guide in solving the rest of the problem. We are not interested in the entire vector electric field, but just the $x$ component, so find a general expression for that. | $$\vec{E} = \int \frac{k\,dq}{r^2}\hat{r}$$ $$E_x = \int \frac{k\,dq}{r^2}\hat{r}\cdot\hat{i} = \int \frac{k\,dq}{r^2}\cos\theta = \int \frac{k\,dq}{r^2}\frac{x}{r}$$ |
| 3. Determine an expression for the charge element $dq$. The differential length $dL$ for an arc is given by $dL = R\,d\phi$. | $dq = \lambda\,dL = \lambda a\,d\phi$ |
| 4. Determine an expression for $r$. | $r = \sqrt{a^2 + x^2}$ |

| 5. Substitute all values into the expression found in step 2, and integrate to find the $x$ component of the electric field. The limits of integration are given by the location of the charged wire. | $$E_x = \int_0^\pi \frac{k\lambda a x \, d\phi}{r^3} = \frac{k\lambda a x}{\left(a^2 + x^2\right)^{3/2}} \int_0^\pi d\phi$$ $$= \frac{\left(8.99 \times 10^9 \text{ N·m}^2/\text{C}^2\right)\left(4 \times 10^{-9} \text{ C/m}\right)\left(0.3\,\text{m}\right)\left(0.4\,\text{m}\right)\pi}{\left[\left(0.3\,\text{m}\right)^2 + \left(0.4\,\text{m}\right)^2\right]^{3/2}}$$ $$= 108 \text{ N/C}$$ |

**Example #4—Interactive.** Find the $x$ component of the electric field at point $P$ on the $x$ axis at $x = 40.0\,\text{cm}$ due to the charge on a thin semicircular plate in the upper half of the $yz$ plane and centered at the origin, as shown in Figure 22-2. The plate has a uniform charge per unit area of $40.0\,\text{nC/m}^2$ and a radius $R$ of $30.0\,\text{cm}$.

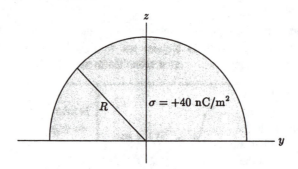

Figure 22-2

**Picture the Problem.** Model your solution after the derivation of the expression for the electric field on the axis of a uniformly charged disk on page 688 of the text. The semicircular surface can be thought of as a set of semicircular surface elements, each of infinitesimal thickness. Using the formula for the electric field on the axis of a uniformly charged semicircular wire obtained in the solution of Example #3, integrate to find the electric field of the uniformly charged semicircular surface. **Try it yourself.** Work the problem on your own, in the spaces provided, to get the final answer.

| 1. Sketch the situation. On the sketch, draw a differential element of charge $dq$ that will be a thin semicircular arc of length $\pi r$ and thickness $dr$ located at a radius of $r$ from the origin. | |
| 2. Using the result from Example #3, write an expression for the $x$ component of the differential electric field created by a thin semicircular arc at a location on the $x$ axis. | |

| 3. Integrate the expression found in step 2. | |
|---|---|
| | $E_x = 226\,\text{N/C}$ |

**Example #5.** A uniform line of charge with linear charge density $\lambda = 4\,\text{nC/m}$ extends from $x = -2\,\text{m}$ to $x = +2\,\text{m}$ on the $x$ axis. Evaluate your results to five significant figures. (*a*) What is the total charge $Q$ on the line? (*b*) Estimate the electric field at point $P$ on the $y$ axis at $y = 120\,\text{m}$ by assuming the entire charge $Q$ is located at the origin. (*c*) Compare this estimate with the actual electric field at $P$ due to the line charge. (*d*) Estimate the electric field at point $P'$ on the $y$ axis at $y = 2\,\text{cm}$ by assuming the line charge is infinitely long with the same linear charge density. (*e*) Compare this estimate with the actual electric field at $P'$ due to the line charge.

**Picture the Problem.** Use the results for the electric field along a perpendicular bisector of a finite line and an infinite line of charge presented in this study guide and derived in the textbook. To obtain results accurate to five significant figures, you should calculate with values accurate to six significant figures or more. Use $k = 8.98755 \times 10^9\,\text{N} \cdot \text{m}^2\,/\,\text{C}^2$.

| 1. (*a*) The total charge on a uniformly charged rod is the charge density multiplied by the length of the rod. | $Q = \int dq = \int \lambda\,dx = \lambda \int dx = \lambda L$ <br><br> $= \left(4 \times 10^{-9}\,\text{C/m}\right)\left(4\,\text{m}\right) = 1.6 \times 10^{-8}\,\text{C}$ |
|---|---|
| 2. (*b*) Assume all of this charge is at the origin. | $\vec{E} = \dfrac{kq}{r^2}\hat{r} = \dfrac{\left(8.98755 \times 10^9\,\text{N} \cdot \text{m}^2\,/\,\text{C}^2\right)\left(1.6 \times 10^{-8}\,\text{C}\right)}{\left(120\,\text{m}\right)^2}\hat{j}$ <br><br> $= 9.9862 \times 10^{-3}\,\text{N/C}\,\hat{j}$ |
| 3. (*c*) Calculate the actual electric field due to the line of charge. | $\vec{E} = \dfrac{2k\lambda}{y}\dfrac{\frac{1}{2}L}{\sqrt{\left(\frac{1}{2}L\right)^2 + y^2}}\hat{j}$ <br><br> $= \dfrac{2\left(8.98755 \times 10^9\,\text{N} \cdot \text{m}^2\,/\,\text{C}^2\right)\left(4 \times 10^{-9}\,\text{C}\right)\left(2\,\text{m}\right)}{\left(120\,\text{m}\right)\sqrt{\left(2\,\text{m}\right)^2 + \left(120\,\text{m}\right)^2}}\hat{j}$ <br><br> $= 9.9848 \times 10^{-3}\,\text{N/C}\,\hat{j}$ |

| 4. ($d$) Estimate the electric field from an infinite line of charge at point $P'$. | $\vec{E} = \dfrac{2k\lambda}{y}\hat{j} = \dfrac{2\left(8.98755\times10^9 \text{ N•m}^2/\text{C}^2\right)\left(4\times10^{-9}\text{ C}\right)}{0.02\,\text{m}}\hat{j}$ <br><br> $= 3595.0\,\text{N/C}\,\hat{j}$ |
|---|---|
| 5. ($e$) Calculate the actual electric field at point $P'$. | $\vec{E} = \dfrac{2k\lambda}{y}\dfrac{\frac{1}{2}L}{\sqrt{\left(\frac{1}{2}L\right)^2 + y^2}}\hat{j}$ <br><br> $= \dfrac{2\left(8.98755\times10^9 \text{ N•m}^2/\text{C}^2\right)\left(4\times10^{-9}\text{ C}\right)\left(2\,\text{m}\right)}{\left(0.02\,\text{m}\right)\sqrt{\left(2\,\text{m}\right)^2 + \left(0.02\,\text{m}\right)^2}}\hat{j}$ <br><br> $= 3954.8\,\text{N/C}\,\hat{j}$ |

**Example #6—Interactive.**    A thin circular disk has a uniform charge per unit area $\sigma = 40\,\text{nC/m}^2$ and a radius of $1\,\text{m}$. Evaluate your results to five significant figures. ($a$) What is the total charge $Q$? ($b$) Estimate the electric field at point $P$ on the axis of the disk a distance of $120\,\text{m}$ from the disk by assuming the entire charge $Q$ is locate at the origin. ($c$) Compare this estimate with the actual electric field at point $P$ due to the charged disk. ($d$) Estimate the electric field at point $P'$ on the axis a distance of $2\,\text{cm}$ from the disk by assuming the disk's radius is infinite and its charge density remains $40\,\text{nC/m}^2$. ($e$) Compare this estimate with the actual electric field at $P'$ due to the charged disk.

**Picture the Problem.**    Use the results for the electric field of a charged disk and an infinite plane of charge presented in this study guide and derived in the textbook. To obtain results accurate to five significant figures, you should calculate with values accurate to six significant figures or more. Use $k = 8.98755\times10^9 \text{ N•m}^2/\text{C}^2$. **Try it yourself.** Work the problem on your own, in the spaces provided, to get the final answer.

| 1. ($a$) The total charge on a uniformly charged disk is the charge density multiplied by the area of the disk. | $Q = 126.66\,\text{nC}$ |
|---|---|
| 2. ($b$) Assume all of this charge is at the origin, and estimate the electric field. | $E = 7.8429\times10^{-2}\text{ N/C}$ |
| 3. ($c$) Calculate the actual electric field due to the disk of charge. | $E = 7.8426\times10^{-2}\text{ N/C}$ |

| | |
|---|---|
| 4. (*d*) Estimate the electric field from an infinite sheet of charge at point *P'*. | $E = 2258.8\,\text{N/C}$ |
| 5. (*e*) Calculate the actual electric field at point *P'*. | $E = 2213.7\,\text{N/C}$ |

**Example #7.** Figure 22-3 shows a prism-shaped surface that is 40.0 cm high, 30.0 cm deep, and 80.0 cm long. The prism is immersed in a uniform electric field of $(500\,\text{N/C})\hat{i}$. Calculate the electric flux out of each of its five faces and the net electric flux out of the entire closed surface.

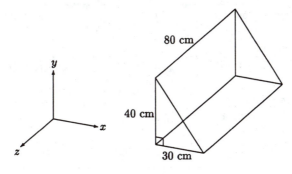

Figure 22-3

**Picture the Problem.** There is no net flux out of any face where $\vec{E} \cdot \hat{n} = 0$. The total flux equals the sum of the fluxes out of each face.

| | |
|---|---|
| 1. Draw a sketch of each of the faces of the prism, including the direction of $\hat{n}$ for each face. | 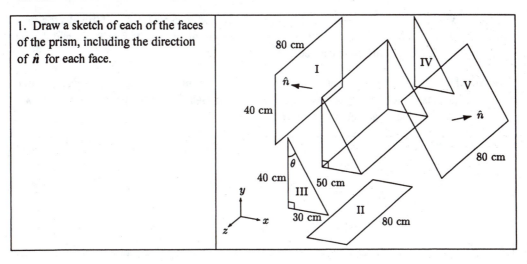 |

| | |
|---|---|
| | |
| 2.  Calculate the electric flux out of surface I. | $\phi_{\text{I}} = \vec{E} \cdot \hat{n}_{\text{I}} A_{\text{I}} = (500 \,\text{N/C}) \hat{i} \cdot (-\hat{i})(0.4 \,\text{m})(0.8 \,\text{m})$ <br> $= -160 \,\text{N} \cdot \text{m}^2 / \text{C}$ |
| 3.  Calculate the electric flux out of surface II. | $\phi_{\text{II}} = \vec{E} \cdot \hat{n}_{\text{II}} A_{\text{II}} = (500 \,\text{N/C}) \hat{i} \cdot (-\hat{j})(0.3 \,\text{m})(0.8 \,\text{m})$ <br> $= 0 \,\text{N} \cdot \text{m}^2 / \text{C}$ |
| 4.  Calculate the electric flux out of surface III. | $\phi_{\text{III}} = \vec{E} \cdot \hat{n}_{\text{III}} A_{\text{III}} = (500 \,\text{N/C}) \hat{i} \cdot (\hat{k}) \tfrac{1}{2}(0.3 \,\text{m})(0.4 \,\text{m})$ <br> $= 0 \,\text{N} \cdot \text{m}^2 / \text{C}$ |
| 5.  Calculate the electric flux out of surface IV. | $\phi_{\text{IV}} = \vec{E} \cdot \hat{n}_{\text{IV}} A_{\text{IV}} = (500 \,\text{N/C}) \hat{i} \cdot (-\hat{k}) \tfrac{1}{2}(0.3 \,\text{m})(0.4 \,\text{m})$ <br> $= 0 \,\text{N} \cdot \text{m}^2 / \text{C}$ |
| 6.  Calculate the electric flux out of surface V.  To compute the dot product, we need cosine of the angle between $\hat{i}$ and $\hat{n}$, which we can get from the figure in step 1. | $\phi_{\text{V}} = \vec{E} \cdot \hat{n}_{\text{V}} A_{\text{V}} = (500 \,\text{N/C}) \hat{i} \cdot (\hat{n})(0.5 \,\text{m})(0.8 \,\text{m})$ <br> $= (200 \,\text{N} \cdot \text{m}^2 / \text{C}) \cos\theta$ <br> $\cos\theta = \dfrac{40 \,\text{cm}}{50 \,\text{cm}} = 0.8$ <br> $\phi_{\text{V}} = 160 \,\text{N} \cdot \text{m}^2 / \text{C}$ |
| 7.  The net electric flux out of the prism is the sum of the fluxes. | $\phi_{\text{net}} = -160 \,\text{N} \cdot \text{m}^2 / \text{C} + 0 + 0 + 0 + 160 \,\text{N} \cdot \text{m}^2 / \text{C}$ <br> $= 0$ |

**Example #8—Interactive.**  The prism of Example #7 is now immersed in a uniform electric field of $(500 \,\text{N/C}) \hat{i} + (400 \,\text{N/C}) \hat{j}$.  Calculate the electric flux out of each of its five faces, and the net electric flux out of the entire closed surface.

**Picture the Problem.**  Use the sketch of each face of the prism from the previous example, and calculate the flux through each face, as before.  **Try it yourself.**  Work the problem on your own, in the spaces provided, to get the final answer.

| | |
|---|---|
| 1.  Calculate the electric flux out of surface I. | <br><br><br><br><br> $\phi_{\text{I}} = -160 \,\text{N} \cdot \text{m}^2 / \text{C}$ |

| | |
|---|---|
| 2. Calculate the electric flux out of surface II. | $\phi_{II} = -96.0 \, \text{N} \cdot \text{m}^2 / \text{C}$ |
| 3. Calculate the electric flux out of surface III. | $\phi_{III} = 0 \, \text{N} \cdot \text{m}^2 / \text{C}$ |
| 4. Calculate the electric flux out of surface IV. | $\phi_{IV} = 0 \, \text{N} \cdot \text{m}^2 / \text{C}$ |
| 5. Calculate the electric flux out of surface V. To compute the dot product, we need cosine of the angle between $\hat{i}$ and $\hat{n}$, which we can get from the figure in step 1 of the previous example. | $\phi_V = +256 \, \text{N} \cdot \text{m}^2 / \text{C}$ |
| 6. The net electric flux out of the prism is the sum of the fluxes. | $\phi_{net} = 0$ |

**Example #9.** A flat slab of thickness $d$, shown in Figure 22-4, has a uniform volume charge density $\rho$. The slab extends indefinitely in its two other dimensions. Use Gauss's law to determine the electric field everywhere.

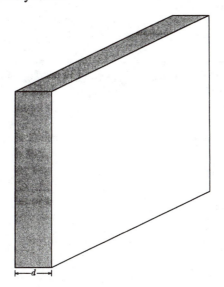

Figure 22-4

**Picture the Problem.**  The slab exhibits planar symmetry, so you can utilize a cylindrical gaussian surface.

| | |
|---|---|
| 1.  First determine the electric field everywhere inside the slab.  Start by sketching a gaussian cylinder inside the slab, centered on the symmetry plane of the slab, with the ends of the cylinder parallel to the symmetry plane.  Sketch both the direction of the electric field and the normal vector of the surface, for each distinct surface on the cylinder.<br><br>Because the charge distribution extends indefinitely, the electric field must be perpendicular to the symmetry plane. | 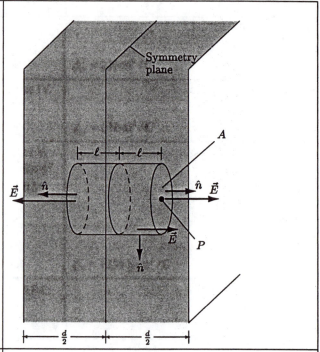 |
| 2.  There are three surfaces to the cylinder.  Start by finding the electric flux through the side of the cylinder.  Because the electric field is perpendicular to $\hat{n}$ everywhere on the side of the cylinder, the electric flux is zero. | $\phi_{\text{side}} = \int \vec{E}_{\text{side}} \cdot \hat{n} \, dA_{\text{side}} = 0$ |
| 3.  Calculate the electric flux on either end of the cylinder.  The electric field is parallel to $\hat{n}$ everywhere on the end of the cylinder.  The electric field is also uniform everywhere on the end of the cylinder, so the result is simply the electric field times the area.  By examining the sketch, you can tell that the flux through each end will be the same. | $\phi_{\text{left}} = \int \vec{E}_{\text{left}} \cdot \hat{n} \, dA_{\text{left}} = \int E \, dA_{\text{end}} = E \int dA = EA_{\text{end}}$<br><br>$\phi_{\text{right}} = \phi_{\text{left}} = EA_{\text{end}}$ |

| 4.  The total flux out of the gaussian cylinder is the sum of the flux through each of the sides. | $\phi_{net} = \phi_{left} + \phi_{right} + \phi_{side} = 2EA$ |
|---|---|
| 5.  Find the total charge enclosed by the cylindrical gaussian surface. Because the charge density is uniform, this will reduce to the charge density times the volume of the cylinder. | $Q_{inside} = \int \rho \, dV = \rho \int dV = \rho V = \rho A(2\ell)$ |
| 6.  Put Gauss's law all together to solve for the electric field inside the slab. | $\phi_{net} = Q_{inside} / \varepsilon_0$ <br> $2EA = 2\rho A\ell / \varepsilon_0$ <br> $E_{inside} = \rho\ell / \varepsilon_0$ |
| 7.  To determine the electric field everywhere outside the slab, we follow the same procedure, only now we extend the gaussian cylinder so the endcaps are outside the slab, as shown. | |
| 8.  The electric field still has the same symmetry, so the same arguments can be made to calculate the electric flux through each of the surfaces. | $\phi_{net} = \phi_{left} + \phi_{side} + \phi_{right} = Q_{inside} / \varepsilon_0$ <br> $\quad = EA + 0 + EA = 2EA$ |

| 9. The gaussian cylinder, however, is not completely filled with charge. There is only charge present in the thickness of the slab, so instead of using $2\ell$ for the length of the cylinder, we need to use $d$ when calculating the total enclosed charge. | $Q_{\text{inside}} = \rho A d$<br><br>$\phi_{\text{net}} = 2EA = \rho A d / \varepsilon_0$<br><br>$E_{\text{outside}} = \rho d / (2\varepsilon_0)$ |
| --- | --- |

**Example #10—Interactive.**   A sphere of radius $a$ has uniform volume charge density $\rho$. Consider a gaussian surface shown in Figure 22-5, consisting of a circular disk and a hemisphere, both of radius $b$, $b > a$, that are concentric with the charged sphere and together form a closed surface. What is the flux out of each segment of the gaussian surface? That is, what is the flux out of the disk and what is the flux out of the hemisphere? Show that the net flux out of the gaussian surface equals the charge inside it divided by $\varepsilon_0$.

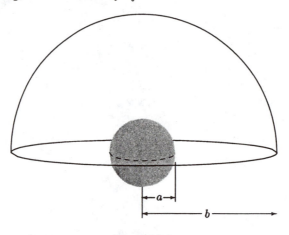

Figure 22-5

**Picture the Problem.**  Use the spherical symmetry of the charge distribution to simplify the flux calculations through each of the surfaces. **Try it yourself.** Work the problem on your own, in the spaces provided, to get the final answer.

| 1. Use symmetry arguments to determine $E_n$ everywhere on the surface of the circle. The electric field outside a uniformly charged solid sphere with charge $Q$ is the same as that of a point charge $Q$ located at the center of the sphere. Use this to calculate the flux through the disk. | |
| --- | --- |
| | $\phi_{\text{disk}} = 0$ |

| 2.  Use the spherical symmetry to determine $E_n$ everywhere on the hemisphere, and use this to calculate the flux through the hemisphere. | $\phi_{\text{hemisphere}} = \dfrac{2\pi a^3 \rho}{3\varepsilon_0}$ |
|---|---|
| 3.  Determine the total flux through the closed surface.  The charge inside the surface is half the total charge of the sphere. | $\phi_{\text{total}} = \dfrac{2\pi a^3 \rho}{3\varepsilon_0}$ |

**Example #11.**  At all points just outside the surface of a 2.00-cm-diameter steel ball bearing, there exists an electric field of magnitude 400 N/C.  Assuming the ball bearing is in electrostatic equilibrium, what is the total charge on the ball?

**Picture the Problem.**  Use Gauss's law to find the total charge of the ball.

| 1.  Imagine a spherical gaussian surface concentric with the ball bearing, with a radius just outside the surface of the ball bearing.  Because of the spherical symmetry, the electric field will be everywhere normal to the gaussian sphere and will have the same magnitude everywhere on the sphere, allowing you to calculate the electric flux through the sphere. | $\phi = \oint \vec{E} \cdot \hat{n}\, dA = \oint E_n\, dA = E_n \oint dA = E_n A$ <br> $= E_n 4\pi r^2$ |
|---|---|
| 2.  Knowing the magnitude of the normal component of the electric field, you can use Gauss's law to calculate the charge on the ball, which is the same as the charge inside the surface.  The problem does not tell us the direction of the electric field, so we do not know whether the charge on the ball bearing is positive or negative. | $E_n 4\pi r^2 = Q_{\text{inside}} / \varepsilon_0$ <br> $Q_{\text{inside}} = E_n 4\pi r^2 \varepsilon_0$ <br> $= (400\,\text{N/C})\, 4\pi (0.01\,\text{m})^2 \left[ 8.85 \times 10^{-12}\,\text{C}^2 / \left( \text{N} \cdot \text{m}^2 \right) \right]$ <br> $= 4.45 \times 10^{-12}\,\text{C}$ |

**Example #12—Interactive.** The electric field in the atmosphere at the surface of the earth is 150 N/C downward. (*a*) What is the total charge on the surface of $1\,m^2$ of level ground? (Assume the earth to be a perfect conductor.) (*b*) What is the total charge on the surface of the earth?

**Picture the Problem.** Knowing the normal component of the electric field, use Gauss's law to find the total charge enclosed by a spherical gaussian surface located just above the surface of the earth. You may need to look up the radius of the earth. **Try it yourself.** Work the problem on your own, in the spaces provided, to get the final answer.

| | |
|---|---|
| 1. Imagine a spherical gaussian surface concentric with the earth, with a radius just outside the surface of the earth. Because of the spherical symmetry, the electric field will be everywhere normal to the gaussian sphere and will have the same magnitude everywhere on the sphere, allowing you to calculate the electric flux through the sphere. | |
| 2. Knowing the magnitude of the normal component of the electric field, you can use Gauss's law to calculate the charge on the earth, which is the same as the charge inside the surface. We know the direction of the electric field, so the sign of the charge can also be determined. | $Q = -6.77 \times 10^5\,C$ |
| 3. From the total charge and the surface area of the earth, calculate the surface charge density of the earth. | $\sigma = -1.32 \times 10^{-9}\,C/m^2$ |

# Chapter **23**

# Electric Potential

## I. Key Ideas

Electrostatic forces are conservative. Thus, changes in potential energy are associated with the work done by these forces. Charges of opposite sign attract each other; so when they are released from rest, they accelerate toward each other, gaining kinetic energy and losing potential energy. The farther apart they are, the *more* potential energy the system of two charges has. Conversely, charges of like sign repel each other; so when they are released from rest, they accelerate away from each other, gaining kinetic energy and losing potential energy. The farther apart they are, the *less* potential energy the system of two charges has.

*Section 23-1.  Potential Difference.* When a conservative force $\vec{F}$ acts on a particle that undergoes a displacement $d\vec{\ell}$, the associated change in potential energy $dU$ is $dU = -\vec{F} \cdot d\vec{\ell}$. The force exerted on a test charge $q_0$ is $\vec{F} = q_0 \vec{E}$; so when the test charge undergoes a displacement $d\vec{\ell}$, the change in electrostatic potential energy is

$$dU = -q_0 \vec{E} \cdot d\vec{\ell}$$

If we divide both sides of this equation by $q_0$, we obtain

$$dV = -\vec{E} \cdot d\vec{\ell} \qquad\qquad \text{Increment of potential difference}$$

Here $dV = dU / q_0$ is defined as the change in electrostatic potential energy per unit charge, which we call the **potential difference.** When a charge undergoes a finite displacement from location $a$ to location $b$, the change in electrostatic potential energy is

$$\Delta U = U_b - U_a = \int_a^b dU = -\int_a^b q_0 \vec{E} \cdot d\vec{\ell} \qquad\qquad \text{Change in electrostatic potential energy}$$

Dividing through this equation by $q_0$, we obtain

$$\Delta V = V_b - V_a = \frac{\Delta U}{q_0} = -\int_a^b \vec{E} \cdot d\vec{\ell} \qquad\qquad \text{Potential difference}$$

where $\Delta V$ is the potential difference of point $b$ relative to point $a$.

*Potential difference defined*

The potential difference $V_b - V_a$ is the negative of the work per unit charge done by the electric field on a positive test charge when the charge moves from point *a* to point *b*.

The potential function $V$ is called the **electric potential** (or sometimes just the **potential**). The SI unit for potential and potential difference is the joule per coulomb, called the volt (V). Because potential difference is measured in volts, it is sometimes called **voltage.**

If the electric potential and the electrical potential energy are chosen to be zero at the same point in space, then they are related according to

$$U = q_0 V \qquad\qquad \text{Relationship between potential energy and potential}$$

When a small stone is released from rest, it accelerates in the direction of the gravitational field. As it falls, its potential energy decreases as its kinetic energy increases. Analogously, when a positively charged particle is released from rest, it accelerates in the direction of the electric field. As it accelerates, its kinetic energy increases as its potential energy decreases. Because the change in potential is the change in potential energy per unit charge, the particle accelerates in the direction of decreasing potential, so the *electric field lines point in the direction of decreasing electric potential.*

When an electron accelerates through a potential difference of 1.00 V, its kinetic energy increases by $1.60 \times 10^{-19}$ J or one **electron volt** (eV).

***Section 23-2. Potential due to a System of Point Charges.*** The electric potential due to a point charge at the origin can be calculated from the electric field, which is given by

$$\vec{E} = \frac{kq}{r^2}\hat{r}$$

If the test charge $q_0$ at a distance *r* from a particle with charge *q*, shown in Figure 23-1, is given a displacement $d\vec{\ell}$, the change in its potential is

$$dV = -\vec{E}\cdot d\vec{\ell} = -\frac{kq}{r^2}\hat{r}\cdot d\vec{\ell} = -\frac{kq}{r^2}dr$$

where $\hat{r}\cdot d\vec{\ell} = d\ell\cos\phi = dr$. Integrating, we obtain

$$V = -\int\frac{kq}{r^2}dr = \frac{kq}{r} + V_0$$

where $V_0$ is the integration constant.

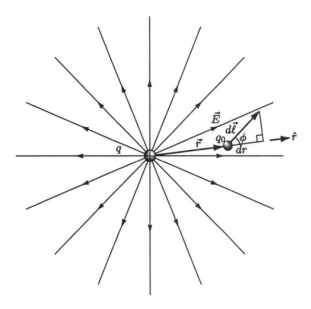

Figure 23-1

It is customary to define the potential to be zero at an infinite distance from the point charge $q$ (that is, at $r = \infty$). Then the constant $V_0 = 0$, and the potential at a distance $r$ from the point charge is

$$V = \frac{kq}{r}$$    Potential for a point charge

The electrostatic potential at a point $P$ due to a system of point charges $q_i$ is the sum of the potentials associated with each charge. That is,

$$V = \frac{kq_1}{r_{10}} + \frac{kq_2}{r_{20}} + \ldots = \sum_i \frac{kq_1}{r_{i0}}$$    Potential for a system of point charges

where $r_{i0}$ is the distance from the $i$th charge to the field point $P$.

***Section 23-3. Computing the Electric Field from the Potential.*** The electric field lines point in the direction of *decreasing* potential. If the potential is known, it can be used to calculate the electric field $\vec{E}$. Consider an arbitrary displacement $d\vec{\ell}$ in an electric field $\vec{E}$. The change in potential is $dV = -\vec{E}\cdot d\vec{\ell} = -E_\ell d\ell$, where $E_\ell$ is the component of the electric field $\vec{E}$ in the direction of the displacement. If we divide by $d\ell$, we have

$$E_\ell = -\frac{dV}{d\ell}$$

If the displacement $d\vec{\ell}$ is perpendicular to the electric field, the potential does not change. A surface on which the potential does not change is called an **equipotential surface**. The greatest change in $V$ occurs when the displacement is either parallel or antiparallel to the field.

If we express the electric field and displacement vectors in terms of their components, we have

$$dV = -\left(E_x\hat{i} + E_y\hat{j} + E_z\hat{k}\right)\cdot\left(dx\hat{i} + dy\hat{j} + dz\hat{k}\right)$$
$$= -E_x dx - E_y dy - E_z dz$$

Therefore,

$$E_x = -\frac{\partial V}{\partial x}, \; E_y = -\frac{\partial V}{\partial y}, \; \text{and } E_z = -\frac{\partial V}{\partial z}$$

or

$$\vec{E} = E_x\hat{i} + E_y\hat{j} + E_z\hat{k}$$
$$= -\left(\frac{\partial V}{\partial x}\hat{i} + \frac{\partial V}{\partial y}\hat{j} + \frac{\partial V}{\partial z}\hat{k}\right) \equiv -\vec{\nabla}V \qquad \text{The gradient of the potential}$$

where the vector $\vec{\nabla}V$ is called the gradient of the potential.

For any charge distribution with point symmetry the potential varies only with $r$ and the electric field is related to the potential by

$$\vec{E} = -\vec{\nabla}V = -\frac{\partial V}{\partial r}\hat{r} \qquad \text{The gradient for charge distributions with point symmetry}$$

*Section 23-4. Calculations of V for Continuous Charge Distributions.* For a continuous charge distribution the potential $V$ at a point is equal to the sum of the potentials due to all elements of charge $dq$. Thus, for a charge distribution that is confined to a finite region of space, the potential is

$$V = \int_Q \frac{k\,dq}{r} \qquad \text{Potential function for a continuous charge distribution}$$

In the text, expressions for the electrical potentials due to a uniformly charged ring and disk are determined by evaluating this integral. The resulting expressions are

$$V = \frac{kQ}{\sqrt{x^2 + a^2}} \qquad \text{V on the axis of a charged ring}$$

$$V = 2\pi k\sigma\left[\left(x^2 + R^2\right)^2 - |x|\right] \qquad \text{V on the axis of a uniformly charged disk}$$

Calculating the potential for charge distributions that are not confined to a finite region of space (such as a uniformly charged infinite plane) requires a different approach. For such charge distributions the potential cannot be zero at infinity. To calculate the potential for these distributions, we first find the electric field and then use the relation

$$V - V_0 = -\int_{P_0}^{r} \vec{E} \cdot d\vec{\ell}$$

where $P_0$ is the reference point at which the potential is $V_0$ and $P$ is the field point. This method is very general (it also works for charge distributions that do not extend to infinity). Also, this definition makes it clear that the potential function is continuous.

In the text, expressions for the electric potentials due to several extended charge distributions are determined by evaluating this integral. The resulting expressions follow.

$$V(x) = V_0 - 2\pi k \sigma |x|$$    V due to a uniformly charged infinite plane

$$V = \frac{kQ}{R} \quad r \leq R$$

$$V = \frac{kQ}{r} \quad r \geq R$$    V inside and outside a uniformly charged spherical shell

$$V = -2k\lambda \ln \frac{r}{a}$$    V due to a uniformly charged infinite line

where $a$ is an arbitrary distance from the line. Because the logarithm of one is zero, when $r = a$, $V = 0$.

***Section 23-5. Equipotential Surfaces.*** All points on any surface which is perpendicular to electric field lines are at the same potential. We call such surfaces **equipotential surfaces.** When a charge moves on such a surface, the electric field does no work on it. Thus the entire body of a conductor in electrostatic equilibrium is an equipotential volume (or just an "equipotential").

Conductors not in contact with one another are usually not at the same potential, and the potential difference between them depends on their sizes, shapes, location, relative orientation, and charges. When conductors are brought into contact, charge is transferred between them as the charge distributions rearrange to reestablish electrostatic equilibrium. In equilibrium, conductors touching each other constitute a single equipotential region. The transfer of charge from one conductor to another during contact is called charge sharing. Consider the special case of a conductor that is completely enclosed by a larger conductor. If the smaller conductor touches the inside surface of the outer conductor, all the charge that was on the inner conductor flows to the outer surface of the outer conductor. Repetition of this process is what charges a Van de Graaf generator to a high potential.

The total amount of charge that can be transferred to a conductor in this way is limited only by **dielectric breakdown** of the air or whatever insulating material surrounds the outer surface of the outer conductor. Dielectric breakdown occurs when an electric field is strong enough to accelerate ions sufficiently to ionize the molecules with which they collide. Dielectric breakdown occurs in dry air when the magnitude of the electric field reaches $E_{max} \approx 3 \times 10^6$ V/m. The electric field strength at which dielectric breakdown occurs for a particular material is called the **dielectric strength** of the material. The electric discharge through the air resulting from dielectric breakdown is called **arc discharge.** Charge tends to concentrate on sharp points or corners on the surface of a conductor where the radius of curvature is very small. Thus, an electric field is strongest just outside the regions of the surface of a conductor where the radius of curvature of the surface is least; so dielectric breakdown occurs most frequently at sharply pointed regions on the surface of a conductor.

## II.   Physical Quantities and Key Equations

### Physical Quantities

volt (V)                               $1 \text{ V} = 1 \text{ J/C}$

electron volt (eV)                     $1 \text{ eV} = 1.60 \times 10^{-19} \text{ J}$

dielectric strength of air             ~3 MV/m

### Key Equations

*Change in electrostatic potential energy*     $\Delta U = U_b - U_a = \int_a^b dU = -\int_a^b q_0 \vec{E} \cdot d\vec{\ell}$

*Increment of potential difference*            $dV = -\vec{E} \cdot d\vec{\ell}$

*Potential difference*                         $\Delta V = V_b - V_a = \dfrac{\Delta U}{q_0} = -\int_a^b \vec{E} \cdot d\vec{\ell}$

*Relationship between potential energy and potential*   $U = q_0 V$

*Potential for a point charge*                 $V = \dfrac{kq}{r} \ ; \ V = 0 \text{ at } r = \infty$

*Potential due to a system of point charges*   $V = \dfrac{kq_1}{r_{10}} + \dfrac{kq_2}{r_{20}} + \ldots = \sum_i \dfrac{kq_1}{r_{i0}}$

*Potential function for a finite charge distribution*   $V = \int_Q \dfrac{k \, dq}{r}$

*V on the axis of a charged ring*              $V = \dfrac{kQ}{\sqrt{x^2 + a^2}}$

*V on the axis of a uniformly charged disk*    $V = 2\pi k\sigma \left[ \left( x^2 + R^2 \right)^{1/2} - |x| \right]$

*V due to an infinite plane of charge: continuity of V*   $V(x) = V_0 - 2\pi k\sigma |x|$

*V inside and outside a uniformly charged spherical shell of charge*
$$V = \frac{kQ}{R} \quad r \le R$$
$$V = \frac{kQ}{r} \quad r \ge R$$

*V due to a uniformly charged infinite line charge*    $V = -2k\lambda \ln\frac{r}{a}$

*The gradient of the potential*    $\vec{E} = -\vec{\nabla} V = -\left( \frac{\partial V}{\partial x}\hat{i} + \frac{\partial V}{\partial y}\hat{j} + \frac{\partial V}{\partial z}\hat{k} \right)$

*The gradient for charge distributions with point symmetry*    $\vec{E} = -\vec{\nabla} V = -\frac{\partial V}{\partial r}\hat{r}$

## III. Potential Pitfalls

Remember that only differences in electrical potential energy, and in electrical potential, are measurable. The position at which the potential is zero is arbitrary, although, as a convention, we usually take the potential of a finite charge distribution to be zero at infinite distance.

Potential is a scalar quantity; it has no direction. Don't confuse it with the electric field, which is a vector; they're different quantities.

The electrostatic force is always in the direction of decreasing potential energy, whereas the electric field is always in the direction of decreasing potential. These directions are not necessarily the same because the electrostatic force on a negative charge is in a direction opposite that of the electric field.

The electrostatic force on a point charge is in the direction of the electric field only if the charge is positive.

## IV. True or False Questions and Responses

**True or False**

_____ 1.  Charges of like sign have more electrostatic potential energy when they are farther apart.

_____ 2.  Electric field lines point in the direction of decreasing potential.

_____ 3.  The volt is a unit of power.

_____ 4.  The term *voltage* refers to a potential difference.

_____ 5.  The electron volt is a unit of energy.

_____ 6.  Electric potential is a scalar quantity.

_____ 7. The potential difference $V_b - V_a$ between $a$ and $b$ is the work per unit charge done by the electric field on a test charge when it moves from $a$ to $b$.

_____ 8. If the expression used for the potential of a point charge $q$ is $kq/r$, this potential approaches zero as $r$ approaches infinity.

_____ 9. Like the electric field, the potential function is discontinuous at a surface charge distribution.

_____ 10. The electric field equals the *negative* of the gradient of the potential.

_____ 11. The surface charge density on a conducting surface is larger where the radius of curvature of the surface is smaller, and vice versa.

_____ 12. Dielectric breakdown in air occurs when the electric field is strong enough to ionize the air molecules.

**Responses to True or False**

1. False. Charges of like sign repel each other. If they are released from rest, they fly apart, their kinetic energy increases and their potential energy decreases. The statement is true for charges of opposite sign.

2. True.

3. False. A volt is a joule per coulomb, which is energy per unit charge. Power is energy per unit of time.

4. True.

5. True, it equals $1.6 \times 10^{-19}$ J. Can you recognize where the numerical value $1.6 \times 10^{-19}$ comes from?

6. True.

7. False. The potential difference is the *negative* of the work per unit charge done by the electric field on a positive test charge when it moves from $a$ to $b$. Imagine moving a positive test charge from a lower $V_a$ to a higher $V_b$. You have to do positive work on it. As it moves from $a$ to $b$, the electric field does negative work on it.

8. True.

9. False. The potential function is always continuous; but its spatial derivatives, and thus $\vec{E}$, can be discontinuous.

10. True.

11. True.

12. True.

# V. Questions and Answers

**Questions**

1. An electron is released from rest in an electric field. Will it move in the direction of increasing or decreasing potential?

2. Is the electric potential at a location associated with a charge at the location or the location itself?

3. In a region where there is an electric field, two nearby points are at the same potential. What is the angle between the line joining the points and the direction of the electric field?

4. If the electric field is zero throughout some volume of space, what can you say about the potential in that volume?

**Answers**

1. Any negatively charged particle, including an electron, will move in a direction opposite to that of the electric field. The electric field always points in the direction of *decreasing* potential, so the electron will move in the direction of *increasing* potential.

2. The electric potential is associated with the location. It is the ratio of electrostatic potential energy to charge that a test charge at the location would have—if a test charge were at the location.

3. The electric field points in the direction of decreasing potential. For the two points to be at the same potential, the line joining the points must be perpendicular to the direction of the electric field.

4. Since no work can be done by a nonexistent field, the potential must have the same value throughout the volume. Any difference in potential equals the negative of the work per unit charge done by the electric field.

# VI. Problems, Solutions, and Answers

**Example #1.** A uniform electric field of $2.00\,\text{kN/C}$ is in the positive-$x$ direction. (*a*) What is the potential difference $V_b - V_a$ when point $a$ is at $x = -30.0\,\text{cm}$ and point $b$ is at $x = +50.0\,\text{cm}$? (*b*) A test charge $q_0 = +2.00\,\text{nC}$ is released from rest at point $a$. What is its kinetic energy when it passes through point $b$?

**Picture the Problem.** Use the relationship between potential and electric field to calculate the potential difference. The change in the charge's potential energy, which can be calculated from the change in potential, will result in a change in the charge's kinetic energy.

| 1. Sketch the physical situation described. | |
|---|---|
| 2. Calculate the potential difference by relating it to the electric field. | $$V_b - V_a = -\int_a^b \vec{E} \cdot d\vec{\ell} = -\int_{x_a}^{x_b} \left(2 \times 10^3 \text{ N/C} \,\hat{i}\right) \cdot \left(dx \,\hat{i}\right)$$ $$= -\left(2 \times 10^3 \text{ N/C}\right) \int_{x_a}^{x_b} dx = -\left(2 \times 10^3 \text{ N/C}\right) x \Big|_{-0.3\,\text{m}}^{0.5\,\text{m}}$$ $$= -\left(2 \times 10^3 \text{ N/C}\right)\left[(0.5\,\text{m}) - (-0.3\,\text{m})\right]$$ $$= -1.60 \times 10^3 \text{ V}$$ |
| 3. Mechanical energy is conserved. Since the initial kinetic energy is zero, the change in kinetic energy will be equivalent to the final kinetic energy. | $$\Delta U + \Delta K = 0$$ $$\Delta K = K_f - K_i = K_f = -\Delta U = -q_0 \, \Delta V$$ $$= -\left(2 \times 10^{-9} \text{ C}\right)\left(-1.6 \times 10^3 \text{ V}\right)$$ $$= 3.2 \times 10^{-6} \text{ J}$$ |

**Example #2—Interactive.** An electric field is given by $\vec{E} = \left(-3.00 \,\text{kV/m}^2\right) x \,\hat{i}$ on the $x$ axis. (*a*) What is the potential difference $V_b - V_a$ when point *a* is at $x = -30.0$ cm and point *b* is at $x = +50.0$ cm? (*b*) How much work is done by an external agent in bringing a test charge $q_0 = +2.00$ nC from rest at *a* to rest at *b*?

**Picture the Problem.** Use the relationship between potential and electric field to calculate the potential difference. The work done by the external agent, plus the work done by the electric field equals the change in the charge's kinetic energy. **Try it yourself.** Work the problem on your own, in the spaces provided, to get the final answer.

| 1. Sketch the physical situation described. | |
|---|---|
| 2. Calculate the change in electric potential due to the electric field as the charge moves from point *a* to point *b*. | $$\Delta V = 240 \text{ V}$$ |

| 3. Calculate the work that must be done by the external agent. This must be equal in magnitude to the work done by the electric field. | |
|---|---|
| | $W_{\text{ext}} = +4.80 \times 10^{-7} \text{ J}$ |

**Example #3.** Consider a 1.00-m³ cube that has +2.00-$\mu$C charges located at seven of its corners. (a) Find the potential at the vacant corner. (b) How much work by an external agent is required to bring an additional +2.00-$\mu$C charge from rest at infinity to rest at the vacant corner?

**Picture the Problem.** The potential at the remaining point is the sum of the potentials due to each of the individual charges. By the work-energy theorem, the work done by an external force will equal the sum of the change in kinetic and potential energy of the charge.

| 1. Sketch the cube with the charges at its corners. | 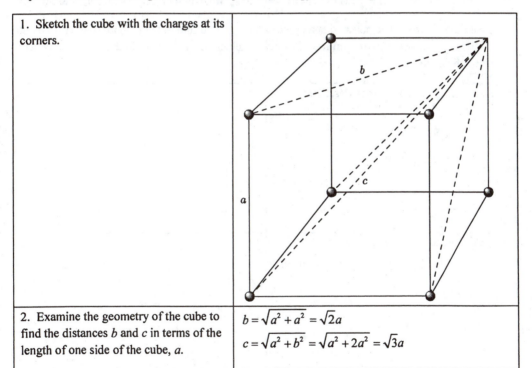 |
|---|---|
| 2. Examine the geometry of the cube to find the distances $b$ and $c$ in terms of the length of one side of the cube, $a$. | $b = \sqrt{a^2 + a^2} = \sqrt{2}a$ <br> $c = \sqrt{a^2 + b^2} = \sqrt{a^2 + 2a^2} = \sqrt{3}a$ |

| 3. The total potential at the remaining corner is the sum of the potential from each of the charges. As can be seen in the figure, there are three charges a distance $a$ from the vacant corner, three charges a distance of $\sqrt{2}a$ from the vacant corner, and one charge a distance of $\sqrt{3}a$ from the corner. | $$V_0 = \sum_i \frac{kq_i}{r_{i0}}$$ $$= 3\frac{kq}{a} + 3\frac{kq}{\sqrt{2}a} + \frac{kq}{\sqrt{3}a} = \frac{kq}{a}\left(3 + \frac{3}{\sqrt{2}} + \frac{1}{\sqrt{3}}\right)$$ $$= \frac{\left(8.99 \times 10^9 \text{ N} \cdot \text{m}^2/\text{C}^2\right)\left(2 \times 10^{-6} \text{ C}\right)}{1\,\text{m}}\left(3 + \frac{3}{\sqrt{2}} + \frac{1}{\sqrt{3}}\right)$$ $$= 1.02 \times 10^5 \text{ V}$$ |
|---|---|
| 4. The work done by an external agent will equal the total change in energy of the external particle as it moves from infinitely far away to the remaining corner. | $$W_{ext} = \Delta K + \Delta U = \Delta U$$ $$= q\Delta V = q\left(V - V_\infty\right)$$ $$= \left(2.00 \times 10^{-6} \text{ C}\right)\left(1.02 \times 10^5 \text{ V} - 0 \text{ V}\right)$$ $$= 0.204 \text{ J}$$ |

**Example #4—Interactive.** A charge of $+2.00\,\mu\text{C}$ is at the origin, and a charge of $-3.00\,\mu\text{C}$ is on the $y$ axis at $y = 40.0\,\text{cm}$. (a) What is the potential at point $a$, which is on the $x$ axis at $x = 40.0\,\text{cm}$? (b) What is the potential difference $V_b - V_a$ when point $b$ is at $(40.0\,\text{cm}, 30.0\,\text{cm})$?

**Picture the Problem.** The potential at a point is the algebraic sum of the potentials due to each of the individual charges. **Try it yourself.** Work the problem on your own, in the spaces provided, to get the final answer.

| 1. Sketch the physical situation described. | |
|---|---|
| 2. Determine the potential at point $a$. | $V_a = -2730 \text{ V}$ |

| 3. Determine the potential at point $b$. | |
|---|---|
| 4. Determine the potential difference. | $V_b - V_a = -26,700\,\text{V}$ |

**Example #5.** The $x$ axis coincides with the symmetry axis of a uniformly charged thin disk with radius $R$ and uniform surface charge density $\sigma$ centered at the origin. ($a$) Make a sketch of $E_x$ versus $x$ for $-4R < x < +4R$. ($b$) Make a sketch of $V(x)$ versus $x$ for $-4R < x < +4R$.

**Picture the Problem.** Use the expressions for the electric field and potential on the axis of a uniformly charged disk.

| | |
|---|---|
| 1. Recall the equation for the magnitude $E$ of the electric field on the axis of a uniformly charged disk. | $E = +2\pi k\sigma\left(1 - \dfrac{\lvert x\rvert}{\sqrt{x^2 + R^2}}\right)$ |
| 2. Sketch $E_x(x)$. The electric field is directed away from the disk on both sides of it. Thus $E_x$ is positive for $x > 0$ and negative for $x > 0$. | 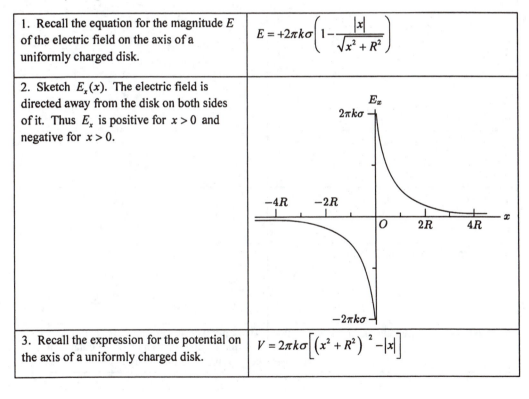 |
| 3. Recall the expression for the potential on the axis of a uniformly charged disk. | $V = 2\pi k\sigma\left[\left(x^2 + R^2\right)^2 - \lvert x\rvert\right]$ |

| | |
|---|---|
| 4.  Sketch $V(x)$.  This function is continuous and always positive. | 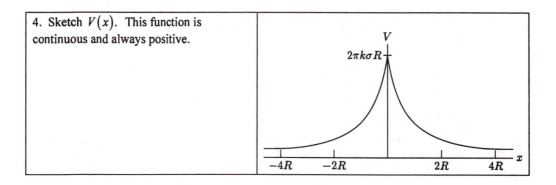 |

**Example #6—Interactive.**  An infinitely long line charge of linear charge density 2.00 nC/m  lies on the $z$ axis.  (*a*) Find the potential difference  $V_c - V_b$  when point $b$ is at  $(40.0\,\text{cm}, 30\,\text{cm}, 0)$  and point $c$ is at  $(200\,\text{cm}, 0, 50.0\,\text{cm})$.  Does your answer depend on your choice of the reference point at which $V = 0$?  (*b*) What are the equipotential surfaces?

**Picture the Problem.**  Use the expression for the potential of a long straight-line charge.  **Try it yourself.**  Work the problem on your own, in the spaces provided, to get the final answer.

| | |
|---|---|
| 1.  Sketch the physical situation described. | |
| 2.  Calculate $V_c$. | |
| 3.  Calculate $V_b$. | |
| 4.  Calculate the potential difference. | $V_c - V_b = -49.9\,\text{V}$ |
| 5.  Determine the equipotential surfaces. | The equipotential surfaces are equidistant from the line of charge.  They are circular cylinders coaxial with the line charge. |

**Example #7.** A uniformly charged sphere has radius $a$ and total charge $Q$. (a) Find the potential $V(r)$ for $0 < r < \infty$. (b) Use your result from part (a) to find an expression for the electric field $\vec{E}$ everywhere.

**Picture the Problem.** Consider the uniformly charged sphere to be a collection of uniformly charged thin spherical shells. Integrate the expression for the potential of a uniformly charged thin spherical shell to obtain the expression for the potential of a uniformly charged sphere. The electric field is the negative gradient of the potential function.

| 1. Sketch the sphere. | |
|---|---|
| | 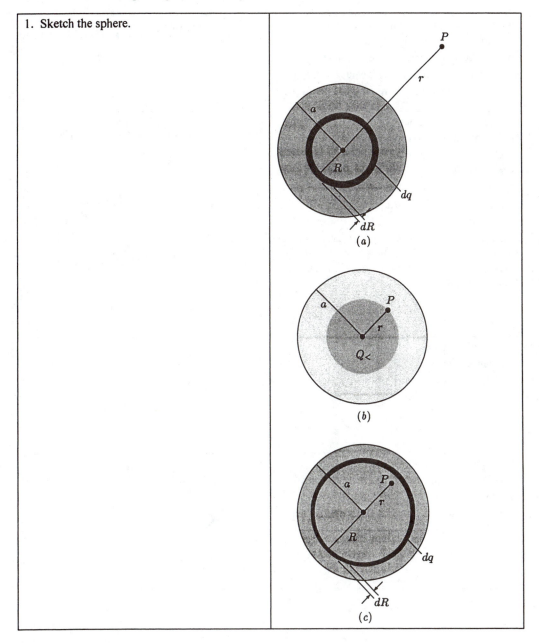 |

| | |
|---|---|
| 2.  Using the expression for the electric potential of a shell of radius $r$, and total charge $Q$, find an expression for the potential $dV$ due to a charge $dq$. | $V = \dfrac{kq}{r}$ <br><br> $dV = \dfrac{k\,dq}{r}$ |
| 3.  For $r > a$, neither $k$ nor $r$ depend on the charge $dq$. Here $Q = \frac{4}{3}\pi a^3 \rho$ is the total charge of the sphere. | $V = \int dV = \displaystyle\int_0^Q \dfrac{k\,dq}{r} = \dfrac{kQ}{r} \qquad r > a$ |
| 4.  To find the potential inside the sphere, we first determine the electric potential $V_<$ due to the charge within some radius $r$.  Since we are only considering the charge inside radius $r$ at this stage, we can use the result from step 3. | $V_< = \dfrac{kQ_<}{r} = \dfrac{k\left(\frac{4}{3}\right)\pi r^3 \rho}{r} = \dfrac{4}{3}\pi k\rho r^2 = \dfrac{kQ}{a}\left(\dfrac{r}{a}\right)^2$ |
| 5.  Now we need to determine the potential at a radius of $r$ due to all the charge at a radius of $R$, such that $r < R < a$ as shown in part (c) of the sketch of step 1.  The charge $dq$ resides in a spherical shell with a radius of $R$ and a thickness $dR$ and hence a total volume of $4\pi R^2 \, dR$. | $dV_> = \dfrac{k\,dq}{R} = \dfrac{k\left(4\pi\rho R^2\,dR\right)}{R} = 4\pi k\rho R\,dR$ <br><br> $V_> = \displaystyle\int_r^a 4\pi k\rho R\,dR = 4\pi k\rho \int_r^a R\,dR$ <br><br> $= 2\pi k\rho\left(a^2 - r^2\right) = \dfrac{3}{2}\dfrac{kQ}{a}\left[1 - \left(\dfrac{r}{a}\right)^2\right]$ |
| 6.  The total potential at some radius $r < a$ is then the sum of the partial potentials calculated in steps 4 and 5. | $V = V_< + V_> = \dfrac{kQ}{a}\left(\dfrac{r}{a}\right)^2 + \dfrac{3}{2}\dfrac{kQ}{a}\left[1 - \left(\dfrac{r}{a}\right)^2\right]$ <br><br> $= \dfrac{1}{2}\dfrac{kQ}{a}\left[3 - \left(\dfrac{r}{a}\right)^2\right] \qquad r < a$ |
| 7.  The electric field is found by taking the gradient of the electric potential.  We will start with the field outside the sphere. | $\vec{E} = -\vec{\nabla}V = -\dfrac{dV}{dr}\hat{r}$ <br><br> $\vec{E} = -\dfrac{d}{dr}\left(\dfrac{kQ}{r}\right)\hat{r} = -kQ\dfrac{d\left(r^{-1}\right)}{dr}\hat{r}$ <br><br> $= \dfrac{kQ}{r^2}\hat{r} \qquad r > a$ |
| 8.  Repeat the process for radii inside the sphere.  The expressions for the electric field found here agree with those in Chapter 22 of the text, demonstrating the consistency of our definitions of electric field and potential. | $\vec{E} = -\dfrac{d}{dr}\left(\dfrac{1}{2}\dfrac{kQ}{a}\left[3 - \left(\dfrac{r}{a}\right)^2\right]\right)\hat{r}$ <br><br> $= -\dfrac{1}{2}\dfrac{kQ}{a}\dfrac{d\left[3 - (r/a)^2\right]}{dr}\hat{r}$ <br><br> $= \dfrac{1}{2}\dfrac{kQ}{a^3}\dfrac{d\left(r^2\right)}{dr}\hat{r} = \dfrac{kQr}{a^3}\hat{r} \qquad r < a$ |

**Example #8—Interactive.** A charge $q_1 = +2q$ is at the origin, and a charge $q_2 = -q$ is on the $x$ axis at $x = a$. (a) Find an expression for the electric potential $V(x)$ on the $x$ axis in the region for $0 < x < a$. (b) Use your result from part (a) to find an expression for the electric field $\vec{E}$ in the same region. Compare your result with the solution to Example #11(b) in Chapter 21 of this Study Guide.

**Picture the Problem.** The potential at a point is the algebraic sum of the potentials due to each of the individual charges. The electric field is the negative of the gradient of the potential. **Try it yourself.** Work the problem on your own, in the spaces provided, to get the final answer.

| | |
|---|---|
| 1. Draw a sketch of the two charges on the $x$ axis. | |
| 2. Determine the potential at an arbitrary coordinate $x$ due to the charge at the origin. | |
| 3. Determine the potential at an arbitrary coordinate $x$ due to the charge at $x = a$. | |
| 4. The potential at any point an the $x$ axis is the sum of the potentials found in steps 2 and 3. | $V = kq\left(\dfrac{2}{x} - \dfrac{1}{a-x}\right)$    $0 < x < a$ |
| 5. The electric field in the same region is the negative gradient of the potential. | $\vec{E} = kq\left(\dfrac{2}{x^2} + \dfrac{1}{(a-x)^2}\right)\hat{i}$    $0 < x < a$ |

**Example #9.** An isolated conducting sphere is to be charged to 100 kV. What is the smallest radius the sphere can have if its electric field is not to exceed the dielectric strength of air?

**Picture the Problem.** All the charge on a conducting sphere resides on its surface. Obtain expressions for both the electric field strength and electric potential. Use these expressions to find an expression for the electric field of a charged conducting sphere in terms of its potential and radius. Solve for $R$.

| 1. Use the expression for the electric field outside a uniformly charged shell. The electric field is greatest right at the surface of the shell. | $\vec{E} = \dfrac{kQ}{r^2}\hat{r}$     $r > R$ <br> $E = \dfrac{kQ}{R^2}$     $r = R$ |
|---|---|
| 2. Do the same for the electric potential just at the surface of the sphere. | $V = \dfrac{kQ}{r}$     $r > R$ <br> $V = \dfrac{kQ}{R}$     $r = R$ |
| 3. Solve the result of step 2 for $kQ$ and substitute into the result from step 1, and solve for $R$. Use the dielectric breakdown strength of air for $E$. | $kQ = VR$ <br> $E = \dfrac{VR}{R^2}$ <br> $R = \dfrac{V}{E} = \dfrac{1 \times 10^5 \text{ V}}{3 \times 10^6 \text{ V/m}} = 3.33 \text{ cm}$ |

**Example #10—Interactive.**   Find the maximum surface charge density $\sigma_{max}$ that a plane conducting surface in air can sustain. **Try it yourself.** Work the problem on your own, in the spaces provided, to get the final answer.

| 1. Draw a sketch of a segment of a conducting surface. | |
|---|---|
| 2. Use Gauss's law to obtain an expression for the electric field just outside the surface in terms of the surface charge density. | |
| 3. The maximum electric field strength equals the dielectric strength of air. | $\sigma_{max} = 2.66 \times 10^{-5} \text{ C/m}^2$ |

# Chapter **24**

# **Electrostatic Energy and Capacitance**

## I.  Key Ideas

A **capacitor** is a practical device for storing charge and electrostatic potential energy. It consists of two separated conductors that carry charges that are equal in magnitude but opposite in sign. The region between the conductors can be empty or can be filled with an insulating material.

*Section 24-1. Electrostatic Potential Energy.* The **electrostatic potential energy** of a system of point charges is the work needed to bring the charges from an infinite separation to their final positions. The work being referred to here is not the work done by the electric field but the work done by external agents. Since the charges start at rest and finish at rest, the net work done is zero, so the work done by external agents equals the negative of the work done by the electric field.

To calculate the electrostatic potential energy of a system of point charges, we start with one of the charges in its final position and then calculate the work required to bring a second charge from infinity to its final position. Then we calculate the work required to bring a third charge from infinity to its final position in the resultant electric field of the first two charges, and so forth. Thus, for a system of charges the electrostatic potential energy is

$$U = q_2 \frac{kq_1}{r_{1,2}} + q_3 \left( \frac{kq_1}{r_{1,3}} + \frac{kq_2}{r_{2,3}} \right) + \ldots = \frac{1}{2} \sum_{i=1}^{n} q_i V_i$$

Electrostatic potential energy of a system of point charges

where $V_i$ is the potential at the location of the $i$th charge due to all of the other charges. This formula is also valid for a system of charged conductors. If we have a set of $n$ conductors with the $i$th conductor at potential $V_i$ and carrying a charge $Q_i$, the electrostatic potential energy is $\frac{1}{2}\Sigma Q_i V_i$.

*Section 24-2.  Capacitance.* The **capacitance** of an air-filled capacitor depends only upon its geometry.  Capacitance is the measure of how much charge a capacitor can store for a given potential difference (coulombs stored per volt).

A capacitor consists of any two conductors that are often referred to as "plates" even when they are not flat plates. In practice, charge is transferred from one plate to the other by a charging device such as a battery. (A battery is a device that supplies electrical energy. An ideal battery maintains a constant potential difference between its terminals.) The amount of charge that accumulates on the capacitor plates depends on the potential difference produced by the charging device, the geometry of the capacitor, and the choice of insulating material. Let $Q$ denote the *magnitude* of the charge on either plate, and let $V$ be the potential difference between the plates. (When we speak of the charge on a capacitor, we mean the magnitude of the charge on either plate. Also, it is conventional to use $V$ for the *magnitude* of the potential difference $\Delta V$ between the plates.) The ratio of $Q/V$ defines the capacitance $C$:

$$C = \frac{Q}{V}$$

Capacitance defined

The SI unit of capacitance is the **farad** (F), thus $1\,\text{F} = 1\,\text{C/V}$. The capacitance of common capacitors is usually given in microfarads ($\mu$F) or picofarads (pF).

The electric field at a point between the plates of a capacitor is proportional to the charge on the capacitor because the potential difference $V = \left| -\int \vec{E} \cdot d\vec{\ell} \right|$ is proportional to the electric field, and because the electric field is in turn proportional to the charge. Thus, the capacitance $\left( C = Q/V \right)$ does not depend on either $Q$ or $V$; that is, if $Q$ increases, $V$ increases proportionately so that the ratio $Q/V$ remains constant.

***The Parallel-Plate Capacitor.*** The common **parallel-plate capacitor** consists of two large parallel conducting plates separated by a narrow gap. Parallel-plate capacitors are traditionally discussed in physics textbooks because of their simplicity. Consider the parallel-plate capacitor shown in Figure 24-1. Electrons with a charge of $-Q$ were transferred from the upper plate to the lower plate by a battery, resulting in an excess of positive charge $(+Q)$ on the upper plate and an excess of negative charge $(-Q)$ on the lower plate. These charges are distributed approximately uniformly on the inner surfaces of the plates, with charge densities of $\sigma = Q/A$ on the upper plate and $-\sigma$ on the lower plate. (Figure 24-1 is not drawn to scale. The plate separation $d$ is assumed to be *very* small compared with the size of the plates. Near the edges of the plates the surface charge density $\sigma$ is not uniform. However, in this presentation, such edge effects are not considered.) The electric field just outside the surface of a conductor, in the region between the plates, is $\sigma/\varepsilon_0$, so $E = \sigma/\varepsilon_0 = Q/(\varepsilon_0 A)$, where $A$ is the area of the inner surface of either of the plates. Since the electric field is constant between the plates, the potential difference is $V = \left| -\int \vec{E} \cdot d\vec{\ell} \right| = Ed = Qd/(\varepsilon_0 A)$. Solving for the ratio $Q/V$ we have

$$C = \frac{\varepsilon_0 A}{d}$$

Capacitance of a parallel-plate capacitor

Note that because the potential difference $V$ is proportional to the charge $Q$, the capacitance $C$ does not depend on either $Q$ or $V$, but only on the geometric factors $d$ and $A$. The SI unit for the permittivity of free space (vacuum) in terms of capacitance, is

$$\varepsilon_0 = 8.85 \times 10^{-12} \text{ F/m} = 8.85 \text{ pF/m}$$

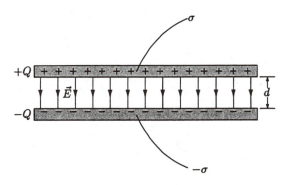

Figure 24-1

*The Cylindrical Capacitor.* The procedure used to obtain the expression for the capacitance of a parallel-plate capacitor can be used to obtain an expression for the capacitance of other capacitors. This procedure is to (1) draw the capacitor with the plates charged; (2) obtain an expression for the electric field between the plates in terms of the charge $Q$ and such geometric parameters as the areas and distances; (3) integrate this expression to obtain an expression for the potential difference $V$; and (4) solve for the ratio $Q/V$. Applying this procedure to a capacitor consisting of an inner conducting cylinder of radius $r_1$ and an outer coaxial conducting cylinder shell of inner radius $r_2$ gives the expression

$$C = \frac{2\pi\varepsilon_0 L}{\ln(r_2/r_1)} \qquad\qquad \text{Capacitance of a cylindrical capacitor}$$

where $L$ is the length of the cylinders.

*Section 24-3. The Storage of Electrical Energy.* The expression for the electrostatic potential energy of a capacitor is $U = \frac{1}{2}QV$. It may seem that the potential energy of a capacitor should be $QV$, not $\frac{1}{2}QV$, but charging a capacitor means transferring charge from one plate to the other. When the capacitor is uncharged, the potential difference is zero. As charge is transferred bit by bit from one plate to the other, the potential difference between the plates increases linearly in proportion to the charge transferred. During this transfer the average potential difference is half the final potential difference $V$. Thus, the final potential energy is $\frac{1}{2}QV$ (the product of total charge transferred and the average potential difference during the transfer). Combining the expression $U = \frac{1}{2}QV$ for potential energy with the expression $C = Q/V$ defining capacitance, we have other equivalent expressions for the energy stored in a capacitor:

$$U = \frac{1}{2}QV = \frac{1}{2}\frac{Q^2}{C} = \frac{1}{2}CV^2 \qquad\qquad \text{Energy stored in a capacitor}$$

*Section 24-4. Capacitors, Batteries, and Circuits.* When capacitors are combined in a circuit, the **equivalent capacitance** of the combination is the capacitance of a single capacitor that could replace them and, for a given potential difference, store the same charge. Capacitors can be combined in a number of ways. Two common ways are in series or in parallel. Figure 24-2a illustrates a combination of **capacitors in parallel**. When capacitors are connected in parallel, the potential drop across each capacitor is the same and the total charge stored is the sum of the charges stored on the individual capacitors. As a consequence, the equivalent capacitance is

$$C_{eq} = C_1 + C_2 + C_3 + \dots \qquad \text{Equivalent capacitance for capacitors in parallel}$$

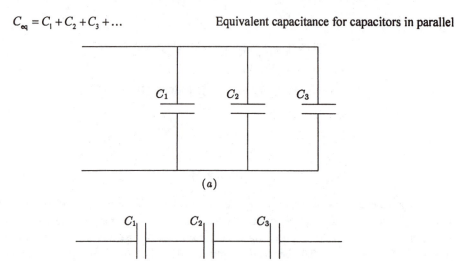

(a)

(b)

Figure 24-2

Figure 24-2*b* illustrates a combination of **capacitors in series**. When capacitors are connected in series, each capacitor has the same charge and the voltage drop across the combination equals the sum of voltage drops across the individual capacitors. As a consequence, the equivalent capacitance is

$$\frac{1}{C_{eq}} = \frac{1}{C_1} + \frac{1}{C_2} + \frac{1}{C_3} + \dots \qquad \text{Equivalent capacitance for capacitors in series}$$

***Section 24-5. Dielectrics.*** An insulating material is called a **dielectric.** When the space between the plates of a capacitor is filled with a dielectric, the capacitance increases by a factor $\kappa$ (lowercase kappa) called the **dielectric constant.** This increase comes about because the dielectric becomes polarized.

Suppose we start with an uncharged capacitor whose plates are separated by empty space. A battery is then connected to the plates and charge $Q$ is transferred from one plate to the other, leaving one plate with charge $+Q$ and the other with charge $-Q$. The battery is then disconnected from the capacitor, an action that results in the charge on each plate remaining fixed. The region between the two plates is then filled with a dielectric. This results in the electric field being reduced by a factor of $1/\kappa$.

Since the electric field is proportional to the potential difference between the plates, the potential difference is reduced by the same factor. This means that the capacitance is increased by a factor of $\kappa$. If, in an isolated charged capacitor, the electric field *without* a dielectric is $\vec{E}_0$, the field *within* the dielectric is

$$\vec{E} = \frac{\vec{E}_0}{\kappa} \qquad\qquad \text{Electric field inside a dielectric}$$

Then the potential difference $V$ between the plates is $V = \left| -\int \vec{E} \cdot d\vec{\ell} \right| = \left| -\int \vec{E}_0 \cdot d\vec{\ell} \right| / \kappa = V_0 / \kappa$ where $V_0 = \left| -\int \vec{E}_0 \cdot d\vec{\ell} \right|$ is the original potential difference without the dielectric. The new capacitance is

$$C = \frac{Q}{V} = \frac{Q}{V_0 / \kappa} = \kappa \frac{Q}{V_0} = \kappa C_0 \qquad\qquad \text{Capacitance with a dielectric}$$

where $C_0 = Q/V_0$ is the capacitance without a dielectric. The capacitance of a parallel-plate capacitor filled with a dielectric with constant $\kappa$ is thus $C = \kappa C_0 = \kappa \varepsilon_0 A / d = \varepsilon A / d$, where

$$\varepsilon = \kappa \varepsilon_0 \qquad\qquad \text{Permittivity defined}$$

is called the **permittivity** of the dielectric.

It is common practice to consider the electrostatic energy stored by the capacitor as being stored in the electric field and in the stress of any dielectric material present. The equation for the electrostatic energy density $u_e$ (the energy per unit volume) is

$$u_e = \tfrac{1}{2} \varepsilon E^2 \qquad\qquad \text{Energy density}$$

***Section 24-6. Molecular View of a Dielectric.*** If a dielectric is placed in an external electric field, the field pulls the positively and negatively charged particles (electrons and protons) within the dielectric in opposite directions. On average the electrons undergo a small displacement directed opposite to the field direction, and the positive nuclei undergo a small displacement in the direction of the field. These displacements result in surface charges on the dielectric called **bound charges.** They are so-named because they are bound to individual atoms or molecules and are not free to move throughout the dielectric as are the free electrons in conductors.

In a parallel-plate capacitor with a dielectric occupying the space between the plates, the bound surface charge density $\sigma_b$ results in an increase in capacitance by producing an electric field $\vec{E}_b$ within the dielectric that is smaller than, and oppositely directed to, the electric field $\vec{E}_f = \vec{E}_0$ produced by the free surface charge density $\sigma_f$ on the surface of the metal plates. The relation between the surface charge densities is

$$\sigma_b = \left( 1 - \frac{1}{\kappa} \right) \sigma_f$$

The way to remember this relationship is to recall that $E_b = -(1/\kappa) E_f$, so *all but* $(1/\kappa)$ of the field lines due to $\sigma_f$ are stopped at the surface of the dielectric. Hence, $\sigma_b$ is *all but* $(1/\kappa)$ of $\sigma_f$. Dielectric constants for some common materials are given in Table 24-1 on page 768 of the textbook. The dielectric constant of most materials is between one (for vacuum) and ten.

# II.   Physical Quantities and Key Equations

## Physical Quantities

*Capacitance*                                    1 farad (F) = 1 C/V

*Permittivity of free space*           $\varepsilon_0 = 8.85 \times 10^{-12}$ F/m = 8.85 pF/m

## Key Equations

*Electrostatic potential energy of a system of point charges*

$$U = q_2 \frac{kq_1}{r_{1,2}} + q_3 \left( \frac{kq_1}{r_{1,3}} + \frac{kq_2}{r_{2,3}} \right) + \dots = \frac{1}{2} \sum_{i=1} q_i V_i$$

*Capacitance defined*                    $C = \dfrac{Q}{V}$

*Capacitance of a plane, parallel-plate capacitor*     $C = \dfrac{\varepsilon_0 A}{d}$

*Capacitance of a cylindrical capacitor*     $C = \dfrac{2\pi\varepsilon_0 L}{\ln(r_2 / r_1)}$

*Energy stored in a capacitor*         $U = \dfrac{1}{2}QV = \dfrac{1}{2}\dfrac{Q^2}{C} = \dfrac{1}{2}CV^2$

*Equivalent capacitance for capacitors in parallel*     $C_{eq} = C_1 + C_2 + C_3 + \dots$

*Equivalent capacitance for capacitors in series*     $\dfrac{1}{C_{eq}} = \dfrac{1}{C_1} + \dfrac{1}{C_2} + \dfrac{1}{C_3} + \dots$

$C$

Capacitor

*Circuit symbol for a capacitor*

*Permittivity defined*                    $\varepsilon = \kappa \varepsilon_0$

*Electric field inside a dielectric*     $\vec{E} = \dfrac{\vec{E}_0}{\kappa}$

*Capacitance with a dielectric*

$$C = \frac{Q}{V} = \frac{Q}{V_0 / \kappa} = \kappa \frac{Q}{V_0} = \kappa C_0$$

*Energy density of an electrostatic field*

$$u_e = \tfrac{1}{2}\varepsilon E^2$$

## III. Potential Pitfalls

The capacitance of any two conductors depends only on the geometry of their configuration, that is, their sizes, shapes, and relative positions. In any case, it does not depend on the charge on the conductors or on the potential difference between them. In circuit analysis we almost always deal with differences in potential rather than potential itself, and so we often write $V$ rather than $\Delta V$ for potential difference. Make sure you know which you're talking about. Another possible confusion is that V is used to stand for "volt," so one encounters equations like $V = 3\,\mathrm{V}$.

The charge of a capacitor refers to the magnitude of the charge on *one* of the plates. The total charge on the two plates is, of course, zero.

Adding a dielectric between the plates of an isolated capacitor weakens the field in the capacitor gap because of the opposing electric field due to the bound charges on the surfaces of the polarized dielectric. This means that a given charge on the plates corresponds to a smaller potential difference, so the capacitance is increased.

## IV. True or False Questions and Responses

**True or False**

____ 1.  The capacitance of a capacitor is the maximum charge it can hold.

____ 2.  The capacitance of a charged capacitor is directly proportional to its charge.

____ 3.  The electrostatic potential energy stored in a capacitor with charge $Q$ and potential difference $V$ is the product $QV$.

____ 4.  The effect of a dielectric between the plates of an isolated capacitor is to diminish the electric field of the capacitor.

____ 5.  Within the dielectric of a capacitor the electric field due to the bound charges is in the same direction as the electric field due to the free charges on the metal plate.

____ 6.  The equivalent capacitance of two capacitors connected in series is less than the capacitance of either capacitor.

____ 7.  Capacitors in parallel necessarily have the same charge.

____ 8.  The expression for electric energy density is $u_e = \tfrac{1}{2}\varepsilon E^2$, where $u_e$ is the energy per unit charge.

**Responses to True or False**

1. False. The capacitance of a capacitor is the ratio of the charge to the potential difference. The charge may increase or decrease, but as it does, this ratio, and therefore the capacitance, does not change.

2. False. Capacitance is the ratio of charge to potential difference. Capacitance is independent of the charge because the electric field, and thus the potential difference, is proportional to the charge. That is, if you double the charge, the potential difference also must double, and the capacitance remains unchanged.

3. False. The electrostatic potential energy of a capacitor is $\frac{1}{2}QV$.

4. True.

5. False. The electric field $\vec{E}_f$ of the free charges causes a small displacement of the positive charges in the dielectric in the direction of this field and a small displacement of the negative charges in the opposite direction. The resulting bound surface charges on the dielectric produce a second electric field $\vec{E}_b$ within the dielectric (Figure 24-3$a$) that is directed opposite to $\vec{E}_f$. The resultant field is shown in Figure 24-3$b$.

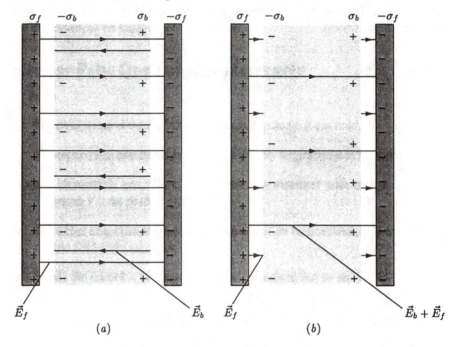

Figure 24-3

6. True. The larger a number is, the smaller the reciprocal of the number. We know that $1/C_{eq}$ is greater than either $1/C_1$ or $1/C_2$ because, for capacitors in series, $1/C_{eq} = 1/C_1 + 1/C_2$. therefore, $C_{eq}$ (the reciprocal of $1/C_{eq}$) is less than either $C_1$ (the reciprocal of $1/C_1$) or $C_2$ (the reciprocal of $1/C_2$).

7. False. However, they must have the same potential difference across each one.

8. False. This is the expression for energy per unit volume.

## V. Questions and Answers

**Questions**

1. Capacitors $A$ and $B$ are identical except that the region between the plates of capacitor $A$ is filled with a dielectric. As shown in Figure 24-4, the plates of these capacitors are maintained at the same potential difference by a battery. Is the electric field intensity in the region between the plates of capacitor $A$ smaller, the same, or larger than the field in the region between the plates of capacitor $B$?

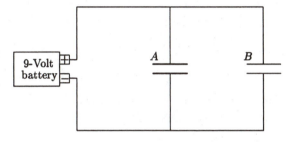

Figure 24-4

2. Using a single 100-V battery, what capacitance do you need to store $10\,\mu\text{C}$ of charge?

3. Using a single 100-V battery, what capacitance do you need to store 1 mJ of electrostatic potential energy?

4. Why is the bound surface charge density on a dielectric always less than the free surface charge density on the capacitor plates?

**Answers**

1. The electric field in the region between the plates is the same for both capacitors. The potential difference $V$ across a capacitor and the electric field $\vec{E}$ in the region between the plates are related by the equation $V = \left| -\int \vec{E} \cdot d\vec{\ell} \right| = Ed$. If the potential difference across each capacitor is the same, then the electric fields in the regions between the plates must also be the same. However, $A$ has the greater charge.

2. $C = Q/V = 10\,\mu\text{C}/100\,\text{V} = 0.1\,\mu\text{F}$

3. $U = \tfrac{1}{2}QV = \tfrac{1}{2}CV^2$
   $C = 2U/V^2 = 2(1\text{ mJ})/(100\text{ V})^2 = 0.2\,\mu\text{F}$

4. If the free surface density and the bound surface charge density were equal, then their electric fields would completely cancel. However, to sustain a bound surface charge requires that the dielectric be polarized, which requires a nonzero electric field. Therefore, the bound surface charge density is always less than the free surface charge density.

## VI. Problems, Solutions, and Answers

**Example #1.** As shown in Figure 24-5, three particles, each with charge $q$, are at different corners of a rhombus with sides of length $a$ and with a diagonal length $a$. (*a*) What is the electrostatic potential energy of this charge distribution? (*b*) How much work by an external agent is required to bring a fourth particle, also of charge $q$, from rest at infinity to rest at the vacant corner of the rhombus? (*c*) What is the total electrostatic potential energy of the four charges?

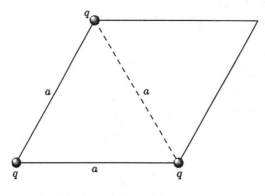

Figure 24-5

**Picture the Problem.** The electrostatic potential energy is the work by an external agent required to assemble the charge distribution.

| | |
|---|---|
| 1. Label the three charges $q_1$, $q_2$, and $q_3$, and the long diagonal $b$. | 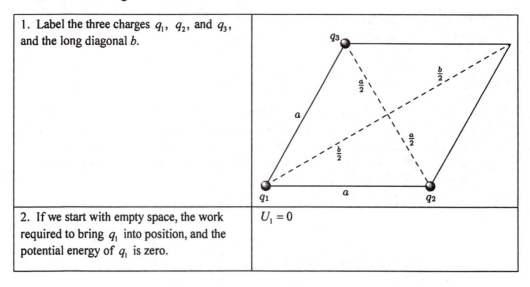 |
| 2. If we start with empty space, the work required to bring $q_1$ into position, and the potential energy of $q_1$ is zero. | $U_1 = 0$ |

| 3. Calculate the change in potential energy required to bring charge $q_2$ into place while $q_1$ is present. | $U_2 = \dfrac{kq_1 q_2}{a}$ |
|---|---|
| 4. Calculate the change in potential energy that result from bringing charge $q_3$ into place in the presence of $q_1$ and $q_2$. The final resting place of $q_3$ is a distance $a$ from each of the other charges. | $U_3 = q_3\left(\dfrac{kq_1}{a} + \dfrac{kq_2}{a}\right)$ |
| 5. The total potential energy of the system is the sum of the potential energies. Remember $q_1 = q_2 = q_3 = q$. | $U = U_1 + U_2 + U_3$ <br><br> $= 0 + q_2\dfrac{kq_1}{a} + q_3\left(\dfrac{kq_1}{a} + \dfrac{kq_2}{a}\right) = \dfrac{3kq^2}{a}$ |
| 6. To find the work required to bring in the fourth charge, we can first find the electric potential at the fourth corner, which is the sum of the electric potential due to the other three charges. | $V = V_1 + V_2 + V_3$ <br><br> $= \dfrac{kq}{b} + \dfrac{kq}{a} + \dfrac{kq}{a} = \dfrac{2kq}{a} + \dfrac{kq}{b}$ <br><br> $\dfrac{b}{2} = \sqrt{a^2 - \left(\dfrac{a}{2}\right)^2}$ <br><br> $b = \sqrt{3}a$ <br><br> $V = \left(2 + \dfrac{1}{\sqrt{3}}\right)\dfrac{kq}{a}$ |
| 7. Use the electric potential to calculate the work required by an external agent to bring the fourth charge into place. | $W_{ext} = \Delta U = q\,\Delta V = qV = \left(2 + \dfrac{1}{\sqrt{3}}\right)\dfrac{kq^2}{a}$ |
| 8. The total potential energy of the four charges is the sum of the results of step 5 and 7. | $U_{tot} = 3\dfrac{kq^2}{a} + \left(2 + \dfrac{1}{\sqrt{3}}\right)\dfrac{kq^2}{a} = \left(5 + \dfrac{1}{\sqrt{3}}\right)\dfrac{kq^2}{a}$ |

**Example #2—Interactive.** Three $+2.00\text{-}\mu C$ charges are located on the $x$ axis at $x = 0.00$, 10.0 cm, and 20.0 cm. (*a*) What is the electrostatic potential energy of this distribution? (*b*) Each of the three charges is replaced by a $-2.00\text{-}\mu C$ charge. Now what is the electrostatic potential energy of the distribution?

**Picture the Problem.** The electrostatic potential energy is the work by an external agent required to assemble the charge distribution. **Try it yourself.** Work the problem on your own, in the spaces provided, to get the final answer.

| 1. Sketch the three charges on the axis. | |
|---|---|
| | |

| | |
|---|---|
| 2. Determine the work required to bring the first charge at $x = 0$ into place. | |
| 3. Determine the work required to bring the second charge at $x = 10\,\text{cm}$ into place in the presence of the first charge. | |
| 4. Determine the work required to bring the final charge into place in the presence of the first two. | |
| 5. The total potential energy is the sum of the work calculated in steps 2-4. | $U = +0.899\,\text{J}$ |
| 6. Repeat steps 2-5 if the charges are negative instead of positive. | $U = +0.899\,\text{J}$ |

**Example #3.** A parallel-plate, air-gap capacitor has a charge of $20.0\,\mu\text{C}$ and a gap width of $0.100\,\text{mm}$. The potential difference between the plates is $200.0\,\text{V}$. (*a*) What is the electric field in the region between the plates? (*b*) What is the surface charge density on the positive plate? (*c*) What is the electrostatic potential energy stored in the capacitor?

**Picture the Problem.** The electric field between the plates of a parallel-plate capacitor is uniform. The surface charge density of each plate contributes equally to the electric field. Use one of the derived relationships to find the potential energy stored in the capacitor.

| | |
|---|---|
| 1. Use the relationship between potential, electric field, and distance to find the electric field, remembering that the electric field is uniform in a parallel-plate capacitor. | $\|V\| = \int \vec{E} \cdot d\vec{\ell} = Ed$ $E = \dfrac{V}{d} = \dfrac{200.0\,\text{V}}{1 \times 10^{-4}\,\text{m}} = 2.00 \times 10^{6}\,\text{V/m}$ |

| 2. The electric field between the plates is due to two thin sheets of charge, one on each of the plates. | $E = E_+ + E_- = \dfrac{\sigma}{2\varepsilon_0} + \dfrac{\sigma}{2\varepsilon_0} = \dfrac{\sigma}{\varepsilon_0}$ <br><br> $\sigma = \varepsilon_0 E = 1.77 \times 10^{-5} \text{ C/m}^2$ |
|---|---|
| 3. Calculate the potential energy stored in the capacitor. | $U = \tfrac{1}{2}QV = \tfrac{1}{2}\left(20.0 \times 10^{-6} \text{ C}\right)\left(200.0 \text{ V}\right)$ <br><br> $= 2.00 \times 10^{-3} \text{ J}$ |

**Example #4—Interactive.** The plates of a parallel-plate capacitor are separated by a 0.500-mm-thick Pyrex sheet. The area of each plate is $3.00 \text{ m}^2$. (*a*) What is the maximum voltage between the plates before dielectric breakdown occurs? (*b*) What is the capacitance? (*c*) What is the maximum amount of electrostatic energy this capacitor can store?

**Picture the Problem.** The potential difference is the product of the electric field and the gap width. Use Table 24-1 on page 768 of the text to find the dielectric constant and dielectric strength of Pyrex. Dielectric breakdown occurs when the electric field exceeds the dielectric strength. The capacitance and potential energy stored can be found using the appropriate equations for a parallel-plate capacitor. **Try it yourself.** Work the problem on your own, in the spaces provided, to get the final answer.

| 1. Lookup the dielectric strength of Pyrex in Table 24-1 of the text. | |
|---|---|
| 2. Relate the voltage on the plates to the dielectric strength of Pyrex and the separation of the plates. | $V = 7 \text{ kV}$ |
| 3. Use the expression for the capacitance of a parallel-plate capacitor filled with a dielectric to determine the capacitance. | $C = 0.297 \,\mu\text{F}$ |
| 4. Using the maximum voltage and capacitance found in steps 2 and 3, calculate the maximum electrical potential energy that can be stored by this capacitor. | $U_{max} = 7.29 \text{ J}$ |

**Example #5.** A 10.0-$\mu$F capacitor, a 40.0-$\mu$F capacitor, and a 100.0-$\mu$F capacitor are connected in series. A 12-V battery is connected across this combination. (*a*) What is the equivalent capacitance of the combination? (*b*) What is the charge on each capacitor? (*c*) What is the potential difference across each capacitor?

**Picture the Problem.** Use the expression for equivalent capacitance of capacitors in series. Capacitors in series all hold the same charge, and that charge is the same as the charge stored by the equivalent capacitance. Use the stored charge and the capacitance of each capacitor to find the potential difference across each capacitor.

| | |
|---|---|
| 1. Sketch the schematic of capacitors to visualize the circuit. | |
| 2. Determine the equivalent capacitance for capacitors in series. | $$\frac{1}{C_{eq}} = \frac{1}{C_1} + \frac{1}{C_2} + \frac{1}{C_3}$$ $$= \frac{1}{10\,\mu F} + \frac{1}{40\,\mu F} + \frac{1}{100\,\mu F} = \frac{13.5}{100\,\mu F}$$ $$C_{eq} = \frac{100\,\mu F}{13.5} = 7.41\,\mu F$$ |
| 3. The charge stored by each capacitor is the same as the charge stored by an equivalent capacitance. | $$Q = C_{eq}V = (7.41\,\mu F)(12\,V) = 88.9\,\mu C$$ |
| 4. This charge is stored by each of the capacitors. Use this to find the potential difference across each capacitor. The sum of all the potential differences is 12 V, within rounding errors. | $$V = \frac{Q}{C}$$ $$V_{10} = \frac{88.9\,\mu C}{10\,\mu F} = 8.89\,V$$ $$V_{40} = \frac{88.9\,\mu C}{40\,\mu F} = 2.22\,V$$ $$V_{100} = \frac{88.9\,\mu C}{100\,\mu F} = 0.889\,V$$ |

**Example #6—Interactive.** A 10.0-$\mu$F capacitor, a 40.0-$\mu$F capacitor, and a 100.0-$\mu$F capacitor are connected in parallel. A 12-V battery is connected across this combination. (*a*) What is the equivalent capacitance of the combination? (*b*) What is the total charge on the three capacitors? (*c*) What is the total charge on each capacitor?

**Picture the Problem.** Use the expression for equivalent capacitance of capacitors in parallel. Use the equivalent capacitance to find the total charge stored by this arrangement of capacitors. Capacitors in parallel all have the same potential difference which can be used to determine the charge stored by each capacitor. **Try it yourself.** Work the problem on your own, in the spaces provided, to get the final answer.

| | |
|---|---|
| 1. Sketch the schematic of capacitors to visualize the circuit. | |

| | |
|---|---|
| 2. Determine the equivalent capacitance for capacitors in parallel. | $C_{eq} = 150 \, \mu F$ |
| 3. Find the charge stored by the equivalent capacitance. | $Q = 1.80 \, mC$ |
| 4. Determine the charge on each capacitor. The total will sum to the total charge stored calculated in step 3. | $Q_{10} = 120 \, \mu C$ ; $Q_{40} = 480 \, \mu C$ ; $Q_{100} = 1.20 \, mC$ |

**Example #7.**  You have a bucketful of capacitors, each with a capacitance of $1.0 \, \mu F$ and a maximum voltage rating of $250$ V. You are to come up with a combination that has a capacitance of $0.75 \, \mu F$ and a maximum voltage rating of $1000$V. What is the minimum number of capacitors you need?

**Picture the Problem.**  Because you want the combination to withstand a maximum voltage four times greater than a single capacitor, you will need to connect several in series. Then add enough of these strings in parallel to get an effective capacitance of $0.75 \, \mu F$.

| | |
|---|---|
| 1. Determine the number of capacitors needed in series to withstand 1000 V. For identical capacitors in series, they will each have the same voltage. Why? | $1000 \, V = x(250 \, V)$ <br> $\therefore x = 4 \, capacitors$ |
| 2. Determine the equivalent capacitance of these four capacitors connected in series. | $\dfrac{1}{C_{series}} = \dfrac{1}{C} + \dfrac{1}{C} + \dfrac{1}{C} + \dfrac{1}{C} = \dfrac{4}{C}$ <br> $C_{series} = \dfrac{C}{4} = \dfrac{1 \, \mu F}{4} = 0.25 \, \mu F$ |
| 3. Determine the number of these strings of capacitors, connected in parallel, needed to reach an equivalent capacitance of $0.75 \, \mu F$. | $C_{eq} = 0.75 \, \mu F = x(C_{series}) = x(0.25 \, \mu F)$ <br> $x = 3 \, chains$ |

| 4. Determine the total number of capacitors needed. | $(3\,\text{chains})(4\,\text{capacitors/chain}) = 12\,\text{capacitors}$ 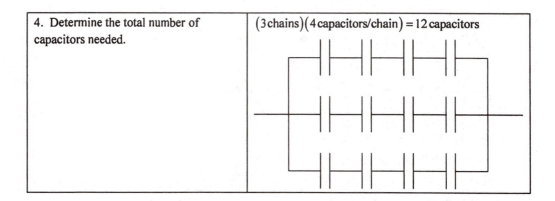 |
| --- | --- |

**Example #8—Interactive.** What is the equivalent capacitance of the combination of three capacitors shown in Figure 24-6.

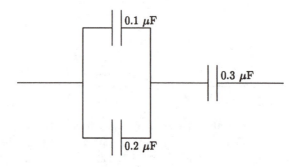

Figure 24-6

**Picture the Problem.** Use the expressions for parallel and series capacitors. **Try it yourself.** Work the problem on your own, in the spaces provided, to get the final answer.

| 1. Calculate the equivalent capacitance of the $0.1\,\mu\text{F}$ and $0.2\,\mu\text{F}$ capacitors which are attached in parallel. | |
| --- | --- |
| 2. The equivalent capacitance calculated in step 1 is in series with the $0.3\,\mu\text{F}$ capacitor. Find that total equivalent capacitance. | $C_{eq} = 0.15\,\mu\text{F}$ |

**Example #9.** A parallel-plate capacitor has area $A$ and separation $d$. How is its capacitance affected if a conducting slab of thickness $d' < d$ is inserted between, and parallel to, the plates as shown in Figure 24-7? Does your answer depend on where, vertically, between the plates the slab is positioned?

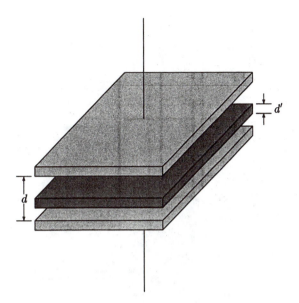

Figure 24-7

**Picture the Problem.** Inserting a conducting slab between the original plates turns the original capacitor into two parallel-plate capacitors in series, with possibly different plate separations.

| | |
|---|---|
| 1. Determine the capacitance of the original capacitor, without the additional plate. | $C = \dfrac{\varepsilon_0 A}{d}$ |
| 2. Determine the total space available between the plates after the conducting plate has been inserted. | $y = d - d'$ |
| 3. If the distance between the top plate and the middle plate is $x$, then the distance between the middle plate and the bottom plate is $y - x$. Use this to calculate the capacitance of the new upper and lower capacitors. | $C_{\text{top}} = \dfrac{\varepsilon_0 A}{x}$ <br><br> $C_{\text{bottom}} = \dfrac{\varepsilon_0 A}{y - x} = \dfrac{\varepsilon_0 A}{d - d' - x}$ |
| 4. Calculate the equivalent capacitance of these two capacitors, which are in series with each other. Since $x$, which governs the relative separation of the plates does not appear in the final expression, the equivalent capacitance does not depend on where the middle plate is positioned. | $\dfrac{1}{C_{\text{eq}}} = \dfrac{1}{C_{\text{top}}} + \dfrac{1}{C_{\text{bottom}}} = \dfrac{x}{\varepsilon_0 A} + \dfrac{d - d' - x}{\varepsilon_0 A} = \dfrac{d - d'}{\varepsilon_0 A}$ <br><br> $C_{\text{eq}} = \dfrac{\varepsilon_0 A}{d - d'}$ |

**Example #10—Interactive.** A modified form of the parallel-plate capacitor consisting of three plates is shown in Figure 24-8. The middle plate is connected to one terminal of the battery and the two outer plates are connected to the other terminal. The area per side of each plate is $A$, the separation between the leftmost pair of plates is $2d$, and between the rightmost pair of plates is $d$. What is the equivalent capacitance of the arrangement?

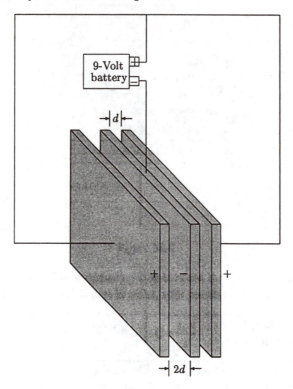

Figure 24-8

**Picture the Problem.** The potential difference between the inner plate and either of the two outer plates is the same. Thus, this arrangement consists of two parallel-plate capacitors in parallel. **Try it yourself.** Work the problem on your own, in the spaces provided, to get the final answer.

| | |
|---|---|
| 1. Find the capacitance of the leftmost capacitor. | |
| 2. Find the capacitance of the righthand capacitor. | |
| 3. Find the equivalent capacitance. | $C_{eq} = \dfrac{3\varepsilon_0 A}{2d}$ |

**Example #11.** An isolated air-gap, parallel-plate capacitor has area $A$, gap width $d$, and charge $Q$. (*a*) Find expressions for the electric field intensity and energy density in the region between the plates. (*b*) Use the expression for energy density to obtain an expression for the electrostatic potential energy of the capacitor. (*c*) The plates are pulled apart, doubling the gap width. Use a work-energy method to show that the force exerted by the electric field on one of the plates is $\frac{1}{2}QE$, where $E$ is the electric field.

**Picture the Problem.** The electric field can be expressed in terms of the surface charge density and the energy density can be expressed in terms of the electric field. The potential energy is the product of the energy density in the gap and the volume of the gap. Determine the potential energy of the capacitor with the gap equal to $2d$. Find an expression for the force by equating the work done with the increase in potential energy.

| | |
|---|---|
| 1. Draw the capacitor and put in the electric field lines. | 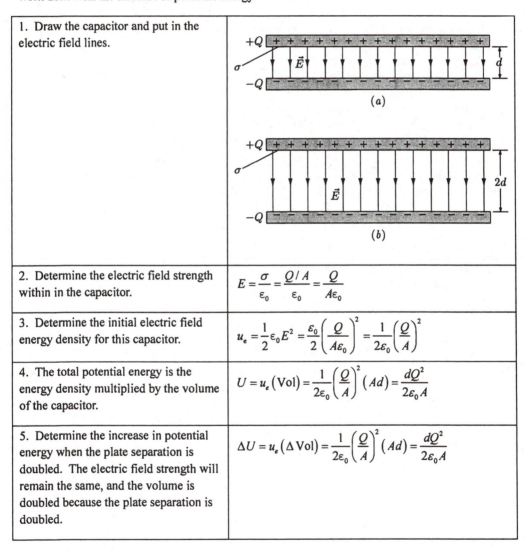 |
| 2. Determine the electric field strength within in the capacitor. | $E = \dfrac{\sigma}{\varepsilon_0} = \dfrac{Q/A}{\varepsilon_0} = \dfrac{Q}{A\varepsilon_0}$ |
| 3. Determine the initial electric field energy density for this capacitor. | $u_e = \dfrac{1}{2}\varepsilon_0 E^2 = \dfrac{\varepsilon_0}{2}\left(\dfrac{Q}{A\varepsilon_0}\right)^2 = \dfrac{1}{2\varepsilon_0}\left(\dfrac{Q}{A}\right)^2$ |
| 4. The total potential energy is the energy density multiplied by the volume of the capacitor. | $U = u_e(\text{Vol}) = \dfrac{1}{2\varepsilon_0}\left(\dfrac{Q}{A}\right)^2(Ad) = \dfrac{dQ^2}{2\varepsilon_0 A}$ |
| 5. Determine the increase in potential energy when the plate separation is doubled. The electric field strength will remain the same, and the volume is doubled because the plate separation is doubled. | $\Delta U = u_e(\Delta\text{Vol}) = \dfrac{1}{2\varepsilon_0}\left(\dfrac{Q}{A}\right)^2(Ad) = \dfrac{dQ^2}{2\varepsilon_0 A}$ |

| 6. This change in potential energy comes from work that must be done by an external force. Assume the force used to separate the plates is constant. | $W = Fd = \Delta U = \dfrac{dQ^2}{2\varepsilon_0 A}$ <br><br> $F = \dfrac{Q^2}{2\varepsilon_0 A}$ |
|---|---|

For a plate to remain at rest after the separation, the net force on the plate must equal zero. Thus, the force exerted by the electric field on the plate is equal in magnitude to the force exerted by the external agent(s). Therefore, the force exerted by the field is also $\frac{1}{2}QE$.

Why is a factor of one-half in this expression? The reason is that the resultant electric field is the superposition of the two fields, one due to the charge $-Q$ on the negative plate and the other due to the charge $+Q$ on the positive plate. The electric force exerted on a charge is $\vec{F} = q\vec{E}_{\text{other}}$, where $\vec{E}_{\text{other}}$ is the electric field at the location of the charge $q$ due to all other charges. When we determine the electric force exerted on the charge $Q$ on one plate of a charged parallel-plate capacitor, $\vec{E}_{\text{other}} = \frac{1}{2}\vec{E} = \frac{1}{2}(\sigma/\varepsilon_0)\hat{i}$, so the electric force on each plate is $\vec{F} = Q\vec{E}_{\text{other}} = \frac{1}{2}Q\vec{E}$.

**Example #12—Interactive.** An air-gap, parallel-plate capacitor with area $A$ and gap width $d$ is connected to a battery that maintains the plates at potential difference $V$. (*a*) Find expressions for the electric field and energy density in the region between the plates and for the charge on the positive plate. (*b*) Use the expression for energy density to obtain an expression for the electrotatic potential energy of the capacitor. (*c*) The plates are pulled apart, doubling the gap width, while they remain in electrical contact with the battery terminals. By what factor does the potential energy of the capacitor change? **Try it yourself.** Work the problem on your own, in the spaces provided, to get the final answer.

| 1  Determine the electric field strength within in the capacitor. | $E = V/d$ |
|---|---|
| 2  Determine the electric field energy density for this specific capacitor. | $u_e = \dfrac{\varepsilon_0}{2}\left(\dfrac{V}{d}\right)^2$ |
| 3. Determine the charge on the positive plate. | $Q = \dfrac{\varepsilon_0 A V}{d}$ |
| 4. The total potential energy is the energy density multiplied by the volume of the capacitor. | $U = \dfrac{\varepsilon_0 A V^2}{2d}$ |
| 5. Determine the change in potential energy when the plate separation is doubled. | Potential energy is halved. |

# Chapter 25

# Electric Current and Direct-Current Circuits

## I.  Key Ideas

An electric current is time rate of the net flow of charge. Currents can occur whenever there are mobile charge carriers.

*Section 25-1. Current and the Motion of Charges.* Electric current is the time rate of flow of electric charge through a surface, such as the cross section of a wire. Thus, if $\Delta Q$ is the charge flowing through a surface in time $\Delta t$, the current $I$ through that area is

$$I = \frac{\Delta Q}{\Delta t}$$
Electric current defined

The SI unit of current is the **ampere** (A):

$$1\ A = 1\ C/s$$

The current $I$ through a cross section of area $A$ of a wire is given by the expression

$$I = nqAv_d$$
Current and drift velocity

where $n$ is the number density of charge carriers (the number of free charge carriers per unit volume); $q$ is the charge of a single charge carrier, and $v_d$, the **drift velocity** of the charge carriers, is the average velocity of the charge carriers. If the current is due to the flow of *positive* charge carriers through a surface, the direction of the current is the direction of the drift velocity of the charge carriers. Conversely, if the current is due to the flow of *negative* charge carriers through a surface, the direction of the current is opposite to the direction of the drift velocity of the charge carriers.  By definition, the direction of current flow is always in the direction of motion of the apparent flow of positive charge carriers.

At normal room temperatures the average *speed* of individual air molecules is about 500 m/s (~1100 mi/hr). However, if the windows of the room you are in are shut so there are no breezes in the air, the average *velocity* of these molecules is zero. On the other hand, if the windows are open and there is a breeze, the bulk motion of the air is quite slow, typically a few meters per second or less. The velocity of this bulk motion of air is the average velocity (the drift velocity) of the air molecules. The motion of free electrons in metals is analogous to the motion of air molecules. At room temperature the average speed of a free electron is much greater than 500 m/s, but the drift velocity, due to an electrical potential difference, of the free electrons in typical household circuits is only a fraction of a millimeter per second.

***Section 25-2. Resistance and Ohm's Law.*** Inside a conductor in electrostatic equilibrium both the electric field and the drift velocity of the charge carriers are zero. However, a conductor in which there is a current is not in electrostatic equilibrium, and neither the drift velocity nor the electric field is zero. In fact, it is forces exerted on the free electrons by these internal electric fields that keep them drifting through the metal. Collisions of free electrons with the atoms of the metal result, on average, in a "drag force" that limits their drift velocity. If the internal electric fields are removed, the "drag" due to these collisions quickly reduces the drift velocity to zero.

The internal electric fields that drive the current in a current-carrying metal wire are directed parallel to the current, opposite to the direction of electron flow. Because an electric field points in the direction of decreasing potential, the current is also in the direction of decreasing potential. The potential drop $V$—the decrease in potential along the direction of the current—across a length $\Delta L$ of the wire is $V = -\Delta V = \int \vec{E} \cdot d\vec{r} = E \Delta L$, because the electric field in a metal wire is approximately constant.

The **resistance** $R$ of a length of any object is defined as

$$R = \frac{V}{I}$$                                             Resistance defined

where $V$ is the potential drop across the object and $I$ is the current in the object. The SI unit of resistance, the volt per ampere, is called the **ohm** ($\Omega$):

$$1\,\Omega = 1\,\text{V/A}$$

It is found that in metals the current is proportional to the potential drop across a given length of wire, that is, $R$ is constant. This empirical result, which holds if the temperature of the metal remains fixed, is known as **Ohm's law:**

$$V = IR, \quad R = \text{constant}$$                          Ohm's law

The resistance of a length $L$ of a conducting object is found to be proportional to the length of the object and inversely proportional to its cross-sectional area $A$:

$$R = \rho\frac{L}{A}$$                                         Resistivity $\rho$

where $\rho$ is the **resistivity,** measured in ohm-meters, of the conducting material. (The reciprocal of the resistivity is called the conductivity $\sigma$).  A **resistor** is any electrical device specifically designed to cause a voltage drop along its length, when a current flows through it.

The resistivity of any given metal varies with temperature. Empirically it is found that the resistivity changes fairly linearly with changes in temperature. The **temperature coefficient of**

**resistivity** $\alpha$, usually given at 20°C, is the ratio of the change in resistivity to the change in temperature, so

$$\rho = \rho_{20}\left[1 + \alpha(t_c - 20°C)\right]$$     Temperature coefficient of resistivity $\alpha$

where $\rho$ and $\rho_{20}$ are the resistivities at the Celsius temperatures of $t_c$ and 20°C, respectively. Table 25-1 on page 792 of the text lists the resistivities and temperature coefficients of several conducting materials.

***Section 25-3. Energy in Electric Circuits.*** In a current-carrying conductor the charge carriers lose potential energy but do not, on average, gain kinetic energy. Instead, they transfer energy to the metal. This is a dissipative process in which the potential energy of the charge carriers is transferred to the thermal energy of the conductor. The rate at which thermal energy is generated equals the rate of loss of potential energy by the charge carriers. That is,

$$P = IV$$     Power

where the power $P$ is the rate at which electrical potential energy is dissipated, $V$ is the potential drop in the direction of the current, and $I$ is the current. If $V$ is in volts (joules/coulomb) and $I$ is in amperes (coulombs/second), then the product $IV$ is in joules/second, or watts.

This heating of a conductor by a current is known as **joule heating.** Using the definition of resistance $\left(R = V / I\right)$, we can express the power dissipated in a resistor alternatively as

$$P = IV = I^2 R = \frac{V^2}{R}$$     Power dissipated in a resistor

When a current flows through a metal wire, the electrical potential energy of the charge carriers is dissipated. Thus, to maintain a current requires a source of electrical potential energy. A device that transforms energy from some other form to electrical potential energy is called a source (or seat) of **electromotive force** or simply a source of **emf.** (The term "electromotive force" is an archaic term and a misnomer in that it is not a force at all. Instead, it is the work per unit charge done by the battery.) Such a device converts chemical energy, mechanical energy, or some other form of energy into electrical potential energy. For example, a battery converts chemical energy into potential energy. A source of emf does work on the charges passing through it, raising their electrical potential energy. The work per unit charge is called the emf $\mathcal{E}$ of the source. When a charge $\Delta Q$ flows through a source of emf, the work done on it is $\mathcal{E}\Delta Q$. The unit of emf is the joule per coulomb, or volt. The power being delivered by a battery, that is, the rate at which electrical potential energy is generated, is

$$P = \frac{\mathcal{E}\Delta Q}{\Delta t} = \mathcal{E}I$$     Power delivered by a source of emf

An **ideal battery** is a source of emf that maintains a constant potential difference between its two terminals, independent of the current through it. The potential difference between the terminals of an ideal battery is equal in magnitude to the emf. Such a device is also called a "voltage source."

Turning on a flashlight connects a battery to the terminals of an incandescent light bulb. (An incandescent light bulb consists of a metal wire, called a filament, in a glass envelope. When there is a current through it, the filament gets hot and radiates heat and light.) As charges pass through a battery, they gain electrical potential energy at the expense of the battery's stored chemical energy. After leaving the battery, the charges pass through the filament where they lose potential energy, which is dissipated as heat. After leaving the filament they return to the battery, where the process is repeated.

This process is analogous to the rainwater cycle near the earth's surface. Energy from the sun evaporates seawater, producing water vapor near the surface of the sea. The water vapor diffuses upward, opposite to the direction of the gravitational force that is acting on them. Thus, this motion results in an increase in the gravitational potential energy of the water vapor. When conditions are favorable, the water vapor condenses into raindrops, which fall to earth. As they fall, they lose gravitational potential energy but do not gain kinetic energy. The drag force exerted by the air on the drops opposing their motion causes them to fall with a constant (terminal) velocity. As they fall, they lose gravitational potential energy, which is transformed into thermal energy of air. After reaching the ground the water returns to the sea, where the process begins again.

A battery consists of two dissimilar metals, called electrodes, separated by a nonmetallic conducting solution called an electrolyte. Electric currents in an electrolyte are due to ionic motion. At the surface of the metals, chemical processes result in the concentration of ions (electrically charged atoms) near the surfaces of the electrodes. These ions then diffuse away from these surfaces, and their motion constitutes an electric current. Just as the water vapor in the previous example diffuses upward, opposite to the direction of the gravitational force acting on it, these ions move (diffuse) opposite to the direction of an electric force that acts on them. This diffusion results in an increase in the electrical potential energy of the ions. Within the electrolyte of a battery, the current, driven by diffusion, is directed opposite to the direction of the electric field; but in a current-carrying conductor outside a battery, the current, here driven by the electric forces, is in the same direction as the electric field. The electric field, both within the electrolyte of a battery and elsewhere, is in the direction of decreasing electric potential.

The potential difference across the terminals of a battery is called the **terminal voltage.** In an **ideal battery,** there are no internal losses of electrical energy, so the terminal voltage always equals the emf. In a **real battery,** the terminal voltage is not equal to the emf except when the current is zero. A real battery has an **internal resistance** $r$, which reduces the terminal voltage when current is present in the circuit. Thus, if, as shown in Figure 25-1$a$, a real battery is connected to an external resistor, the terminal voltage is

$$V_a - V_b = \mathcal{E} - Ir \qquad\qquad\qquad \text{Terminal voltage of a real battery}$$

where $a$ and $b$ are, respectively, the $+$ and $-$ terminals of the battery. The potential drop across the external resistor is also $V_a - V_b$, so

$$V_a - V_b = IR$$

where $R$ is the load resistance, the resistance of the external resistor. Figure 25-1$b$ is a schematic diagram of the same circuit. Equating the right sides of these equations and solving for the current, we have

$$I = \frac{\mathcal{E}}{R+r}$$

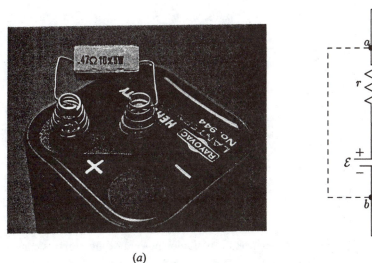

(a)

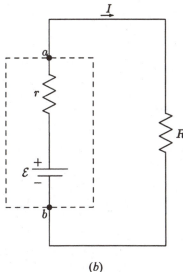

(b)

Figure 25-1

***Section 25-4. Combinations of Resistors.*** Resistors can be combined in various ways. Each resistor connected in series in a circuit carries the same current, and the potential drop across a combination of resistors connected in series is the sum of the potential drops across each of the individual resistors in the combination.  The **equivalent resistance** of a combination of resistors is the resistance of a single resistor that would give the same potential drop $V$ as the combination when carrying the same current $I$, so the equivalent resistance is

$$R_{eq} = R_1 + R_2 + R_3 + \ldots$$                    Resistors in series

Two or more resistors are said to be connected in series when they are connected as shown in Figure 25-2$a$.

Each resistor connected in parallel has the same potential drop across it, and the total current through the combination equals the sum of the currents through each of the individual resistors in the combination; so the equivalent resistance is

$$\frac{1}{R_{eq}} = \frac{1}{R_1} + \frac{1}{R_2} + \frac{1}{R_3} + \ldots$$                    Resistors in parallel

Two or more resistors are said to be connected in parallel when they are connected as shown in Figure 25-2$b$. The reciprocal of the resistance is called the conductance. For resistors in parallel, the equivalent conductance is the sum of the individual conductances.

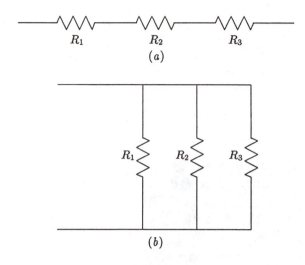

Figure 25-2

***Section 25-5. Kirchhoff's Rules.*** **Kirchhoff's rules** are the very foundation of circuit theory.

**1.** When any closed circuit loop is traversed, the algebraic sum of the changes in potential must equal zero.

**2.** At any junction point in a circuit where the current can divide, the sum of the currents entering the junction must equal the sum of the currents leaving the junction.

Kirchhoff's first rule, called the **loop rule,** follows from the presence of a conservative electric field. For the most part, the net charge in any region inside a conductor is zero; however, the nonzero charge distributions on the surfaces of the wires, resistors, capacitors, and other components create a conservative electric field $\vec{E}$. Because it is conservative, we have

$$\oint \vec{E} \cdot d\vec{\ell} = 0$$

where the integral is taken around any closed path. The potential difference between two points $a$ and $b$ is defined as

$$\Delta V = V_b - V_a = -\int_a^b \vec{E} \cdot d\vec{\ell}$$

To say that the algebraic sum of the changes in potential must equal zero when any closed-circuit loop is traversed is equivalent to stating $-\oint \vec{E} \cdot d\vec{\ell} = 0$. That is, Kirchhoff's loop rule is a consequence of the fact that the field is conservative.

Kirchhoff's second rule, the **junction rule,** follows from the conservation of charge and the fact that very large surface areas are needed to accumulate a meaningful amount of charge. *Current* is the rate of flow of charge through a cross section of a wire. If three or more wires join at a junction, some of the wires may be carrying charge to the junction while others carry it away from the junction. The rate at which the charge accumulates at the junction equals the sum of the currents entering the junction less the sum of the currents leaving the junction. Kirchhoff's junction rule, that the sum of the currents entering the junction equals the sum of the currents leaving the junction, assumes that the rate of accumulation of charge at a junction is zero.

The procedure for applying Kirchhoff's rules follows:

**1.** To reduce the complexity of the algebra, replace any series or parallel combinations of resistors with their equivalent resistances.

**2.** Make an arbitrary choice of the positive direction for the current in each branch of the circuit. A single branch of a circuit spans an entire distance between two junctions. Indicate this direction by drawing a small arrow on the circuit diagram. Label the current in each branch with labels such as $I_1$, $I_2$,....

The current through a resistor is in the direction of decreasing potential. Assuming that the current is positive (that it actually flows in the direction indicated by your arrow), put a plus sign at the high-potential end and a minus sign at the low-potential end of each resistor. Also mark the high-potential and low-potential terminals of each source of emf with + and −.

**3.** Apply the junction rule to some but not all of the junctions. Application of the junction rule to all of the junctions results in redundant information that will most likely complicate your algebra.

**4.** Apply the loop rule once for each interior loop (that is, loops that have no loops interior to themselves). Additional applications of the loop rule only result in redundant information.

**5.** Solve the equations to obtain values for the unknowns. There are various methods of doing this. If you have only three unknowns, you can use one equation to obtain an expression for one of the unknowns in terms of the other two; then substitute this expression into the other two equations. This reduces your algebra to the task of solving two simultaneous equations for the other two unknowns. If you need a more formal method for solving simultaneous equations, look up Cramer's rule in any elementary linear algebra text.

**6.** Check your results by verifying that they satisfy the original equations.

*Analysis of Circuits by Symmetry.* Using symmetry sometimes greatly simplifies certain circuit problems, because in some circuits symmetry considerations indicate that the currents in two resistors are the same or that the potential drops across two resistors are the same.

For example, if a circuit consists of two 4-$\Omega$ resistors, two 7-$\Omega$ resistors, and a 17-$\Omega$ resistor connected as shown in Figure 25-3, you can see from symmetry that the currents in the two 4-$\Omega$ resistors are equal; so the potential drops across them must be equal. Consequently, we know that the potentials at *a* and *b* are equal, and so the potential difference across the 17-$\Omega$ resistor is zero. If the potential difference across the 17-$\Omega$ resistor is zero, then we know that the current in that resistor must also be zero.

*Ammeters, Voltmeters, and Ohmmeters.* Ammeters, voltmeters, and ohmmeters are instruments for measuring current, potential difference (voltage), and resistance. The heart of each of these instruments is the **galvanometer**, a device that indicates the current passing through it. A typical galvanometer has a resistance $R_g$ of $20\,\Omega$ and requires a current $I_g$ of 0.5 mA for a full-scale reading. Galvanometers are delicate ammeters that are useful for measuring currents on the order of 1 mA or less.

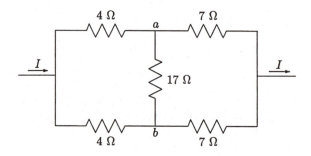

Figure 25-3

To construct an ammeter capable of measuring larger currents, a resistor called a shunt resistor (shunt) is placed in parallel with the galvanometer, as in Figure 25-4. The resistance $R_p$ of the shunt resistor is chosen to be much less than $R_g$ so that most of the current through the ammeter flows through the shunt.

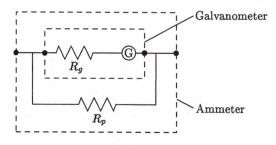

Figure 25-4

To construct a voltmeter, a current-limiting resistor $R_s$ is placed in series with the galvanometer, as in Figure 25-5. (The subscripts p and s stand for parallel and series, respectively.)

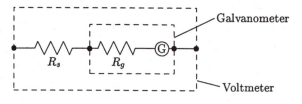

Figure 25-5

The workings of an ohmmeter are shown in Figure 25-6. The current in the circuit is determined by the external resistance to be measured (not shown) connected across the terminals of the instrument.

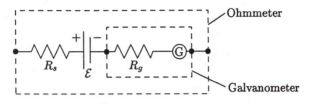

Figure 25-6

***Section 25-6. RC Circuits.*** To apply Kirchhoff's loop rule to a circuit that contains a capacitor, we first arbitrarily label the charge on one plate of the capacitor $+Q$; the other plate, of course, has charge $-Q$. If it is given that a particular plate is positively charged, then it seems appropriate to label that charge $+Q$. If the sign of the charge is not given, a judicious choice is to label as $+Q$ the charge on the plate that becomes more positively charged when the current $I$ enters it, because the current and the charge are then related as shown in Figure 25-7*a*. Otherwise, the charge $Q$ and the current $I$ are related as shown in Figure 25-7*b*.

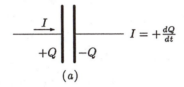

These relations, in combination with Kirchhoff's loop rules, result in equations containing both $Q$ *and* $dQ/dt$. To solve for the charge and the current involves integrating these equations. After the switch is closed in the circuit shown in Figure 25-8, the charge on the capacitor $Q$ is related to the resistance, the capacitance, and the elapsed time by the following equation:

$$Q(t) = Q_0 e^{-t/RC} = Q_0 e^{-t/\tau}$$         Capacitor discharging through a resistor

where $\tau = RC$, called the **time constant,** is the time it takes for the charge to decrease to $1/e$ of its original value $Q_0$.

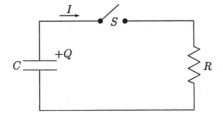

Figure 25-8

If a current $I$ discharges the capacitor, then $I$ is the negative of the time rate of the change of $Q$. Thus,

$$I = -\frac{dQ}{dt} = \frac{Q_0}{RC}e^{-t/RC} = \frac{V_0}{R}e^{-t/\tau} = I_0 e^{-t/\tau}$$

where $V_0$ and $Q_0$ are the initial voltage across and charge on the capacitor. From the definition of capacitance $(C = Q/V)$ we know that $V_0 = Q_0 / C$.

The capacitor in Figure 25-9 is initially uncharged. After the switch is closed, the current charges the capacitor. By applying Kirchhoff's loop rule and solving for the charge, we have

$$Q(t) = C\mathcal{E}\left(1 - e^{-t/RC}\right) = Q_f\left(1 - e^{-t/\tau}\right) \qquad \text{Capacitor charging through a resistor}$$

where $Q_f$ is the final charge as $t \rightarrow \infty$. For this circuit $I$ is the rate of *increase* of the charge $Q$. Thus,

$$I = \frac{dQ}{dt} = \frac{\mathcal{E}}{R}e^{-t/RC} = I_0 e^{-t/\tau}$$

where $I_0 = \mathcal{E}/R$ is the value of the current immediately after the switch is closed at $t \approx 0$.

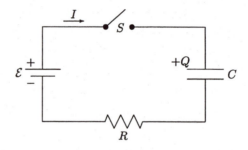

Figure 25-9

# I . Physical Quantities and Key Equations

### Physical Quantities

| | |
|---|---|
| *Current* | 1 ampere (A) $= 1\,\text{C/S}$ |
| *Resistance* | 1 ohm ($\Omega$) $= 1\,\text{V/A}$ |
| *EMF* | 1 V |

### Key Equations

| | |
|---|---|
| *Electric current defined* | $I = \dfrac{\Delta Q}{\Delta t}$ |

| | |
|---|---|
| *Current and drift velocity* | $I = nqAv_d$ |
| *Resistance defined* | $R = \dfrac{V}{I}$ |
| *Ohm's law* | $V = IR \quad R = \text{constant}$ |
| *Resistivity $\rho$* | $R = \rho \dfrac{L}{A}$ |
| *Conductivity $\sigma$* | $\sigma = \dfrac{1}{\rho}$ |
| *Temperature coefficient of resistivity $\alpha$* | $\rho = \rho_{20}\left[1 + \alpha\left(t_C - 20°C\right)\right]$ |
| *Power* | $P = IV = I^2 R = \dfrac{V^2}{R}$ |
| *Terminal voltage of a real battery* | $V_a - V_b = \mathcal{E} - Ir$ |
| *Resistors in series* | $R_{eq} = R_1 + R_2 + R_3 + \ldots$ |
| *Resistors in parallel* | $\dfrac{1}{R_{eq}} = \dfrac{1}{R_1} + \dfrac{1}{R_2} + \dfrac{1}{R_3} + \ldots$ |
| *Capacitor discharging through a resistor* | $Q(t) = Q_0 e^{-t/RC} = Q_0 e^{-t/\tau}$ |
| *Capacitor charging through a resistor* | $Q(t) = C\mathcal{E}\left(1 - e^{-t/RC}\right) = Q_f\left(1 - e^{-t/\tau}\right)$ |

*Circuit symbols*

Battery           Resistor

## III. Potential Pitfalls

When a battery is connected to a resistor, the current through the resistor is in the direction of the electric field within it, which means it is in the direction of decreasing potential. However, within the battery the electric field opposes the current, which means the current through it is in the direction of increasing potential.

Remember that the rules for the equivalent resistance of resistors in series and parallel are exactly opposite to the rules for combining capacitors.

The rules for the equivalent resistance of series and parallel combinations of resistors apply only for those exact combinations. It's easy to look at two resistors in a circuit, for instance, and think they're in series when, actually, another branch forks off between them.

When a capacitor is charged through a resistor, not all of the potential energy delivered by the battery goes into charging the capacitor; some of it is dissipated in the resistor.

When using $P = IV = I^2R = V^2/R$ for determining the power dissipated in a resistor, be sure that $I$ is the current through that resistor and $V$ is the voltage drop across it in the direction of the current.

Resistance depends both on geometry and on a material property, the resistivity. For example, the resistance of a wire with a uniform cross-sectional area $A$ is $R = \rho L/A$.

When you calculate the equivalent resistance of a parallel combination by adding the inverse resistances, don't forget that you have to invert the result to obtain the equivalent.

When you apply Kirchhoff's loop rule, be careful to remember that the potential decreases as you go past a resistor in the direction of the current but increases if you are going "upstream." Likewise, the voltage increases as you go through a source of emf from the negative to the positive terminal, and vice versa. It's easy to make sign errors. It will help to ask yourself if you are moving in the direction of the electric field or not. The electric field always points in the direction of decreasing potential.

You have to assign a direction to the current in each branch of a circuit to apply Kirchhoff's rules. It doesn't matter at all if the actual direction of one or more currents is opposite to the direction you assign it. When you solve the problem, that current will come out negative.

Similarly, if the actual polarity of a capacitor is opposite to the polarity that you assign, when you solve for the charge (or voltage), it will come out negative.

A voltmeter measures the potential difference between its terminals. Its terminals are connected to the two points between which you want to know the potential difference. An ammeter measures the current through it. It must be connected so that the current you want to measure flows through the meter.

Voltmeters are connected in parallel with the potential difference being measured. To minimize their effect on the circuit, they should draw as little current as possible, and so a good voltmeter has a very high resistance. Ammeters are connected in series with the circuit element for which the current is to be measured. Because the full current flows through an ammeter, it should have a small resistance so that it does not significantly reduce the current being measured.

To choose the correct sign in the relation $I = \pm dQ/dt$ you must consider whether a particular current $I$ increases or decreases the charge $Q$ on a capacitor.

## IV.  True or False Questions and Responses

**True or False**

_____  1.  In a metal the drift velocity of the free electrons is in the direction of the current.

_____  2.  A conductor is a material that contains free charge carriers.

_____ 3.  The drift velocity of the free electrons in a current-carrying wire refers to the average speed of the free electrons.

_____ 4.  For an ohmic material the drift velocity is proportional to the electric field.

_____ 5.  For an ohmic material the resistance is independent of the current.

_____ 6.  For an ohmic material the resistance is independent of the temperature.

_____ 7.  The equation $V = IR$ can only be applied to resistors made with ohmic materials.

_____ 8.  In a metallic resistor the free electrons move from low potential to high potential.

_____ 9.  To charge a rechargeable battery requires that, during the recharging cycle, its terminal voltage is maintained at a level in excess of its emf.

_____ 10.  Kirchhoff's rules apply only to circuits that contain resistors but not capacitors.

_____ 11.  If a battery is shorted out by connecting its terminals with a wire of zero resistance, the terminal voltage is zero.

_____ 12.  The resistivity of most metals decreases as the temperature increases.

_____ 13.  When the formula $P = IV$ is applied to a resistor, $V$ is the increase in potential in the direction of the current, and $P$ is the rate at which electrical potential energy is dissipated as thermal energy in the resistor.

_____ 14.  Kirchhoff's loop rule follows directly from the fact that a conservative electric field is present.

_____ 15.  Kirchhoff's rules apply to circuits that contain only ohmic materials.

_____ 16.  The time constant associated with the discharge of a capacitor through a resistor is $R/C$.

_____ 17.  An ideal ammeter has a very large internal resistance.

_____ 18.  An ideal voltmeter has a very large internal resistance.

_____ 19.  A galvanometer is a device that quantitatively indicates small currents.

_____ 20.  An ohmmeter contains a battery and a capacitor.

**Responses to True or False**

1. False. Because the free electrons carry a negative charge, the current direction is opposite to the direction of the drift velocity.

2. True.

3. False. The drift velocity is the average velocity of the free electrons, not the average speed.

When the current is zero, the average velocity of the free electrons is zero; but at room temperature the average speed is very large.

4.  True.

5.  True.

6.  False. A material is said to be ohmic if the resistance at a given temperature does not vary with current.

7.  True.  This is the definition of an ohmic material at constant temperature.

8.  True. The electric field is in the direction of decreasing potential. Because the electrons carry a negative charge, the force exerted on them by the electric field is directed opposite to the electric field and in the direction of increasing potential.

9.  True.

10. False. Kirchhoff's rules apply to all circuits except those that operate at very high frequencies.

11. True.

12. False. The resistivity of most metals increases as the temperature increases.

13. False. If the formula $P = IV$ is applied to a resistor, $V$ is the *potential drop* in the direction of the current and $P$ is the rate at which electrical potential energy is dissipated as thermal energy in the resistor.

14. True.

15. False. Kirchhoff's rules apply to all circuits except for those operating at very high frequencies.

16. False. The time constant is the product $RC$. Check the units using the definitions of resistance $(R = V / I)$ and capacitance $(C = Q / V)$.

17. False. An ammeter has a low-resistance shunt resistor in parallel with a galvanometer. An ammeter with a large resistance would significantly decrease the current in a circuit—which, of course, is undesirable if you want to know what the current is when the meter is not present.

18. True. If a voltmeter had a low resistance, it would short out any device it is placed in parallel with.

19. True.

20. False. An ohmmeter consists of a battery, a galvanometer, and a series resistor. The purpose of the series resistor is to protect the galvanometer by limiting the current. The current is maximum when the terminals of the ohmmeter are externally shorted. Under this condition the current would become excessive if the series resistor were not in place.

## V.   Questions and Answers

**Questions**

1.  We justified a number of electrostatic phenomena by the argument that there can be no electric field in a conductor. Now we say that the current in a conductor is driven by a potential difference and thus there is an electric field in the conductor. Is there a contradiction here?

2.  The average (drift) velocity of electrons in a wire carrying a steady current is constant even though the electric field within the wire is doing work on the electrons. What happens to this energy?

3.  When 120 V is applied to the filament of a 75-W light bulb, the current drawn is 0.63 A. When a potential difference of 3 V is applied to the same filament, the current is 0.086 A. Is the filament made of an ohmic material?

4.  Under ordinary conditions the drift speed of electrons in a metal wire is about 6 mm/min, or less. When you flip on the wall switch, why doesn't it take an hour or so before a light bulb several meters away begins to light up?

5.  Today, ordinary strings of Christmas-tree lights contain eight or so bulbs connected in parallel across a 110-V line. Forty years ago most strings contained eight bulbs connected in series across the line. What would happen if you put one of the old-style bulbs into a modern Christmas-tree light set? (The light sockets are made differently to prevent this.)

6.  Consider the circuit in Figure 25-10. Can you simplify it by replacing a series or parallel combination of resistors with a resistor of equivalent resistance?

7.  Two wires, $A$ and $B$, have the same physical dimensions but are made of different materials. If $A$ has twice the resistance of $B$, how do their resistivities compare?

8.  A variable resistor is connected across the terminals of an ideal battery. Will the rate at which electrical energy is dissipated in the resistor increase or decrease as the resistor's resistance decreases?

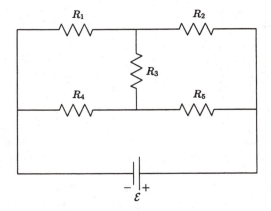

Figure 25-10

9. Does the time required to charge a capacitor through a given resistor with a battery depend on the emf? Does it depend on the total amount of charge to be placed on the capacitor?

10. Give a simple physical explanation of why the charge on a capacitor in an *RC* circuit can't be changed "instantaneously."

11. In order to modify an electrical circuit you need to know the resistance of a specific resistor that is installed in the original circuit. To measure this resistance with an ohmmeter, you must first remove the resistor from the circuit. Why?

12. On a hot day a co-worker of mine needed to measure the voltage at a wall outlet. He set his multimeter (a combination voltmeter, ohmmeter, and ammeter) to measure voltage and connected its terminals to a wall outlet. The voltage reading was about 112 V. He then asked "I wonder what the current is?" and changed the setting of the meter to measure current. What do you think happened when the meter became an ammeter?

**Answers**

1. There is no contradiction. If a conductor is in electrostatic equilibrium, the electric field within it must be zero. However, any conductor that is carrying a current is definitely not in electrostatic equilibrium.

2. The energy is dissipated, heating up the wire. If an electron collides with the ion lattice, it transfers energy to the lattice. In this way the work done by the electric field on the electrons shows up as an increase in the thermal energy of the lattice.

3. Just calculating the resistance of the filament $( R = V / I )$ gives 190 Ω when 120 V is applied and 35 Ω when 3 V is applied. Noting that the resistance varies with the current might lead you to conclude that the filament is made of a nonohmic material. This conclusion would not be justified. The temperature of the filament with 120 V across it is much greater than its temperature with only 3 V across it. Most lamp filaments are made from the metal tungsten. Like most metals, tungsten is ohmic, with a resistivity that increases with temperature. The operating temperature of a typical light bulb filament is about 2800 K, so we expect the resistance of the filament to be higher when 120 V are across it even though tungsten is ohmic. If the temperature of a tungsten filament were kept constant, its resistance would be independent of current.

4. It is the electric field within the wire that causes the electrons to drift and create a current. This electric field is set up throughout the wire almost instantaneously after the switch is turned on, so the light bulb lights up at once. This is somewhat analogous to the question of how long it takes a sprinkler on the other end of a hose to start sprinkling after you open the water valve. The answer depends on whether or not the hose is full of water at the time the valve is opened. If it is full of water, then the sprinkler starts sprinkling almost immediately. An electric wire is like a hose that is always full of water. The wire is always full of free electrons.

5. The voltage drop across each bulb in the old series string was about one-eighth of 110 V, or 14 V. The modern parallel connection puts the full 110 V across each bulb. Placing 110 V across

one of the old bulbs, designed to operate at 14 V, would result in excessive current in the filament, which would burn out the bulb at once, perhaps spectacularly.

6. No. There are no series or parallel combinations of resistors in this circuit.

7. The equation for resistivity is $R = \rho L / A$. If the geometric factors $L$ and $A$ are the same, then if the resistance is twice as large, the resistivity is also twice as large.

8. The terminal voltage $V_t$ of an ideal battery is constant. Thus, lowering the resistance results in an increased current $I$. The rate of joule heating in the resistor is equal to the product $IV_t = V_t^2 / R$, so the rate of heating will increase.

9. This depends on what you mean. The time required to charge the capacitor to a given fraction (say 99%) of its final charge depends only on the time constant $RC$. However, the time to charge the capacitor to, say, 5 V with a 6-V battery is longer than the time to charge it to 5 V with a 12-V battery.

10. It cannot be changed instantaneously because the resistor in the circuit limits the current (the rate of flow of charge). Because the current is finite, it requires time for the charge to flow on and off the capacitor plates.

11. The ohmmeter measures the current in the galvanometer (see Figure 25-6); so if the resistor remains in the circuit, there may be parallel paths for the current to flow around it. Also, the circuit may contain batteries or charged capacitors that could supply additional current.

12. If you remember that an ammeter is a low-resistance device, then you will realize that when 112 V is placed across it, the current will be very large. In this case there was a loud pop like a firecracker, some smoke, and a very startled co-worker. The 112 V delivered a mortal blow to the meter.

## VI. Problems, Solutions, and Answers

**Example #1.** A piece of 14-gauge copper wire 0.163 cm in diameter is 14.0 m long. (*a*) What is its resistance? (*b*) If a potential difference of 1.00 V is applied across the wire, what current flows in it? (*c*) What is the electric field in the wire? The resistivity of copper is $1.7 \times 10^{-8}$ $\Omega$·m.

**Picture the Problem.** Calculate the resistance of the wire, and use Ohm's law to find the current in the wire. The electric field in the wire is the voltage drop per unit length.

| | |
|---|---|
| 1. Calculate the resistance of the wire. | $R = \dfrac{\rho L}{A} = \dfrac{\rho L}{\pi (d/2)^2} = \dfrac{\left(1.7 \times 10^{-8} \ \Omega \text{·m}\right)\left(14.0 \text{ m}\right)}{(\pi/4)\left(1.63 \times 10^{-3} \text{ m}\right)^2} = 0.11 \Omega$ |
| 2. Use Ohm's law to find the current in the wire. | $I = V / R = (1.00 \text{ V})/(0.11 \Omega) = 9.1 \text{ A}$ |
| 3. Find the electric field. | $E = V / L = (1.00 \text{ V})/(14.0 \text{ m}) = 0.0714 \text{ V/m}$ |

It takes a very small field to produce a significant current in a wire.

**Example #2—Interactive.**  A power transmission line is made of copper $1.80\,cm$ in diameter. If the resistivity of copper is $1.7\times10^{-8}\,\Omega\cdot m,$ find the resistance of one mile of this line.

**Picture the Problem.**  Convert the length and diameter to meters, and use these values to find the resistance. **Try it yourself.**  Work the problem on your own, in the spaces provided, to get the final answer.

| | |
|---|---|
| 1.  Convert the wire length to meters. | |
| 2.  Convert the diameter to meters and find the cross-sectional area of the wire. | |
| 3.  Calculate the resistance directly from the geometry of the wire and its resistivity. | $R = 0.11\Omega$ |

**Example #3.**  An electric heater consists of a single resistor connected across a 110-V line. It is used to heat $200.0\,g$ of water in a cup (to make instant coffee) from $20.0°C$ to $90.0°C$ in 2.70 min. Assuming that 90% of the energy drawn from the power source goes into heating the water, what is the resistance of the heater?

**Picture the Problem.**  The heat gained by the water is equal to 90% of the energy dissipated by the resistor.  We can express the power dissipated by the resistor in terms of resistance and voltage, and express the heat gained by the water in terms of its change in temperature.

| | |
|---|---|
| 1.  Find an expression for the power dissipated by the resistor. | $P = V^2/R$ |
| 2.  The total energy dissipated by the resistor is the power multiplied by the time over which that energy is dissipated. | $E = Pt = \left(V^2/R\right)t$ |
| 3.  Find an expression for the heat gained by the water.  You may want to refer back to Chapter 18. | $Q = m_{water}c_{water}\,\Delta T$ |

| 4.  90% of the energy dissipated by the resistor is equal to the heat gained by the water.  Solve for the resistance of the heater. | $0.9\left(V^2/R\right)t = m_{water}c_{water}\,\Delta T$ <br><br> $R = \dfrac{0.9V^2t}{m_{water}c_{water}\,\Delta T}$ <br><br> $= \dfrac{0.9(110\,\text{V})^2(2.7\,\text{min})(60\,\text{s}/1\,\text{min})}{(0.2\,\text{kg})\left(4180\,\text{J}/(\text{kg}\cdot{}^\circ\text{C})\right)(90.0^\circ\text{C}-20.0^\circ\text{C})}$ <br><br> $= 30.1\,\Omega$ |
|---|---|

**Example #4—Interactive.** For a silver wire 0.100 inch in diameter and 100 feet long, carrying a current of 25.0 A, find (*a*) the resistance (*b*) the potential difference between the ends of the wire, (*c*) the electric field in it, and (*d*) the rate at which heat is generated in the wire.

**Picture the Problem.** Convince yourself you should follow the steps in Example #2 for part (*a*), then use the steps in Example #1 for parts (*b*) and (*c*), and use the power relationship for part (*d*). You will need to look up the resistivity of silver in Table 25-1 on page 792 of the main text. **Try it yourself.** Work the problem on your own, in the spaces provided, to get the final answer.

| 1.  Convert the wire length to meters. | |
|---|---|
| 2.  Convert the diameter to meters and find the cross-sectional area of the wire. | |
| 3.  Calculate the resistance directly from the geometry of the wire and its resistivity. | $R = 0.096\,\Omega$ |
| 4.  Use Ohm's law to find the potential difference between the ends of the wire. | $V = 2.4\,\text{V}$ |
| 5.  The electric field is the voltage drop per unit length of the wire. | $E = 0.079\,\text{N/C}$ |
| 6.  Determine the power dissipated in this wire. | $P = 60\,\text{W}$ |

**Example #5.** An automotive battery has a terminal voltage of 12.5 V when it is delivering 30.0 A to the starter motor of a car. Under different load conditions, it delivers 80.0 A and its terminal voltage is 10.7 V. Find the constant internal resistance and the emf of the battery.

**Picture the Problem.** In each case, the circuit diagram consists of the emf of the battery, the internal resistance of the battery, and the resistance of the external load. The terminal voltage of the battery is equal to its emf minus the voltage drop across its internal resistance.

| | |
|---|---|
| 1. Draw a sketch of the circuit. Here, $\mathcal{E}$ is the emf of the battery, and $r$ is its internal resistance. | |
| 2. Find an expression for the terminal voltage, and substitute in values for the two cases. | $V_t = \mathcal{E} - Ir$<br>$12.5\,\text{V} = \mathcal{E} - (30.0\,\text{A})r$<br>$10.7\,\text{V} = \mathcal{E} - (80.0\,\text{A})r$ |
| 3. Subtract the two expressions obtained in step 2 to find the internal resistance. | $1.8\,\text{V} = (50.0\,\text{A})r$<br>$r = 0.0360\,\Omega$ |
| 4. Substitute the value for the internal resistance into one of the expressions obtained in step 2 to find the emf of the battery. | $10.7\,\text{V} = \mathcal{E} - (80.0\,\text{A})(0.0360\,\Omega)$<br>$\mathcal{E} = 13.6\,\text{V}$ |

**Example #6—Interactive.** A potential of 3.60 V is applied between points $a$ and $b$ in the circuit shown in Figure 25-11. Find the current in each of the resistors and the total current through the combination.

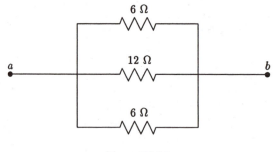

Figure 25-11

**Picture the Problem.** The given voltage is the same across all three resistors, so Ohm's law can be used to find the current through each resistor. The total current may be found using Kirchhoff's junction rule. **Try it yourself.** Work the problem on your own, in the spaces provided, to get the final answer.

| 1. Use Ohm's law to determine the current through each resistor. | |
|---|---|
| | $I_{6,top} = 0.600\,\text{A}, I_{12} = 0.300\,\text{A}, I_{6,bottom} = 0.600\,\text{A}$ |
| 2. Use Kirchhoff's junction rule to find the total current. | |
| | $I_{total} = 1.5\,\text{A}$ |

**Example #7.** A potential difference of 7.5 V is applied between points $a$ and $c$ in the circuit shown in Figure 25-12. Find the difference in potential between points $b$ and $c$.

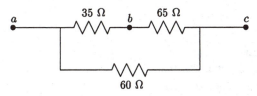

Figure 25-12

**Picture the Problem.** The given voltage is across the combination of the 35$\Omega$ and 65$\Omega$ resistors, which are in series. Once you know the current through the equivalent series resistance, you can use Ohm's law a second time to find the voltage drop across the 65$\Omega$ resistor alone.

| 1. Find the equivalent resistance of the two resistors in series. | $R_{eq} = R_1 + R_2 = (35\,\Omega) + (65\,\Omega) = 100\,\Omega$ |
|---|---|
| 2. Use Ohm's law to determine the current through the equivalent resistance required to maintain the given potential difference. | $I = V / R = (7.5\,\text{V}) / (100\,\Omega) = 0.075\,\text{A}$ |
| 3. The current through the equivalent resistance is the same as the current through each resistor is series. Use the current from step 2 to find the potential difference across the 65$\Omega$ resistor. | $V = IR = (0.075\,\text{A})(65\,\Omega) = 4.88\,\text{V}$ |

**Example #8—Interactive.** In the circuit of Figure 25-13 a potential difference of 5.00 V is applied between points $a$ and $b$. Find ($a$) the equivalent total resistance, ($b$) the current in each resistor, and ($c$) the power being dissipated in each resistor.

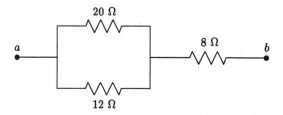

Figure 25-13

**Picture the Problem.**  You will need to find two equivalent resistances, one for the two resistors in parallel, and one for all three resistors.  Remember that the voltage across resistors in parallel is the same for all resistors, and the current is the same through all resistors in series.  **Try it yourself.**  Work the problem on your own, in the spaces provided, to get the final answer.

| | |
|---|---|
| 1.  Find the equivalent resistance of the two resistors in parallel. | |
| 2.  The equivalent resistance calculated in step 1 is in series with the $8\Omega$ resistor.  Calculate the total equivalent resistance of this series combination. | $R_{eq,total} = 15.5\Omega$ |
| 3.  Using Ohm's law, find the total current through the equivalent resistance.  This total current is the same current which flows through the $8\Omega$ resistor. | $I_{total} = 0.323\,A$ |
| 4. Find the potential difference across the two resistors in parallel, using their equivalent resistance calculated in step 1, and the total current just calculated. | |
| 5.  Use this potential difference to determine the current flowing through the two parallel resistors. | $I_{20} = 0.121\,A;\ I_{12} = 0.202\,A$ |
| 6.  Knowing the resistances and the current through each resistor, you can calculate the power dissipated in each resistor. | $P_{20} = 0.293\,W,\ P_{12} = 0.490\,W,\ P_{8} = 0.835\,W$ |

**Example #9.** For the circuit shown in Figure 25-14, find (*a*) the current in each resistor, (*b*) the power supplied by each source of emf, and (*c*) the power dissipated in each resistor.

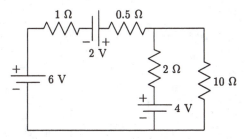

Figure 25-14

**Picture the Problem.** There are three unknown currents, so we need three equations. Therefore, apply the junction rule on one of the junctions, and apply the loop rule to the two interior loops to solve for the current in each segment of the circuit. Once you know the current, you can calculate the power dissipated by each resistor and provided by each source of emf.

| | |
|---|---|
| 1. Redraw the circuit, with labels for the current in each branch, and arrows indicating your guess for the direction of current flow in each branch. Also place plus and minus signs indicating the polarity for each resistor based on your directions for the current flow. | ![circuit diagram with labels $I_1$, $1\,\Omega$, $0.5\,\Omega$, $I_3$, $2\,V$, $6\,V$, $2\,\Omega$, $I_2$, $10\,\Omega$, $4\,V$] |
| 2. Apply the junction rule to the top junction. | $I_1 + I_2 = I_3$ |
| 3. Apply the loop rule to the leftmost loop, starting at the lower left corner and moving clockwise. | $+(6\,\text{V}) - (1\,\Omega) I_1 + (2\,\text{V}) - (0.5\,\Omega) I_1$ <br> $+(2\,\Omega) I_2 - (4\,\text{V}) = 0$ |
| 4. Apply the loop rule to the rightmost loop, starting at its lower left corner and moving clockwise. Reduce this expression by dividing by 2. | $+(4\,\text{V}) - (2\,\Omega) I_2 - (10\,\Omega) I_3 = 0$ <br> $+(2\,\text{V}) - (1\,\Omega) I_2 - (5\,\Omega) I_3 = 0$ |
| 5. Solve the equation in step 2 for $I_1$. | $I_1 = I_3 - I_2$ |
| 6. Substitute this value into the equation from step 3, and simplify. We multiply the entire equation by 2 to assist in the next step. | $+(6\,\text{V}) - (1\,\Omega)(I_3 - I_2) + (2\,\text{V}) - (0.5\,\Omega)(I_3 - I_2)$ <br> $+(2\,\Omega) I_2 - (4\,\text{V}) = 0$ <br> $(4\,\text{V}) + (3.5\,\Omega) I_2 - (1.5\,\Omega) I_3 = 0$ <br> $(8\,\text{V}) + (7.0\,\Omega) I_2 - (3.0\,\Omega) I_3 = 0$ |

| | |
|---|---|
| 7.  Multiply the result from step 4 by 7, and add it to the result from step 6, solving for $I_3$. | $(22\,\text{V}) - (38\,\Omega)\,I_3 = 0$ <br><br> $I_3 = 0.579\,\text{A}$ |
| 8.  Substitute this value back into the result from step 4 to solve for $I_2$. The negative sign simply means that the current $I_2$ really flows opposite to the direction we initially chose. | $+(2\,\text{V}) - (1\,\Omega)\,I_2 - (5\,\Omega)(0.579\,\text{A}) = 0$ <br><br> $I_2 = -0.895\,\text{A}$ |
| 9.  Substitute values for $I_2$ and $I_3$ into the junction rule to solve for $I_1$. | $I_1 = (0.579\,\text{A}) - (-0.895\,\text{A}) = 1.474\,\text{A}$ |
| 10.  Find the power supplied by each source of emf.  If the 4-V battery is a rechargeable battery, it is storing charge, rather than providing it.  If the battery is not rechargeable, then 3.58 W is dissipated as heat. | $P = \mathcal{E}I$ <br><br> $P_{6\text{V}} = (6\,\text{V})\,I_1 = (6.00\,\text{V})(1.474\,\text{A}) = 8.84\,\text{W}$ <br><br> $P_{2\text{V}} = (2\,\text{V})\,I_1 = (2.00\,\text{V})(1.474\,\text{A}) = 2.95\,\text{W}$ <br><br> $P_{4\text{V}} = (4\,\text{V})\,I_2 = (4.00\,\text{V})(-0.895\,\text{A}) = -3.58\,\text{W}$ |
| 11.  Determine the power dissipated by each resistor.  The total power dissipated by the resistors is equal to the total power provided by the batteries. | $P = I^2 R$ <br><br> $P_{1\Omega} = I_1^2 (1\,\Omega) = (1.474\,\text{A})^2 (1.00\,\Omega) = 2.17\,\text{W}$ <br><br> $P_{0.5\Omega} = I_1^2 (0.5\,\Omega) = (1.474\,\text{A})^2 (0.500\,\Omega) = 1.09\,\text{W}$ <br><br> $P_{2\Omega} = I_2^2 (2\,\Omega) = (-0.895\,\text{A})^2 (2.00\,\Omega) = 1.60\,\text{W}$ <br><br> $P_{10\Omega} = I_3^2 (10\,\Omega) = (0.579\,\text{A})^2 (10.0\,\Omega) = 3.35\,\text{W}$ |

**Example #10—Interactive.**  Find the current in each of the three resistors of the circuit shown in Figure 25-15.

**Picture the Problem.**  Use the same steps used to find the currents in Example #9.  **Try it yourself.**  Work the problem on your own, in the spaces provided, to get the final answer.

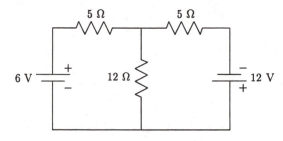

Figure 25-15

| | |
|---|---|
| 1.  Redraw the circuit, with labels for the current in each branch, and arrows indicating your guess for the direction of current flow in each branch.  Also place plus and minus signs indicating the polarity for each resistor based on your directions for the current flow. | |
| 2.  Apply the junction rule to the top junction. | |
| 3.  Apply the loop rule to the leftmost loop. | |
| 4.  Apply the loop rule to the rightmost loop. | |
| 5.  Solve the three equations for the three unknown currents. | $I_{5\Omega,\text{left}} = 1.74\,\text{A}$, to the right<br><br>$I_{5\Omega,\text{right}} = 1.95\,\text{A}$, to the right<br><br>$I_{12\Omega} = 0.207\,\text{A}$, upward |

**Example #11.**  A 12-V battery with an internal resistance of $0.600\,\Omega$ is used to charge a $0.200\text{-}\mu\text{F}$ capacitor through a $5.00\text{-}\Omega$ resistor.  Find (*a*) the initial current drawn from the battery, (*b*) the time constant of the circuit, and (*c*) the time required to charge the battery to 99% of the final charge.

**Picture the Problem.**  Initially, the uncharged capacitor behaves like a short circuit.  With no charge on it, there is no potential difference across the plates.  Use the relationship for the time dependence of the charge on the capacitor to solve for part (*c*).

| 1. Draw the circuit. Label the current direction, and the plate of the capacitor with charge Q. | |
|---|---|
| 2. Apply Kirchhoff's loop rule to the circuit. | $\mathcal{E} - Ir - IR - Q/C = 0$ |
| 3. Initially, there is no charge on the capacitor. Solve for the initial current $I$. | $I = \dfrac{\mathcal{E}}{r+R} = \dfrac{12\,\text{V}}{(0.6\,\Omega)+(5\,\Omega)} = 2.14\,\text{A}$ |
| 4. Calculate the time constant. The internal resistance adds to the equivalent resistance of the circuit, and impacts the time constant. | $\tau = R_{eq}C = (5.6\,\Omega)(0.200\,\mu\text{F}) = 1.12 \times 10^{-6}\,\text{s}$ |
| 5. Use the expression for the time-dependence of the charge to solve for the time to reach 99% of the final charge. | $Q = Q_f\left(1 - e^{-t/\tau}\right)$ <br><br> $e^{-t/\tau} = 1 - \left(Q/Q_f\right)$ <br><br> $-t/\tau = \ln\left(1 - \left(Q/Q_f\right)\right)$ <br><br> $t = -\tau \ln\left(1 - \left(Q/Q_f\right)\right)$ <br><br> $= -\left(1.12 \times 10^{-6}\,\text{s}\right)\ln\left(1 - \left(0.99 Q_f / Q_f\right)\right) = 5.16 \times 10^{-6}\,\text{s}$ |

**Example #12—Interactive.** A 10.0-$\mu$F capacitor has an initial charge of 80.0 $\mu$C. (*a*) If a resistance of 25.0 $\Omega$ is connected across it, what is the initial current through the resistor? (*b*) What is the time constant of the circuit?

**Picture the Problem.** As this capacitor discharges, the potential difference across it will always be the same as the potential difference across the resistor. This can be used to determine the initial current through the circuit. **Try it yourself.** Work the problem on your own, in the spaces provided, to get the final answer.

| 1. Draw the circuit. Label the current direction, and the plate of the capacitor with charge Q. | |
|---|---|
| | |

| | |
|---|---|
| 2. Determine the initial voltage across the capacitor, which is the same as the initial voltage across the resistor. | |
| 3. Use Ohm's law to determine the initial current through the resistor. | $I_0 = 0.320\,\text{A}$ |
| 4. Calculate the time constant. | $\tau = 2.5 \times 10^{-4}\,\text{s}$ |

**Example #13.** For the circuit shown in Figure 25-16, find (*a*) the current through the battery just after the switch is closed, (*b*) the steady-state current through the battery after the switch has been closed a long time, and (*c*) the maximum charge on the capacitor.

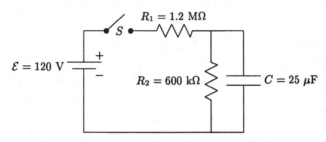

Figure 25-16

**Picture the Problem.** Initially, all the current goes into the capacitor, and none through $R_2$ because there will be no charge, and hence no potential difference across the capacitor or $R_2$. After a long time, the capacitor will be fully charged, so no more charge will flow into it.

| | |
|---|---|
| 1. Sketch the circuit, labeling the current in each segment, and indicating its direction. Label the polarity of the capacitor. | 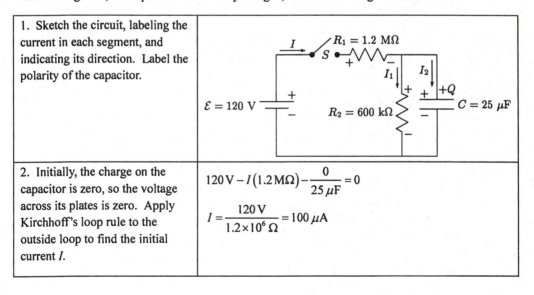 |
| 2. Initially, the charge on the capacitor is zero, so the voltage across its plates is zero. Apply Kirchhoff's loop rule to the outside loop to find the initial current *I*. | $120\,\text{V} - I(1.2\,\text{M}\Omega) - \dfrac{0}{25\,\mu\text{F}} = 0$ <br><br> $I = \dfrac{120\,\text{V}}{1.2 \times 10^6\,\Omega} = 100\,\mu\text{A}$ |

| 3. In the steady state, no current flows through the capacitor, which means that $I = I_1$. Apply Kirchhoff's loop rule to the left hand loop to find the steady-state current provided by the battery. | $(120\,\text{V}) - I(1.2\,\text{M}\Omega) - I_1(600\,\text{k}\Omega) = 0$<br><br>$(120\,\text{V}) - I(1.2\,\text{M}\Omega) - I(600\,\text{k}\Omega) = 0$<br><br>$I = \dfrac{120\,\text{V}}{1.8 \times 10^6\,\Omega} = 66.7\,\mu\text{A}$ |
|---|---|
| 4. Determine the potential difference across $R_2$ in the steady-state condition. This is equal to the voltage across the capacitor when it is fully charged. | $V_C = V_{R2} = IR = \left(66.7 \times 10^{-6}\,\text{A}\right)\left(600 \times 10^3\,\Omega\right)$<br><br>$= 40.0\,\text{V}$ |
| 5. Use the final voltage across the capacitor to determine the final charge on the capacitor. | $Q = CV = (25.0\,\mu\text{F})(40.0\,\text{V}) = 1\,\text{mC}$ |

**Example #14—Interactive.** For the circuit shown in Figure 25-17, just after the switch is closed find (*a*) the current through the battery and (*b*) the current through the 200-k$\Omega$ resistor. (*c*) Find the steady-state charge on the capacitor after the switch has been closed a long time.

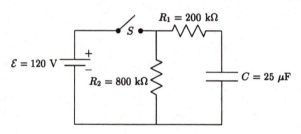

Figure 25-17

**Picture the Problem.** This problem is very similar to Example #13. Follow the solution presented there. **Try it yourself.** Work the problem on your own, in the spaces provided, to get the final answer.

| 1. Sketch the circuit, labeling the current in each segment, and indicating the direction of current flow. Label the polarity of the capacitor. | |
|---|---|
| 2. Initially, the charge on the capacitor is zero, so the voltage across its plates is zero. Apply Kirchhoff's loop rule to the outside loop to find the initial current through the capacitor and $R_1$. | |
| | $I_{R1} = 0.600\,\text{mA}$, to the right |

| 3. Apply Kirchhoff's loop rule to the leftmost loop to find the initial current through $R_2$. | |
|---|---|
| 4. Use Kirchhoff's junction rule to find the total current provided by the battery. | $I_{total} = 0.750\,\text{mA}$, upward |
| 5. In the steady state, no current flows through the capacitor, so no current will flow through $R_1$, either. Apply Kirchhoff's loop rule to the outside loop to find the final charge on the capacitor. | $Q = 3.00\,\text{mC}$ |

**Example #15.** In a certain galvanometer, full-scale deflection corresponds to a current of 0.200 mA. The galvanometer's internal resistance is $50.0\,\Omega$. (a) What resistance must be placed in parallel with it to make an ammeter that reads 3.00 mA full scale? (b) What resistor must be place in series with it to make a voltmeter that reads 1.00 V full scale?

**Picture the Problem.** For part (a) design a circuit so that the combination passes 3.00 mA, while only 0.200 mA passes through the galvanometer branch. For part (b), when the galvanometer resistance passes 0.200 mA, the voltage across the series combination must be 1.00 V.

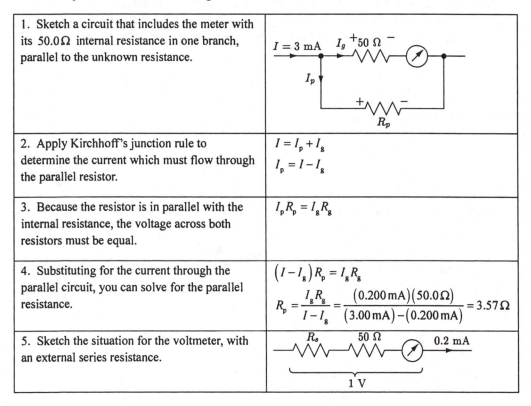

| 1. Sketch a circuit that includes the meter with its $50.0\,\Omega$ internal resistance in one branch, parallel to the unknown resistance. | |
|---|---|
| 2. Apply Kirchhoff's junction rule to determine the current which must flow through the parallel resistor. | $I = I_p + I_g$<br>$I_p = I - I_g$ |
| 3. Because the resistor is in parallel with the internal resistance, the voltage across both resistors must be equal. | $I_p R_p = I_g R_g$ |
| 4. Substituting for the current through the parallel circuit, you can solve for the parallel resistance. | $\left(I - I_g\right) R_p = I_g R_g$<br>$R_p = \dfrac{I_g R_g}{I - I_g} = \dfrac{(0.200\,\text{mA})(50.0\,\Omega)}{(3.00\,\text{mA}) - (0.200\,\text{mA})} = 3.57\,\Omega$ |
| 5. Sketch the situation for the voltmeter, with an external series resistance. | |

| 6. The current through both resistors will be the maximum value when the total potential difference across both resistors and the meter is equal to 1.00 V. | $V = I_g R_{eq} = I_g(r + R)$ $R = \dfrac{V}{I_g} - r = \dfrac{1.00\,\text{V}}{2 \times 10^{-4}\,\text{A}} - (50.0\,\Omega) = 4950\,\Omega$ |
|---|---|

**Example #16—Interactive.** The meter $M$ in Figure 25-18 is calibrated with readings from 0.00 to 1.00, but no units are given. When resistor $R$ is $10\,\Omega$, the meter reads 0.70; when $R$ is $20\,\Omega$, the meter reads 0.42  Find the internal resistance of the meter and the current required for full-scale deflection.

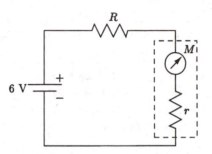

Figure 25-18

**Picture the Problem.** If $I_{max}$ is the current corresponding to full-scale deflection, then the series combination of $10\,\Omega$ and the meter resistance draws a current of $0.70I_{max}$ from the battery. A similar relationship can be determined for the $20\,\Omega$ resistor. These equations can be solved for the desired quantities. **Try it yourself.** Work the problem on your own, in the spaces provided, to get the final answer.

| 1. Relate the current with the $10\,\Omega$ series resistor to the maximum current. | |
|---|---|
| 2. Relate the current with the $20\,\Omega$ series resistor to the maximum current. | |
| 3. Solve these two equations for the two unknowns. | $I_{max} = 0.571\,\text{A}, r = 5.00\,\Omega$ |

# Chapter 26

# The Magnetic Field

## I.  Key Ideas

Magnetism is a fundamental property of matter. Naturally occurring magnetic forces have been known since ancient times. Every magnet has two regions called poles where the magnetic field is strongest. The poles are designated "north" and "south" since a freely suspended magnet orients itself in approximately a north–south direction because the earth itself is a magnet. The north poles of magnets point toward the north geographic pole of the earth. Like poles of two magnets repel each other, and opposite poles attract, with a force that is inversely proportional to the square of the distance between them. Magnetic poles exist only in equal and opposite pairs; and the observation of a single isolated pole, a monopole, has never confirmed. The *source* of the **magnetic field** is really one or more electric charges in motion, that is, electric current.

*Section 26-1. The Force Exerted by a Magnetic Field.* A compass needle is an arrow-shaped magnet, with its tail as the south pole and its tip as the north pole. The existence of a magnetic field $\vec{B}$ at any point in space can be demonstrated by suspending a compass needle there. The magnetic field will exert a torque on the needle, aligning it so that its north end points in the direction of the field.

A magnetic field exerts a force on a charged particle only if the particle is in motion. The magnetic force $\vec{F}$ on a particle with charge $q$ that is moving with velocity $\vec{v}$ in a region with magnetic field $\vec{B}$ is

$$\vec{F} = q\vec{v} \times \vec{B} \qquad \text{Magnetic force on a moving charge}$$

Thus, the direction of the magnetic force is perpendicular to both the velocity of the charged particle and the direction of the field. This equation defines the **magnetic field** $\vec{B}$ in terms of the force exerted on a moving charge. The SI unit of magnetic field is the **tesla** (T), where

$$1 \text{ T} = 1 \text{ N} \cdot \text{s}/(\text{C} \cdot \text{m}) = 1 \text{ N}/(\text{A} \cdot \text{m})$$

Another commonly used unit of magnetic field is the gauss (G), where

$$1 \text{ G} = 10^{-4} \text{ T}$$

The magnetic field strength of the earth is slightly less than $10^{-4}$ T or, equivalently, just less than 1 G.

The magnetic force on a current-carrying wire in a magnetic field is the sum of the forces on all the moving charge carriers in the wire. The net magnetic force on a very small segment of current-carrying wire of length $d\ell$ is

$$d\vec{F} = I \, d\vec{\ell} \times \vec{B} \qquad\qquad \text{Magnetic force on a current element}$$

where the vector $d\vec{\ell}$ is parallel to the direction of the current. The quantity $I \, d\vec{\ell}$ is called a **current element.**

Just as the electric field $\vec{E}$ can be represented by electric field lines, the magnetic field $\vec{B}$ can be represented by magnetic field lines. Whereas electric field lines originate on positive charges and terminate on negative charges, magnetic field lines have no beginnings or ends. They are continuous, and wrap back around onto themselves.

*Section 26-2. Motion of a Point Charge in a Magnetic Field.* The magnetic force on a moving charged particle always acts at right angles to the particle's velocity. Thus, the magnetic force does no work on the particle and has no effect on the particle's kinetic energy. A magnetic force can change the direction of the velocity of a particle but not its magnitude.

Consider an otherwise free charged particle moving with its velocity perpendicular to a uniform magnetic field. The magnetic force has a constant magnitude and acts at right angles to the velocity of the particle. In accordance with Newton's 2nd law $\left(\Sigma\vec{F} = m\vec{a}\right)$, such a particle will move in uniform circular motion. Applying Newton's 2nd law to such a particle results in the equation

$$qvB = m\frac{v^2}{r} \qquad\qquad \text{Circular motion of a charged particle in a uniform magnetic field}$$

where $q$ is the charge, $v$ the speed, $B$ the magnetic field, and $r$ the radius of the circle. The circumference of the circle equals the speed times the time for one revolution (the period $T$). That is, $2\pi r = vT$. The frequency $f$ of the motion is the reciprocal of the period. Thus, combining the last two equations and solving for the reciprocal of the period, we have

$$f = \frac{1}{T} = \frac{v}{2\pi r} = \frac{qB}{2\pi m} \qquad\qquad \text{Cyclotron frequency}$$

Note that the frequency does not depend on either the speed of the particle or the radius of the orbit. In the early 1930s, E. O. Lawrence and M. S. Livingston built a particle accelerator called the **cyclotron** that exploited this result. Consequently, this frequency and period are called the **cyclotron frequency** and **cyclotron period.**

The cyclotron is a machine used to produce high-speed charged particles. In a cyclotron, the particles move in nearly circular orbits in a uniform magnetic field. At two points in the orbit, the particles pass through an electric field produced by the charge distribution on two large D-shaped conductors called dees. The force of the electric field accelerates the particles and they move in orbits of larger and larger radius.

If the velocity of a charged particle is not perpendicular to the magnetic field, the particle moves in a helical, corkscrewlike, path. We can resolve the velocity into a component $v_{\parallel}$ that is parallel to the field and a component $v_{\perp}$ perpendicular to the field. The motion due to $v_{\perp}$ is the same as that just discussed. In accordance with the equation $\vec{F} = q\vec{v} \times \vec{B}$, the magnetic force acts perpendicular to $\vec{v}$, so $v_{\parallel}$ is unaffected by the magnetic field.

Certain instruments require a beam of charged particles in which each particle in the beam moves at the same velocity. A **velocity selector** is a device that can select charged particles moving at a specific velocity while deflecting those moving either faster or slower. It consists of a region of space that contains a mutually perpendicular electric field $\vec{E}$ and magnetic field $\vec{B}$. As shown in Figure 26-1, a charged particle enters this region with a velocity perpendicular to both $\vec{E}$ and $\vec{B}$. The net force on this particle is zero only when the electric and magnetic forces are equal in magnitude and oppositely directed. Equating the magnitudes of these forces, we have $qE = qvB$; so $v = E/B$. Charged particles with this speed pass through the region undeflected. Particles moving either more quickly or more slowly are deflected because the magnetic force is not equal in magnitude to the electric force.

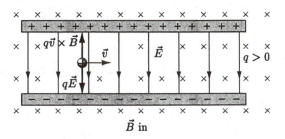

Figure 26-1

A **mass spectrometer** is used to measure the masses of charged particles, such as molecules and ions. In a mass spectrometer a particle with known charge $q$ moving with known speed $v$ enters a region of uniform magnetic field $\vec{B}$, where it moves in uniform circular motion. The mass of the particle determines the radius $r$ of its circular trajectory. Solving the equation for circular motion of a charged particle for the mass $m$ we have $m = qrB/v$.

***Section 26-3. Torques on Current Loops and Magnets.*** A **current loop** is one or more complete turns of a current-carrying wire. The magnetic dipole moment $\vec{\mu}$ of a current loop is

$$\vec{\mu} = NIA\hat{n}$$                                    Magnetic dipole moment of a current loop

where $N$ is the number of turns, $I$ the current in the wire, $A$ the area of the plane surface bounded by a single turn of the wire, and $\hat{n}$ a unit vector normal to the surface. $\hat{n}$ points in the direction of the thumb when the fingers of the right hand curl in the direction of the current. When a current loop is placed in a region containing a uniform magnetic field, the magnetic forces exerted on the loop produce a torque. The magnetic forces exerted by a uniform magnetic field on a current loop result in a *couple,* that is, a pair of oppositely directed offset forces of equal magnitude (for a discussion of couples see page 377 of the text). The torque $\vec{\tau}$ exerted by the field $\vec{B}$ on a current loop is a vector.

$$\vec{\tau} = \vec{\mu} \times \vec{B}$$                    Torque on a current loop

A current loop in a magnetic field has some potential energy that depends on the orientation of the magnetic dipole moment relative to $\vec{B}$.  This energy is at a minimum when they are oriented in the same direction, and at a maximum when they are oriented in opposite directions.

$$U = -\vec{\mu} \cdot \vec{B}$$                    Potential energy of a magnetic dipole

Permanent magnets may also be described in terms of a magnetic dipole moment.  Consequently, the discussions of torque and energy presented here can be applied not only to current loops, but to permanent magnets as well.

*Section 25-4. The Hall Effect.* When a current-carrying wire is placed in a magnetic field $\vec{B}$, the magnetic force exerted on the charge carriers in the wire causes them to drift to one side of the wire. This results in a separation of charge across the wire, called the **Hall effect.** There is a potential difference $V_H$, called the **Hall voltage,** across the width of the wire due to the electric field $\vec{E}$ associated with this charge separation.

$$V_H = Ew = v_d Bw$$                    Hall voltage

where $w$ is the width of the wire and $v_d$ the drift velocity of the charge carriers.

For a given magnetic field strength, the Hall voltage across a wire is proportional to the drift velocity of the charge carriers. Thus, the Hall effect can be used to measure magnetic field strengths. For a known current, magnetic field strength, and wire width, the Hall effect can be used to determine the sign and number density of the charge carriers.

## II.  Physical Quantities and Key Equations

**Physical Quantities**

1 tesla (T) $= 1 \text{ N} \cdot \text{s} / (\text{C} \cdot \text{m}) = 1 \text{ N} / (\text{A} \cdot \text{m})$

1 gauss (G) $= 10^{-4} \text{ T}$

**Key Equations**

*Magnetic force on a moving charge*                    $\vec{F} = q\vec{v} \times \vec{B}$

*Magnetic force on a current element*                    $d\vec{F} = I \, d\vec{\ell} \times \vec{B}$

| | |
|---|---|
| *Cyclotron frequency* | $f = \dfrac{1}{T} = \dfrac{v}{2\pi r} = \dfrac{qB}{2\pi m}$ |
| *Motion of a charged particle in a uniform magnetic field* | $qv_\perp B = m\dfrac{v_\perp^2}{r}$ |
| *Torque on a current loop* | $\vec{\tau} = \vec{\mu} \times \vec{B}$ |
| *Magnetic dipole moment of a current loop* | $\vec{\mu} = NIA\hat{n}$ |
| *Potential energy of a magnetic dipole in a magnetic field* | $U = -\vec{\mu} \cdot \vec{B}$ |
| *Hall voltage* | $V_{\mathrm{H}} = Ew = v_d Bw$ |

## III. Potential Pitfalls

The behaviors of electric and magnetic fields are fundamentally different. The electric field is always parallel to—and the magnetic field always perpendicular to—the direction of the force they exert on a charged particle. Interestingly, both fields exert torques on their respective dipoles that tend to align the dipole moments with the fields. That is, the electric field exerts torques on electric dipoles that tend to align their electric dipole moments with the electric field, and the magnetic field exerts torques on magnetic dipoles (such as compass needles and current loops) that tend to align their magnetic dipole moments with the magnetic field.

What we call the north pole of a magnet is more appropriately called a north-seeking pole because it seeks the geographic north pole of the earth when the magnet is suspended. Because opposite poles attract and like poles repel, it is a south magnetic pole near the geographic north pole that attracts the north pole of the compass.

Remember that the tangent to the magnetic field lines points in the direction of the magnetic field $\vec{B}$, not in the direction of the magnetic force $\vec{F}$ exerted on a moving electric charge. (Magnetic fields cannot be defined to be parallel to the magnetic force because the force depends on the direction of motion of the charge.)

The magnetic force on a moving charged particle is directed perpendicularly to both the field direction and the direction of the velocity of the charge. Therefore, no work is ever done by magnetic forces.

The magnetic force acts only on charges that have a component of velocity perpendicular to the magnetic field. A charge at rest or moving along the field direction experiences no magnetic force.

The general motion of a charged particle in a uniform magnetic field is in a helical path with the axis of the helix in the field direction. In the special case in which the velocity of a particle is directed perpendicularly to the field direction, the motion is uniform circular motion. The period of this motion depends on the magnetic field strength and the charge and mass of the particle. It does not depend on the speed of the particle or the radius of the circular trajectory; if the speed varies, the radius will change proportionally.

In a uniform magnetic field a net torque, but no net force, is exerted on either a permanent magnetic dipole or a current loop.

# IV.  True or False Questions and Responses

**True or False**

_____ 1.   The Earth's south magnetic pole is near its north geographic pole.

_____ 2.   If a bar magnet is broken in half, each half is a bar magnet with both a north and a south pole.

_____ 3.   If the direction of a charged particle's velocity is the same as the direction of the magnetic field, the magnetic force does work on the particle.

_____ 4.   The tesla is an SI unit.

_____ 5.   The magnitude of the earth's magnetic field at the surface of the earth is slightly less than one tesla.

_____ 6.   For a given current, the magnetic force on a current-carrying wire is independent of the charge and drift velocity of the charge carriers in the wire.

_____ 7.   The torque exerted on a current loop by a uniform magnetic field tends to orient the plane of the loop at right angles to the magnetic field.

_____ 8.   Magnetic field lines are drawn so that at any point their direction is that in which a small compass needle would tend to line up.

_____ 9.   In a cyclotron, as the speed of the particle increases, the ratio of the speed to the circumference of the orbit remains constant.

_____ 10. Magnetic field lines always begin on the north pole of a magnet and end on the south pole.

**Responses to True or False**

1. True.

2. True. As far as we know, single isolated magnetic poles do not exist.

3. False. Magnetic forces are always perpendicular to the velocities of the particles they act on. Thus, they never do any work on the particles.

4. True. Although the gauss is not an SI unit, it is commonly used.

5.  False. It is slightly less than 1 G ($10^{-4}$ T ).

6.  True.

7.  True.

8.  True.

9.  True. For this reason, the period of the orbit does not vary with either the speed or the radius of the orbit.

10. False. Outside the magnet, the field lines appear to leave the north pole and end at the south pole but, in actuality, they do not. Magnetic field lines never begin or end; they are continuous through the magnet.

## V.  Questions and Answers

**Questions**

1.  A current-carrying wire is in a region where there is a magnetic field, but there is no magnetic force acting on the wire. How can this be?

2.  Both electric and magnetic fields can exert a force on a moving charge. In a particular case, how could you tell whether it is an electric or a magnetic force that is causing a moving charge to deviate from a straight-line path?

3.  A velocity selector consists of crossed electric and magnetic fields, with the $\vec{B}$ field directed straight up. A beam of positively charged particles passing through the velocity selector from left to right is undeflected by the fields. (*a*) In what direction is the electric field? (*b*) The direction of the particle beam is reversed so that it travels from right to left. Is it deflected? If so, in what direction? (*c*) A beam of electrons (negatively charged) moving with the same speed is passed through from left to right. Is it deflected? If so, in what direction?

4.  Physicists refer to crossed $\vec{E}$ and $\vec{B}$ fields as a velocity selector. In the same sense, the deflection of charged particles in a strong magnetic field perpendicular to their motion can be thought of as a momentum selector. Why is this?

**Answers**

1.  The wire is aligned with the magnetic field.

2.  The simplest way I can think of is to reverse the direction of the velocity of the particle. If the force is magnetic, then reversing the direction of the velocity will result in reversing the direction of the force. However, if the force is electric, then reversing the direction of the velocity will leave the force unaffected.

3.  (*a*) The electric field is perpendicular to both the particle velocity and the magnetic field. Using the right-hand rule, we know that the magnetic force exerted on the positively charged

particles moving to the right is toward us. (We are viewing the beam from the side.) The electric field must be away from us if these particles are to pass undeflected. (*b*) Yes; reversing the direction of the velocity reverses the direction of the magnetic force, but not that of the electric force. Thus, the two forces are now in the same direction and the positively charged particles are deflected in the direction of the electric field. (*c*) Changing the sign of the charge results in both forces reversing direction. They still add to zero so there is no deflection.

4.  Applying Newton's 2nd law to a charged particle whose velocity is at right angles to a uniform magnetic field $\vec{B}$ results in the equation $qvB = mv^2/r$ (or $r = mv/qB$). Thus, as long as the particles have the same charge, the radius of their orbits depends on their momentum, not their velocity alone.

## VI. Problems

**Example #1.**  A horizontal, current-carrying wire of mass 40.0 g is free to slide withouth friction along two vertical conducting rails spaced 80.0 cm apart as in Figure 26-2. A uniform magnetic field of 1.20 T is directed into the plane of the drawing. What magnitude and direction must the current have if the force exerted by the magnetic field on the wire is just sufficient to balance the force of gravity on the wire?

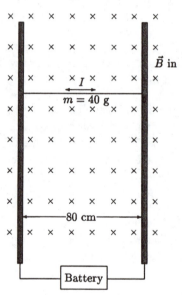

Figure 26-2

**Picture the Problem.**  The net force on the wire is zero.

| 1. Draw a sketch which includes the forces acting on the wire. For the net force on the wire to be zero, the magnetic force must be directed upward. Use the right-hand-rule to determine the direction of current flow that results in an upward magnetic force. | |
|---|---|
| 2. Set up the vector form of Newton's 2nd law, and solve for the current in the wire. | $\vec{w} + \vec{F}_{mag} = 0$ $$mg\left(-\hat{j}\right) + I\vec{L} \times \vec{B} = 0$$ $$mg\left(-\hat{j}\right) + ILB\sin 90° \,\hat{j} = 0$$ $$I = \frac{mg}{LB} = \frac{\left(0.040\,\text{kg}\right)\left(9.81\,\text{m/s}^2\right)}{\left(0.8\,\text{m}\right)\left(1.2\,\text{T}\right)} = 0.409\,\text{A}$$ |

**Example #2—Interactive.** A straight segment of wire 35.0 cm long carrying a current of 1.40 A is in a uniform magnetic field. The segment makes an angle of 53° with the direction of the magnetic field. If the force on the segment is 0.200 N, what is the magnitude of the magnetic field?

**Picture the Problem.** Use the expression for the magnetic force on a segment of current-carrying wire of length $L$. **Try it yourself.** Work the problem on your own, in the spaces provided, to get the final answer.

| 1. Sketch the situation to properly visualize all the directions involved. | |
|---|---|
| 2. Write the vector expression for the magnetic force on a segment of current-carrying wire of length $L$. | |
| 3. Determine the magnitude of the magnetic force from the expression in step 1. Rewrite the equation to solve for the magnitude of the magnetic field. | $B = 0.511\,\text{T}$ |

**Example #3.** There is a uniform magnetic field of 1.20 T in the positive $z$ direction. (*a*) A particle with a charge of $-2.00\,\mu\text{C}$ moves with a speed of 2.20 km/s in the $yz$ plane. Find the magnetic force on the particle when the direction of the velocity makes an angle of 55° with the positive $y$ direction and 35° with the negative $z$ direction. (*b*) A second particle, also with a charge of $-2.00\,\mu\text{C}$, moves with a speed of 2.20 km/s in the $xy$ plane. Find the magnetic force on the particle when the direction of the velocity makes an angle of 55° with the positive $y$ direction and an angle of 35° with the positive $x$ direction.

| | |
|---|---|
| 1. Draw a sketch for each situation, to make sure you have the angle between the field direction and the velocity direction correct.<br><br>By using the right-hand-rule, you find that for part (*a*) the cross product $\vec{v} \times \vec{B}$ is in the positive $x$ direction, so the magnetic force on the negatively charged particle will be in the negative $x$ direction.<br><br>For part (*b*), the right-hand-rule tells us that both $\vec{v}' \times \vec{B}$ and the resultant force will lie in the $xy$ plane. | 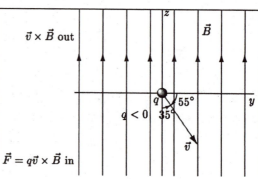<br>The positive-$x$ axis is out of the figure.<br>(*a*)<br><br>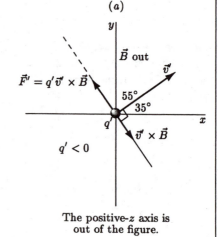<br>The positive-$z$ axis is out of the figure.<br>(*b*) |
| 2. Write the expression for the force on a moving charged particle, and solve for the magnitude of the force for part (*a*). | $\vec{F} = q\vec{v} \times \vec{B}$<br>$F = \|qvB\sin\theta\|$<br>$\quad = \left\|\left(2 \times 10^{-6}\,\text{C}\right)\left(2.2 \times 10^{3}\,\text{m/s}\right)\left(1.2\,\text{T}\right)\sin 145°\right\|$<br>$\quad = 3.03 \times 10^{-3}\,\text{N}$ |

| 3. Write the expression for the force on a moving charged particle, and solve for the magnitude of the force for part (b). Because the motion is in the $xy$ plane, the velocity will also be perpendicular to the direction of $\vec{B}$. | $\vec{F} = q\vec{v} \times \vec{B}$ <br><br> $F = \|qvB \sin\theta\|$ <br><br> $\quad = \left\|\left(2\times10^{-6}\,\text{C}\right)\left(2.2\times10^{3}\,\text{m/s}\right)\left(1.2\,\text{T}\right)\sin90°\right\|$ <br><br> $\quad = 5.28\times10^{-3}\,\text{N}$ |
|---|---|

**Example #4—Interactive.** An electron moves with a velocity of $3.00\times10^{6}$ m/s in the $xy$ plane, making an angle of $40°$ with the positive $x$ direction and $50°$ with the positive $y$ direction. There is a magnetic field of $3.00\,\text{T}$ in the positive $y$ direction. Find the magnitude and direction of the magnetic force on the electron.

**Picture the Problem.** This is very similar to Example #3. Use the right-hand-rule to get the direction of the cross product, and then the force, and calculate the magnitude of the force. **Try it yourself.** Work the problem on your own, in the spaces provided, to get the final answer.

| 1. Draw a sketch of the situation, to make sure you have the angle between the field direction and the velocity direction correct. <br><br> By using the right-hand-rule find the direction of the magnetic force. | <br><br><br><br> Force is in the negative $z$ direction |
|---|---|
| 2. Write the expression for the force on a moving charged particle, and solve for the magnitude of the force. | <br><br><br><br><br><br> $F = 1.10\,\text{pN}$ |

**Example #5.** There is a uniform magnetic field of $2.20\,\text{G}$ in the positive $x$ direction. A compass needle, consisting of a thin 2.00-cm-long iron rod with a cross-sectional area of $0.100\,\text{cm}^{2}$, is suspended from its center of mass so that it is free to rotate in the $xy$ plane. The needle is released from an initial position where it points in a direction making an angle of $135°$ with the positive $x$ axis. As the needle rotates through the position where it is momentarily parallel with the $y$ axis, its angular acceleration is $0.400\,\text{rad/s}^{2}$. What is its magnetic dipole moment?

**Picture the Problem.** The angular acceleration is due to the magnetic torque exerted on the magnet by the magnetic field. Use Newton's 2nd law for rotations, in addition to the expression for magnetic torque, to determine the magnetic dipole moment.

| 1.  Write out Newton's 2nd law for rotations, and substitute the expression for magnetic torque.  Also determine the moment of inertia for a rod rotated about its middle from Table 9-1 on page 274 of the text. | $\vec{\tau} = I\vec{\alpha}$ $$\vec{\mu} \times \vec{B} = \left(\tfrac{1}{12} ML^2\right)\vec{\alpha}$$ $$\left|\mu B \sin\theta\right| = \tfrac{1}{12} ML^2 \left|\alpha\right|$$ |
|---|---|
| 2.  Determine the mass of the rod.  You may need to look up the density in Table 13-1 on page 396 of the text. | $M = \rho AL$ $$\rho = 7.96 \times 10^3 \text{ kg/m}^3$$ |
| 3.  Solve for the magnetic dipole moment.  When the needle is parallel with the $y$ axis, the magnetic moment is perpendicular to the magnetic field. | $$\mu = \frac{\rho AL^3 \alpha}{12B \sin 90°}$$ $$= \frac{\left(7.96 \times 10^3 \text{ kg/m}^2\right)\left(1 \times 10^{-5} \text{ m}^2\right)\left(0.02 \text{ m}\right)^3 \left(0.4 \text{ rad/s}\right)}{12\left(2.2 \times 10^{-4} \text{ T}\right)}$$ $$= 9.65 \times 10^{-5} \text{ A} \cdot \text{m}^2$$ |

**Example #6—Interactive.**  A small 20-turn current loop with a 4.00-cm  diameter is suspended in a region with a magnetic field of $1000 \, \text{G}$,  with the plane of the loop parallel with the magnetic field direction.  What is the current in the loop when the torque exerted by the magnetic field on the loop is $4.00 \times 10^{-5} \, \text{N} \cdot \text{m}$?

**Picture the Problem.**  The torque on the current loop is related to its magnetic dipole moment. The magnetic dipole moment is due to current running through the current loop.  When the plane of the current loop is parallel with the magnetic field, then the magnetic dipole moment of the current loop is perpendicular to the magnetic field.  **Try it yourself.**  Work the problem on your own, in the spaces provided, to get the final answer.

| 1.  Write an expression relating the torque to the magnetic dipole moment. | |
|---|---|
| 2.  Write an expression for the dipole moment of the coil in terms of the current, loop area, and number of turns. | |
| 3.  Substitute the result from step 2 into the expression from step 1.  Solve for the current in the loop. | $I = 15.9 \, \text{mA}$ |

**Example #7.** A cyclotron used to accelerate protons has a uniform magnetic field of 1.10 T. (*a*) What is the frequency at which the potential difference between the dees must oscillate? (*b*) If the inner radius of the dees is 30.0 cm, what is the maximum kinetic energy (in MeV) attainable by the protons? (*c*) If the protons gain 72.0 keV of kinetic energy per revolution, what is the potential difference between the dees as the protons transit from one dee to the other?

**Picture the Problem.** The oscillation frequency of the potential difference between the dees should be the same as the cyclotron frequency. Use Newton's 2nd law to relate the radius of the orbit with the speed, and hence kinetic energy, of the protons. Remember that the protons get sped up each time they cross the gap between the dees, which is twice per revolution.

| | |
|---|---|
| 1. Calculate the cyclotron frequency, which is the same as the frequency of the potential difference between the dees. Look up the mass of the proton on the back cover of the text. | $f = \dfrac{qB}{2\pi m} = \dfrac{eB}{2\pi m_p}$ <br><br> $= \dfrac{\left(1.60\times10^{-19}\,\text{C}\right)\left(1.1\,\text{T}\right)}{2\pi\left(1.67\times10^{-27}\,\text{kg}\right)}$ <br><br> $= 16.8\,\text{MHz}$ |
| 2. Apply Newton's 2nd law to the rotational motion of the proton to find the speed of the proton. | $F = ma_c$ <br><br> $evB = mv^2/R$ <br><br> $v = \dfrac{eBR}{m_p}$ |
| 3. Use the speed to determine the kinetic energy of the proton. | $K = \tfrac{1}{2}m_p v^2 = \dfrac{\left(eBR\right)^2}{2m_p}$ <br><br> $= \dfrac{\left[\left(1.60\times10^{-19}\,\text{C}\right)\left(1.1\,\text{T}\right)\left(0.3\,\text{m}\right)\right]^2}{2\left(1.67\times10^{-27}\,\text{kg}\right)} \dfrac{1\,\text{eV}}{1.60\times10^{-19}\,\text{J}}$ <br><br> $= 5.22\,\text{MeV}$ |
| 4. The increase of 72 keV happens twice each revolution, so there must be an increase of 36 keV across the gap. | $\Delta K = q\,\Delta V$ <br><br> $36,000\,\text{eV} = e\,\Delta V$ <br><br> $\Delta V = 36,000\,\text{V}$ |

**Example #8—Interactive.** A beam of 2.50-MeV particles with charge $q = -2e$ is deflected by the magnetic field of a bending magnet as in Figure 26-3. The radius of curvature of the beam is 20.0 cm, and the strength of the magnetic field is 1.50 T. What is the mass of the particles making up the beam?

**Picture the Problem.** Apply Newton's 2nd law to the particles, which will relate the radius, speed, charge, magnetic field, and mass. Find the speed from the kinetic energy of the particles. **Try it yourself.** Work the problem on your own, in the spaces provided, to get the final answer.

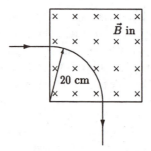

Figure 26-3

| | |
|---|---|
| 1.  Use the kinetic energy to find the speed of the particles in terms of the energy and the mass. | |
| 2.  Apply Newton's 2nd law to the circular motion of the particles. | |
| 3.  Substitute your result for the speed from step 1 into the equation from step 2, and solve for the mass. | $m = 1.15 \times 10^{-26}$ kg $= 6.94$ u (Li$^{2-}$ ions) |

**Example #9.**  As shown in Figure 26-4, the current is downward in a 4.00-cm-wide conductor with a rectangular cross section.  A uniform magnetic field of 1.50 T is directed into the figure.  The potential difference $V_b - V_a$ is $+12.0\,\mu$V.  (*a*) Are the charge carriers positively or negatively charged? (*b*) What is the drift velocity of the charge carriers?

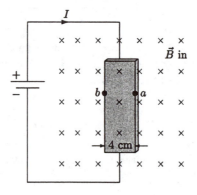

Figure 26-4

**Picture the Problem.** As charge carriers move through the rectangular wire, they experience a magnetic force which causes charge to migrate to the sides of the wire. These separated charges set up an electric field such that in the steady state, the electric field produced by these charges exerts a force on the subsequent charge carriers that directly cancels the magnetic force felt by the charges as they flow through the wire.

| | |
|---|---|
| 1. $V_b$ is larger than $V_a$, which means that positive charges must have accumulated along side $b$ as shown. Positive charge will flow "down" through the conductor, and negative charge will flow "up" through the conductor. Use the right-hand-rule to determine which charges must be moving for positive charges to accumulate on side $b$. | 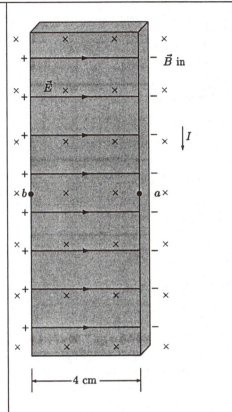 |
| | The charge carriers are negative. |
| 2. Use the expression for the Hall voltage to determine the drift velocity of the charge carriers. | $V_H = v_d w B$ $$v_d = \frac{V_H}{wB} = \frac{V_b - V_a}{wB} = \frac{2\times10^{-6}\,\text{V}}{(0.04\,\text{m})(1.5\,\text{T})}$$ $$= 3.33\times10^{-5}\,\text{m/s}$$ |

**Example #10—Interactive.** A segment of 3.00-mm-diameter copper wire has a narrow region where the diameter is only 1.00 mm. This segment of current-carrying wire is in a region where there is a uniform magnetic field perpendicular to the wire. If the Hall voltage across the wire in the thin region is $V_0$, what is it across the thicker region?

**Picture the Problem.** Because it is all part of the same wire, the current and the number density of charge carriers throughout the wire is the same. Consequently, the drift velocity varies inversely with the cross-sectional area of the wire.

| | |
|---|---|
| 1. Use the expression for the Hall voltage to solve for the drift velocity of the charge carriers. | |
| 2. Determine the drift velocity in the thicker segment of wire. | |
| 3. Use the value obtained in step 2 to find the Hall voltage across the thicker wire segment. | $V_{\mathrm{H}} = V_0 / 9$ |

# Chapter 27

# Sources of the Magnetic Field

## I.  Key Ideas

The earliest known sources of magnetism were permanent magnets. In the 1820's Oersted observed that a current-carrying wire deflected a compass needle and that the electric current was a source of magnetic field $\vec{B}$. During that decade the relationship between the current and the magnetic field was quantified.

*Section 27-1. The Magnetic Field of Moving Point Charges.* When a point charge $q$ moves with velocity $\vec{v}$, it produces a magnetic field at point $P$ given by

$$\vec{B} = \frac{\mu_0}{4\pi}\frac{q\vec{v}\times\hat{r}}{r^2} \qquad (v << c)$$      Magnetic force of a moving charge

where $\hat{r}$ is a unit vector that points from the charge $q$ toward the field point $P$ and $\mu_0$ is a constant called the magnetic **permeability of free space,** which has the value

$$\mu_0 = 4\pi\times10^{-7}\ \text{T}\cdot\text{m/A} = 4\pi\times10^{-7}\ \text{N/A}^2$$      Permeability of free space

The direction of $\vec{B}$, which is perpendicular to both $\vec{v}$ and $\hat{r}$, can be obtained by the standard right-hand-rule for cross products. The field lines are roughly circular around the line of motion. The direction of $\vec{B}$ can also be obtained by imagining that you grab the charge with your right hand with your thumb in the direction of $\vec{v}$. If the charge is positive, the field lines circulate in the direction of your fingers; if it is negative, the field lines are directed opposite to the direction of your fingers. The expression for the magnetic field due to a moving charge is mathematically analogous to Coulomb's law for the electric field due to a point charge,

$$\vec{E} = \frac{1}{4\pi\varepsilon_0}\frac{q}{r^2}\hat{r}$$

with $q\hat{r}$ replaced by $q\vec{v}\times\hat{r}$ and $1/\varepsilon_0$ replaced by $\mu_0$.

**Section 27-2. The Magnetic Field of Currents: The Biot–Savart Law.** A current element $I\,d\vec{\ell}$ associated with a short segment of current-carrying wire is related to the free charges $q_i$ in the segment and their velocities $\vec{v}_i$, by

$$I\,d\vec{\ell} = \sum q_i \vec{v}_i = \left( \sum q_i \right) \vec{v}_d$$

where $\vec{v}_d$ is the average drift velocity of the free charges. Thus, the magnetic field $d\vec{B}$ produced by a current element $I\,d\vec{\ell}$ is given by

$$d\vec{B} = \frac{\mu_0}{4\pi} \frac{I\,d\vec{\ell} \times \hat{r}}{r^2} \qquad \text{Magnetic field due to a current element (Biot–Savart law)}$$

This is analogous to Coulomb's law for the electric field due to a small segment of charge $dq$ with $dq\,\hat{r}$ replaced by $I\,d\vec{\ell} \times \hat{r}$ and with $\varepsilon_0$ replaced by $1/\mu_0$.

The field $\vec{B}$ due to an extended circuit is found by summing the vector contributions (the $d\vec{B}$s ) of all the current elements. The magnitude of the magnetic fields for several circuits with simple geometry are

$$B_x = \frac{\mu_0\, 2\pi R^2 I}{4\pi \left( x^2 + R^2 \right)^{3/2}} \qquad \text{$B$ on the axis of a current loop}$$

where $R$ is the radius of the loop, and $x$ is the distance from the loop along its axis.

$$B = \frac{1}{2}\mu_0 n I \left( \frac{b}{\sqrt{b^2 + R^2}} + \frac{a}{\sqrt{a^2 + R^2}} \right) \qquad \text{$B$ on the axis of a solenoid}$$

where $n = N/L$, $N$ is the number of turns in the solenoid, $L$ is its length, and $L = a + b$ as shown in Figure 27-1.

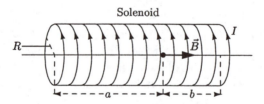

Figure 27-1

$$B = \mu_0 n I \qquad \text{$B$ well inside a long solenoid}$$

$$B = \frac{\mu_0}{4\pi} \frac{I}{R} \left( \sin\theta_1 + \sin\theta_2 \right) \qquad \text{$B$ due to a straight wire segment}$$

where $\theta_1$ and $\theta_2$ are the magnitudes of the angles shown in Figure 27-2.

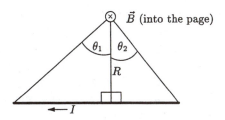

Figure 27-2

$$B = \frac{\mu_0 I}{2\pi R}$$    B due to a long straight wire

The directions of these magnetic fields are obtained by applying the right-hand rule.

*Definition of the Ampere*. If two very long parallel wires one meter apart carry equal currents, the current in each wire is defined as one ampere when the magnetic force per unit length on each wire is $2 \times 10^{-7}$ N/m.

Two parallel current-carrying wires exert equal and opposite forces on each other because the magnetic field due to the current in one wire exerts a force on the current in the other wire and vice versa.

*Section 27-3.  Gauss's Law for Magnetism.*  Gauss's law for magnetism is a mathematical statement that no isolated magnetic monopoles exist.

$$\phi_{m,net} = \oint_s \vec{B} \cdot d\vec{A} = 0$$    Gauss's law for magnetism

*Section 27-4. Ampère's Law.* Ampère's law states that in steady-state situations,

$$\oint_C \vec{B} \cdot d\vec{\ell} = \mu_0 I_C \quad C, \text{ any closed curve}$$    Ampère's law

where $I_C$ is the net current through any surface bounded by the curve $C$. Like Gauss's law, Ampère's law is most useful when applied to highly symmetric current distributions. If, by using symmetry arguments, you can establish that the tangential component $B_\parallel$ of a magnetic field is constant on a curve $C$, then $\oint_C \vec{B} \cdot d\vec{\ell} = B_\parallel \ell$.

Ampère's law and the Biot–Savart law can only be used to determine magnetic fields produced by circuits in which both the current and the charge distribution are constant in time. These laws do provide close approximations to actual magnetic fields produced by circuits in which time-varying changes take place sufficiently slowly. Magnetic fields produced by circuits in which the current or the charge distribution changes in time will be studied in Chapter 28.

Ampère's law can be used to show that the magnetic field inside a tightly wound toroid of inner radius $a$, outer radius $b$, and $N$ turns is

$$B = \frac{\mu_0 N I}{2\pi r} \quad a < r < b$$    B due to a tightly wound toroid

at a distance *r* from the center of the toroid.

In many ways electric dipoles and magnetic dipoles are analogous. For example, an electric field exerts a torque on an electric dipole that tends to align it with the electric field. Similarly, a magnetic field exerts a torque on a magnetic dipole that tends to align it with the magnetic field. However, electric and magnetic dipoles are not completely analogous. The electric field due to an electric dipole and the magnetic field due to a magnetic dipole are geometrically similar externally, but they are exactly opposite internally. The direction of the electric field within an electric dipole is opposite the direction of the electric dipole moment, whereas the direction of the magnetic field within a magnetic dipole is along the direction of the magnetic dipole moment.

***Section 27-5. Magnetism in Matter.*** Most materials fall into one of three main magnetic categories—ferromagnetic, paramagnetic, or diamagnetic—according to the manner in which their atoms or molecules align themselves in an external magnetic field. Most atoms and molecules have an intrinsic dipole moment that is due to both the orbital motion and the spin of their electrons; the atoms in these molecules tend to align their magnetic dipole moments with magnetic fields. Materials consisting of these atoms are said to be either **paramagnetic** or **ferromagnetic.** Some molecules lack intrinsic magnetic moments; for such molecules a magnetic field induces a dipole moment that is directed oppositely to the field direction. Materials consisting of such molecules are said to be **diamagnetic.**

***Magnetization and Magnetic Susceptibility.*** The degree to which a material is magnetized depends on the individual strengths of the molecular magnetic dipole moments and the degree to which they are aligned. The magnetization $\vec{M}$ of a material is the net magnetic dipole moment per unit volume of the material:

$$\vec{M} = \frac{d\vec{\mu}}{dV} \qquad\qquad \text{Magnetization defined}$$

Magnetic dipoles can be thought of as current loops. In a magnetized material in which the magnetic dipole moments are aligned, these current loops are distributed throughout the material. If the material is uniformly magnetized, the currents in neighboring current loops cancel. However, at the edge of the material there are no neighboring current loops on one side, and so there is a net current, called an **amperian current,** on the outside edge of the material as in Figure 27-3. The amperian current is due to the motions of bound rather than free charges.

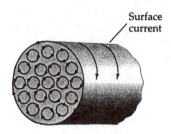

Figure 27-3

The amperian current in a uniformly magnetized, long solid cylinder shown in Figure 27-4 is directed around the perimeter of the cylinder. The geometry of this current is identical to the current in a long, tightly wound solenoid, so the geometry of the magnetic field due to the amperian current is the same as the geometry of the magnetic field due to the current in the solenoid. The magnetic field inside the solenoid is $B = \mu_0 nI$, where $n$ is the number of turns per unit length of the solenoid, and the magnetic field inside the cylinder is

$B = \mu_0 M$                                                             Magnetic field inside a long bar magnet

An argument showing that the magnetism $M$ corresponds to $nI$ is given on pages 875-876 of the textbook.

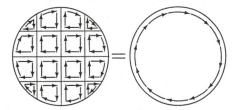

Figure 27-4

Consider an unmagnetized cylinder of material that is located within a solenoid. When a current flows through the windings of the solenoid, a magnetic field $\vec{B}_{app}$ fills the region within it. This applied field of the solenoid magnetizes the material of the cylinder so that it has magnetization $\vec{M}$. There are now two magnetic fields in the material, an applied field $\vec{B}_{app}$ due to the current in the solenoid and a second field due to the magnetization of the material. The net magnetic field $\vec{B}$ is the sum of the two. That is,

$$\vec{B} = \vec{B}_{app} + \mu_0 \vec{M}$$

If the atoms constituting the cylinder have intrinsic magnetic dipole moments, $\vec{B}_{app}$ exerts torques that tend to align them with it. For these ferromagnetic and paramagnetic materials, $\vec{M}$ is parallel with $\vec{B}_{app}$. If the atoms of the cylinder do not have intrinsic magnetic dipole moments, $\vec{B}_{app}$ induces a dipole moment that is opposite in direction to it. For these diamagnetic materials, $\vec{M}$ is opposite to $\vec{B}_{app}$. For both paramagnetic and diamagnetic materials, the magnetization is proportional to the applied field, Thus,

$$\vec{M} = \chi_m \left( \frac{\vec{B}_{app}}{\mu_0} \right)$$                              Magnetic susceptibility defined

where the dimensionless constant $\chi_m$, called the magnetic susceptibility, is a property of the material. Substituting this expression for $\vec{M}$ in the previous equation, we have

$$\vec{B} = \vec{B}_{app} + \mu_0 \vec{M} = \vec{B}_{app} \left( 1 + \chi_m \right)$$

For paramagnetic materials, $\chi_m$ is a small positive number that depends on temperature. For diamagnetic materials, it is a small negative number that is independent of temperature. The magnetic susceptibility of various materials are listed in Table 27-1 on page 876 of the textbook.

*Atomic Magnetic Moments.* The magnetic dipole moment $\vec{\mu}$ of a particle with mass $m$ and charge $q$ moving in a circular orbit can be shown to be

$$\vec{\mu} = \frac{q}{2m}\vec{L} \qquad\qquad \text{Magnetic moment for a charged particle in a circular orbit}$$

where $\vec{L}$ is the angular momentum of the particle about the center of the circle. For an electron, $m = m_e$ and $q = -e$, so the magnetic moment of the electron due to its orbital motion is

$$\vec{u}_\ell = \frac{e\hbar}{2m_e}\frac{\vec{L}}{\hbar} = -\mu_B\frac{\vec{L}}{\hbar} \qquad\qquad \text{Magnetic moment due to orbital motion of an electron}$$

where

$$\mu_B = \frac{e\hbar}{2m_e} = 9.27\times10^{-24}\ \text{A}\bullet\text{m}^2 = 9.27\times10^{-24}\ \text{J/T} \qquad\qquad \text{Bohr magneton}$$

is the quantum unit of magnetic moment, called a **Bohr magneton.**

The fundamental constant $h$ (with units of angular momentum), called Planck's constant, has a value of

$$h = 6.67\times10^{-34}\ \text{J}\bullet\text{s} \qquad\qquad \text{Planck's constant}$$

and the combination $h/2\pi$, which appears frequently, is designated by $\hbar$ (read "$h$ bar"):

$$\hbar = \frac{h}{2\pi} = 1.05\times10^{-34}\ \text{J}\bullet\text{s}$$

The magnetic moment of an electron due to its intrinsic spin angular momentum $\vec{S}$ is

$$\vec{\mu}_s = -2\frac{e\hbar}{2m_e}\frac{\vec{S}}{\hbar} = -2\mu_B\frac{\vec{S}}{\hbar} \qquad\qquad \text{Magnetic moment due to electron spin}$$

Although the calculation of the magnetic moment of any atom is a complicated problem in quantum theory, the result for all atoms, according to both theory and experiment, is that the magnetic moment is of the order of a few Bohr magnetons. For atoms with zero net angular momentum, the net magnetic moment is zero. Angular momentum is found only in certain discrete values; that is, angular momentum is quantized.

Magnetization is greatest when all the atomic magnetic dipole moments are maximally aligned. For this case, the magnetization is called the **saturation magnetization** $M_s$:

$$M_s = n\mu$$

where $n$ is the number of molecules per unit volume and $\mu$ the magnetic dipole moment of each molecule.

*Paramagnetism.* **Paramagnetic materials** have a very small, positive magnetic susceptibility $\chi_m$. The atoms or molecules in a paramagnetic material have an intrinsic magnetic dipole moment. An external magnetic field applied to a paramagnetic material exerts torques on the molecules that tend to align the dipole moments with the field. Not all of the molecules are perfectly aligned with the field because thermal agitation keeps jarring them out of alignment.

The potential energy $U$ of a magnetic dipole moment $\vec{\mu}$ at an angle with a uniform magnetic field $\vec{B}$ is $-\vec{\mu}\cdot\vec{B} = -\mu B\cos\theta$. If $\vec{\mu}$ is parallel with $\vec{B}$, this energy is $-\mu B$; and if it is antiparallel, it is $+\mu B$. Thus, an energy of $2\mu B$ is required to reverse the direction of a magnetic dipole after it is aligned with the field. Typical thermal energies are of the order of $kT$, where $k$ is the Boltzmann constant and $T$ the absolute temperature. Calculations show that for molecular magnetic moments of 1 Bohr magneton, for magnetic fields on the order of 1 tesla, and for temperatures of about 300 K, $kT$ is about 200 times greater than $2\mu B$. Under such conditions alignment is very weak, which is why typical values for the magnetic susceptibility of paramagnetic solids are on the order of $10^{-5}$. The magnetization in weak magnetic fields varies inversely with temperature according to the equation

$$M = \frac{1}{3}\frac{mB_{app}}{kT}M_s \qquad\qquad \text{Curie's law}$$

*Ferromagnetism.* **Ferromagnetic materials** have very large, positive values of magnetic susceptibility $\chi_m$. In these materials, a small applied external magnetic field can produce a very high degree of alignment of the atomic magnetic dipole moments. This is because, unlike the atoms of paramagnetic materials, the magnetic dipole moments of the atoms of ferromagnetic materials interact strongly and form regions, called **magnetic domains,** in which they align with each other even when no applied field is present. At temperatures above a critical temperature called the **Curie temperature,** thermal agitation is sufficient to break up the domains and to prevent new ones from forming.

When an unmagnetized ferromagnetic material is located within a solenoid, the magnetization of the material increases as the solenoid current, and thus the applied magnetic field, increases. The magnetization, however, does not increase linearly with the applied field because the magnetization eventually approaches saturation for the material. If, after saturation is reached, the applied magnetic field is returned to zero, the material remains magnetized with some residual magnetization. This effect is called **hysteresis.** The magnetic field $B_r$ due to the residual magnetism is called the **remnant field.**

To measure the magnetic susceptibility of a ferromagnetic material, we place the unmagnetized ferromagnetic material in a solenoid and increase the solenoid current and, thus, the applied magnetic field. The magnetic susceptibility $\chi_m$ of the material is then given by

$$\chi_m = \frac{M\mu_0}{B_{app}}$$

and

$$\vec{B} = \vec{B}_{app} + \mu_0 \vec{M} = \vec{B}_{app}\left(1 + \chi_m\right) = \vec{B}_{app} K_m$$

where $K_m = 1 + \chi_m$ is called the **relative permeability** of the material. The magnetic **permeability** $\mu$ is defined by

$$\mu = K_m \mu_0 = \left(1 + \chi_m\right)\mu_0 \qquad \text{Permeability defined}$$

Ferromagnetic materials in which hysteresis effects are small are said to be **magnetically soft.** Magnetically soft materials are desirable as transformer cores. For permanent magnets, large hysteresis effects are desirable; materials in which hysteresis effects dominate behavior are said to be **magnetically hard.**

*Diamagnetism.* **Diamagnetic materials** have very small, negative values of magnetic susceptibility $\chi_m$. An applied magnetic field induces currents in the atoms and molecules of any material. In accordance with Lenz's law, in a diamagnetic material these currents produce a magnetic flux that opposes the increase of the applied field as we shall see in Chapter 28. Therefore, the induced magnetic dipole moments are anti-aligned with the applied field. The atoms or molecules of diamagnetic materials have no intrinsic magnetic dipole moment, so the magnetization of these materials is due entirely to the induced magnetic dipole moments. Thus, the magnetic susceptibility of diamagnetic materials is negative.

A superconductor is a perfect diamagnetic material. An applied magnetic field induces currents in a superconductor that decrease the net magnetic field inside the superconductor to zero. Thus, the magnetic susceptibility $\chi_m$ of a superconductor is $-1$.

## II.  Physical Quantities and Key Equations

### Physical Quantities

*Permeability of free space*        $\mu_0 = 4\pi \times 10^{-7} \text{ T} \cdot \text{m/A} = 4\pi \times 10^{-7} \text{ N/A}^2$

*Bohr magneton*        $\mu_B = \dfrac{e\hbar}{2m_e} = 9.27 \times 10^{-24} \text{ A} \cdot \text{m}^2 = 9.27 \times 10^{-24} \text{ J/T}$

*Planck's constant*        $h = 6.67 \times 10^{-34} \text{ J} \cdot \text{s}$

$\hbar = \dfrac{h}{2\pi} = 1.05 \times 10^{-34} \text{ J} \cdot \text{s}$

### Key Equations

*Magnetic field of a moving charge*        $\vec{B} = \dfrac{\mu_0}{4\pi}\dfrac{q\vec{v} \times \hat{r}}{r^2} \qquad \left(v \ll c\right)$

*Magnetic field due to a current element (Biot–Savart law)*      $\vec{B} = \dfrac{\mu_0}{4\pi} \dfrac{I\, d\vec{\ell} \times \hat{r}}{r^2}$

*B on the axis of a current loop*      $B_x = \dfrac{\mu_0}{4\pi} \dfrac{2\pi R^2 I}{\left(x^2 + R^2\right)^{3/2}}$

*B on the axis of a solenoid*      $B = \dfrac{1}{2}\pi_0 n I \left( \dfrac{b}{\sqrt{b^2 + R^2}} + \dfrac{a}{\sqrt{a^2 + R^2}} \right)$

where $n = N/L$ , $L$ is the length of the solenoid, and $a$ and $b$ are as shown in Figure 27-1.

*B well inside a long solenoid*      $B = \mu_0 n I$

*B due to a straight wire segment*      $B = \dfrac{\mu_0}{4\pi} \dfrac{I}{R}\left( \sin\theta_1 + \sin\theta_2 \right)$

where $I$, $R$, $\theta_1$, and $\theta_2$ are shown in Figure 27-2.

*B due to a long straight wire*      $B = \dfrac{\mu_0 I}{2\pi R}$

*Gauss's Law for Magnetics*      $\oint_S \vec{B} \cdot d\vec{A} = 0$

*Ampère's law*      $\oint_C \vec{B} \cdot d\vec{\ell} = \mu_0 I_C$  C, any closed curve

*B due to a tightly wound toroid*      $B = \dfrac{\mu_0 N I}{2\pi r}$   $a < r < b$

*Magnetization defined*      $\vec{M} = \dfrac{d\vec{\mu}}{dV}$

*Magnetic field inside a long bar magnet*      $B = \mu_0 M$

*Susceptibility defined*      $\vec{M} = \chi_{\mathrm{m}} \left( \dfrac{\vec{B}_{\mathrm{app}}}{\mu_0} \right)$

*Magnetic moment for a charged particle in a circular orbit*      $\vec{\mu} = \dfrac{q}{2m}\vec{L}$

*Curie's law*      $M = \dfrac{1}{3}\dfrac{mB_{\mathrm{app}}}{kT} M_{\mathrm{s}}$

*Permeability defined*      $\mu = K_{\mathrm{m}}\mu_0 = \left(1 + \chi_{\mathrm{m}}\right)\mu_0$

## III. Potential Pitfalls

Most students have difficulty visualizing the three-dimensional aspect of magnetism. You can help your visualization by making drawings of every situation. Most teachers practice these drawings before coming to class, and you may want to do the same. If appropriate, make several drawings, each from a different point of view. It is often helpful to use different colors to illustrate cross products.

You can find the directions of the magnetic forces that two moving charges exert on each other, as follows. First, at the location of one charge, find the direction of the magnetic field due to the other charge; and repeat the process at the location of the other charge. Second, find the directions of the forces exerted by these magnetic fields. Use the same two steps to find the directions of the forces exerted by two current elements on each other.

Ampère's law and the Biot–Savart law can only be used for circuits in which the current and the charge distribution are constant in time. These laws provide close approximations to the magnetic field produced by circuits in which changes in the current and the charge distribution take place sufficiently slowly.

The term *magnetization* doesn't sound like a density, but it is. It is the magnetic dipole moment per unit volume—that is, the magnetic dipole moment density.

The induced magnetic field far exceeds the applied field in ferromagnetic materials.

The transition from a ferromagnetic material to a paramagnetic material when the temperature crosses the Curie point is not a gradual one. Rather, it is similar to the transition of water from liquid to gas when the temperature crosses the steam point. Such abrupt transitions are called *phase transitions*.

Atoms and molecules of diamagnetic materials have induced magnetic dipole moments that are always opposed to the field direction. Atoms and molecules of paramagnetic and ferromagnetic materials have permanent magnetic dipole moments that are not 100% aligned except at saturation.

## IV. True or False Questions and Responses

**True or False**

_____ 1. On the surface of a sphere centered on a current element, the magnitude of the magnetic field due to the current element is constant.

_____ 2. In a given direction from a current element, the magnitude of its magnetic field varies inversely with the square of the distance from it.

_____ 3. Imagine a current element at the center of a geographer's globe of the earth, pointed along the globe's axis toward its north pole. The magnetic field lines on the surface of the globe, which are due only to the current element, are along the lines of constant longitude.

_____ 4. If a moving bullet acquires a negative charge, the magnetic field lines due to this moving charge are directed counterclockwise as viewed by an observer watching it move directly away from him.

_____ 5. Ampère's law is valid if the net current passing through a surface bounded by the closed integration path is constant.

_____ 6. The magnetic field lines associated with the magnetic field due to the steady current in a long straight wire are circles centered on the wire.

_____ 7. For a given current $I$ and total number of turns $N$, the magnetic field inside a long, tightly wound solenoid is independent of its length $L$.

_____ 8. The magnetic field produced by a circuit in which the current varies in time can, in principle, be calculated using the Biot–Savart law.

_____ 9. In principle, Ampère's law can be used only in situations of sufficient symmetry that the direction of the magnetic field can be deduced from symmetry arguments.

_____ 10. The magnetic field lines inside a long, tightly wound solenoid are circles coaxial with the solenoid.

_____ 11. The magnetic field lines due to a tightly wound solenoid and the magnetic field lines due to a uniformly magnetized bar magnet of the same size and shape are identical both inside and outside the devices.

_____ 12. Materials whose molecules do not possess intrinsic magnetic dipole moments are called paramagnetic.

_____ 13. Magnetic fields exert torques on magnetic dipoles that tend to align them with the field.

_____ 14. When the magnetic dipole moment of a current loop is aligned in an external magnetic field, the magnetic field that is due to the current loop and that is aligned along the loop's axis is also aligned with the external field.

_____ 15. The magnetic susceptibility is −1 for superconducting materials.

_____ 16. The magnetic susceptibility approaches +1 for most ferromagnetic materials.

_____ 17. The intrinsic magnetic dipole moment of an atom is typically on the order of one Bohr magneton.

_____ 18. The magnetic dipole moment of an atom that is induced by the field of a strong laboratory magnet is typically on the order of one Bohr magneton.

_____ 19. Ferromagnetic material suitable for forming permanent magnets should be magnetically hard.

_____ 20. A diamagnetic material is attracted by either pole of a magnet.

_____ 21. The susceptibility of a paramagnetic material increases with temperature.

_____ 22. The magnetization of a ferromagnetic material depends not only on the applied magnetic field but also on what has happened to the material previously.

_____ 23. At temperatures below a critical temperature called the Curie point, the magnetic dipole moments of all the atoms of a ferromagnetic material align to form a single magnetic domain.

**Responses to True or False**

1. False. The magnetic field on the surface varies as $\sin\theta$, where $\theta$ is the angle between the current element vector and the vector from the particle to the field point on the surface of the sphere. The angle $\theta$ ranges between zero and $\pi$.

2. True.

3. False. The magnetic field due to a current element is always perpendicular to the direction of the current element and to the radius vector from the center of the globe. The field lines on the globe are along lines of constant latitude.

4. True. Imagine yourself, the observer, grabbing the bullet with your right hand with your thumb pointing away from you. Your fingers, as viewed by you, then curl clockwise. But, because the charge on the bullet is negative, the magnetic field lines around the bullet are directed counterclockwise.

5. True. Ampère's law can be used only in steady-state situations. This means not only that the total current must be everywhere constant in time, but also that the current distribution must not change with time.

6. True.

7. False. The expression for the magnetic field inside a long, tightly wound solenoid is $B = \mu_0 nI$, where $n = N / L$, and $L$ is the length of the solenoid.

8. True. To apply Ampère's law, however, requires the circuit to be steady state.

9. False. However, it is only for situations with sufficient symmetry that Ampère's law is very practical for determining the magnitude of the field.

10. False. They are straight lines parallel with the axis of the solenoid.

11. True.

12. False. Materials whose molecules do not possess intrinsic magnetic dipole moments are called diamagnetic.

13. True.

14. True.

15. True.

16. False. The magnetic susceptibility of ferromagnetic materials is usually on the order of several thousand.

17. True.

18. False. It is much less than one Bohr magneton.

19. True. It is desirable for a permanent magnet to retain magnetization.

20. False. Because magnetization in a diamagnetic material is directed oppositely to the applied field, a diamagnetic material is repelled by either pole of a magnet.

21. False. The susceptibility of a paramagnetic material decreases with temperature because thermal motion increases with temperature.

22. True.

23. False. At temperatures above the Curie point, magnetic domains are broken up and prevented from forming; but below the Curie point, a single atomic domain occurs only at magnetic saturation. A ferromagnetic sample that is below saturation typically contains many domains.

## V.  Questions and Answers

**Questions**

1. Two flat, square, current-carrying loops lying in the plane of the page are shown in Figure 27-5. Show how the following statements about magnetic forces apply to this case: Like poles repel whereas opposite poles attract. Parallel currents attract each other whereas antiparallel currents repel.

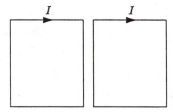

Figure 27-5

2. Think of a long solenoid of $N$ total turns carrying current $I$ as a helical spring. If the solenoid is stretched along its length without changing $N$ or $I$, does the field inside it change? How about its magnetic moment?

3. In Figure 27-6, a mass hangs in equilibrium on the end of a spring. If a current is passed through the spring, which way does the mass move?

Figure 27-6

4. What are the similarities and differences between the fields of electric and magnetic dipoles? Sketch the fields of an electric dipole and a current loop to illustrate your responses.

5. The hysteresis curve of a sample of some ferromagnetic material is shown in Figure 27-7. On the diagram, identify the saturation magnetic field and the remnant field. Is the scale on the $B$ axis the same as the scale on the $B_{app}$ axis?

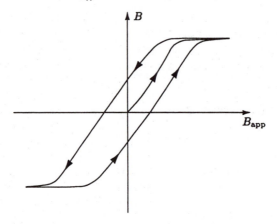

Figure 27-7

6. Explain why a magnet will pick up an unmagnetized iron nail but not an otherwise identical aluminum nail.

7. A permanent magnet may lose much of its magnetization if it is dropped or banged against something. Why?

**Answers**

1. Because the currents in the two loops are both clockwise, the two magnetic dipole moments are parallel. If you think of these loops as bar magnets, you can visualize that they're aligned with north pole next to north pole and south pole next to south pole. Thus, they repel each other. In terms of the currents, the strongest force is between the two currents that are closest together. These are oppositely directed, and oppositely directed currents repel each other.

2. The number of turns per unit length decreases as the coil is stretched, so the magnetic field inside, which is $B = \mu_0 nI = \mu_0 (N/L)I$, decreases. The magnetic dipole moment $\mu = NIA$ remains the same.

3. Attractive magnetic forces between adjacent turns of the spring (think of them as parallel wires) cause the spring to compress longitudinally. The spring contracts and raises the mass. This force gets stronger as the turns get closer together, so the spring continues to collapse. This effect can be quite strong, and the construction of high-field laboratory solenoids has to be very rugged to keep them from imploding.

4. The two fields far from the dipoles look identical (Figure 27-8). In this figure the direction of both the magnetic dipole moment of the current loop and the electric dipole moment of the electric dipole is toward the right. The main difference is that the direction of the field "inside" the current loop is the same as the direction of the magnetic dipole moment (to the right), and the direction of the field inside the electric dipole is opposite to the direction of the electric dipole moment (also to the right).

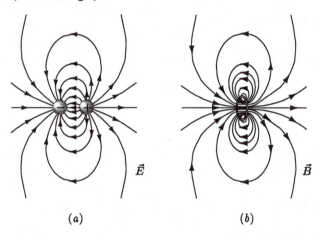

$\vec{E}$          $\vec{B}$

(a)          (b)

Figure 27-8

5. The saturation field is the maximum value of $B$. The remnant field is the value of $B$ that remains when $B_{app}$ becomes zero *after* saturation. The relative permeability of iron, as listed in Table 27-2 on page 882 of the textbook, is 5500; therefore, near the origin, the slope of the curve from origin to saturation is about 5500. On the figure, the slope is about one; so the scale for the $B$ axis is very different from the scale for the $B_{app}$ axis. If the graph were drawn with the same scale for both axes, its height would have to be 5500 times greater than its width.

6. When the field of the magnet is applied to the iron nail, the magnetization of the iron in the nail is very large. When the field is applied to the aluminum nail, the magnetization of the aluminum, which is paramagnetic, is many thousands of times smaller. The force of the nonuniform magnetic field of the magnet on the strong magnetic dipole moment of the iron is much greater than its force on the weak magnetic dipole moment of the aluminum.

7. The ferromagnetic domains may get knocked out of alignment when a permanent magnet is jarred.

# VI.  Problems, Solutions, and Answers

**Example #1.**  A particle with a charge of $+5.00\,\text{nC}$ travels through the origin with a velocity of $\left(3.00\times10^6\,\text{m/s}\right)\hat{k}$.  Find the magnitude and direction of the magnetic field due to this charged particle at points $a=\left(0,0,5.00\,\text{cm}\right)$, $b=\left(0,0,-5.00\,\text{cm}\right)$, $c=\left(0,5.00\,\text{cm},0\right)$, $d=\left(5.00\,\text{cm},0,0\right)$, $e=\left(3.00\,\text{cm},4.00\,\text{cm},0\right)$, and $f=\left(3.00\,\text{cm},0,4.00\,\text{cm}\right)$.

**Picture the Problem.**  Sketch the situation, and use the formula for the magnetic field due to a charged particle to find the field in each case.

| | |
|---|---|
| 1.  Make a sketch of the moving particle and the locations where the magnetic field is to be located.  The right-hand-rule can be used to determine the direction of the magnetic field and to check your calculations. | 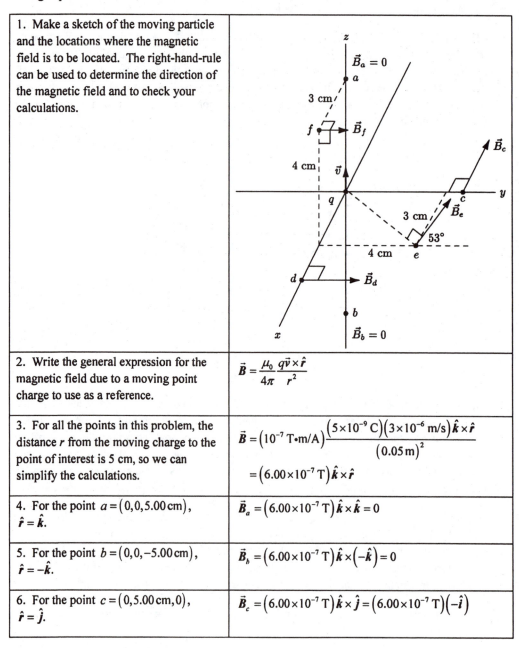 |
| 2.  Write the general expression for the magnetic field due to a moving point charge to use as a reference. | $$\vec{B}=\frac{\mu_0}{4\pi}\frac{q\vec{v}\times\hat{r}}{r^2}$$ |
| 3.  For all the points in this problem, the distance $r$ from the moving charge to the point of interest is 5 cm, so we can simplify the calculations. | $$\vec{B}=\left(10^{-7}\,\text{T}\cdot\text{m/A}\right)\frac{\left(5\times10^{-9}\,\text{C}\right)\left(3\times10^{-6}\,\text{m/s}\right)\hat{k}\times\hat{r}}{\left(0.05\,\text{m}\right)^2}$$ $$=\left(6.00\times10^{-7}\,\text{T}\right)\hat{k}\times\hat{r}$$ |
| 4.  For the point $a=\left(0,0,5.00\,\text{cm}\right)$, $\hat{r}=\hat{k}$. | $$\vec{B}_a=\left(6.00\times10^{-7}\,\text{T}\right)\hat{k}\times\hat{k}=0$$ |
| 5.  For the point $b=\left(0,0,-5.00\,\text{cm}\right)$, $\hat{r}=-\hat{k}$. | $$\vec{B}_b=\left(6.00\times10^{-7}\,\text{T}\right)\hat{k}\times\left(-\hat{k}\right)=0$$ |
| 6.  For the point $c=\left(0,5.00\,\text{cm},0\right)$, $\hat{r}=\hat{j}$. | $$\vec{B}_c=\left(6.00\times10^{-7}\,\text{T}\right)\hat{k}\times\hat{j}=\left(6.00\times10^{-7}\,\text{T}\right)\left(-\hat{i}\right)$$ |

| 7. For the point $d = (5.00\,\text{cm}, 0, 0)$, $\hat{r} = \hat{i}$. | $\vec{B}_d = (6.00 \times 10^{-7}\,\text{T})\hat{k} \times \hat{i} = (6.00 \times 10^{-7}\,\text{T})\hat{j}$ |
|---|---|
| 8. For the point $e = (3\,\text{cm}, 4\,\text{cm}, 0)$, $\hat{r} = 0.6\hat{i} + 0.8\hat{j}$. | $\vec{r} = 3\hat{i} + 4\hat{j}$  ;  $r = \sqrt{3^2 + 4^2} = 5$ <br><br> $\hat{r} = \dfrac{\vec{r}}{r} = 0.6\hat{i} + 0.8\hat{j}$ <br><br> $\vec{B}_e = (6.00 \times 10^{-7}\,\text{T})\hat{k} \times (0.6\hat{i} + 0.8\hat{j})$ <br><br> $= (6.00 \times 10^{-7}\,\text{T})(0.6\hat{j} - 0.8\hat{i})$ <br><br> or $6.00 \times 10^{-7}\,\text{T}$ in the direction shown. |
| 9. For the point $f = (3\,\text{cm}, 0, 4\,\text{cm})$, $\hat{r} = 0.6\hat{i} + 0.8\hat{k}$. | $\vec{B}_f = (6.00 \times 10^{-7}\,\text{T})\hat{k} \times (0.6\hat{i} + 0.8\hat{k})$ <br><br> $= (3.6 \times 10^{-7}\,\text{T})\hat{j}$ |

**Example #2—Interactive.** Particle 1 with a charge of +5.00 nC, travels through the origin with a velocity of $(1.40 \times 10^6\,\text{m/s})\hat{k}$. At the same instant, particle 2, with a charge of +2.00 nC, travels through the point $(5.00\,\text{cm}, 0, 0)$ with a velocity of $(-3.00 \times 10^6\,\text{m/s})\hat{i}$. At that instant, find (a) the magnetic force exerted by the field of particle 1 on particle 2, and (b) the magnetic force exerted by the field of particle 2 on particle 1.

**Picture the Problem.** Note that these are *not* Newton's 3rd law force pairs. The particles are not exerting forces on each other. The particles create fields, and it is these fields which exert a force on each particle. First find the field created by one of the particles, and then determine the magnetic force experienced by the second moving particle while under the influence of the field of the first particle. **Try it yourself.** Work the problem on your own, in the spaces provided, to get the final answer.

| | |
|---|---|
| 1. Draw separate sketches for parts (a) and (b). | |
| 2. Determine the magnetic field created by particle 1 at the position of particle 2. | |

| | |
|---|---|
| 3. Determine the force on particle 2 when it experiences the magnetic field calculated in step 2. | $\vec{F} = \left(-1.68 \times 10^{-9}\,\text{N}\right)\hat{k}$ |
| 4. Determine the magnetic field created by particle 2 at the position of particle 1. | |
| 5. Determine the magnetic force on particle 1 when it experiences the field calculated in step 4. | $\vec{F} = 0$ |

In order to make sense of this apparent paradox, we need to consider the momentum that is carried by the electric and magnetic fields involved in this problem. We will see how to think properly about this concept in Chapter 30.

**Example #3.** Two long, straight, current-carrying wires in the $xy$ plane are parallel to the $y$ axis and intersect the $x$ axis at $x = +4.00\,\text{cm}$ and at $x = -4.00\,\text{cm}$. Each wire carries a constant current of 20.0 A in the positive $y$ direction. Find the magnetic field at point $P$, which is at $z = 3.00\,\text{cm}$ on the $z$ axis.

**Picture the Problem.** Use the formula for the field due to a long straight wire to find the contribution from each wire to the magnetic field at point $P$. Add the magnetic field vector of each contribution to get the net magnetic field at $P$.

| | |
|---|---|
| 1. Sketch the physical situation. | 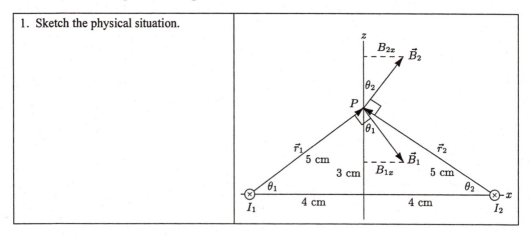 |

| 2. Determine the magnitude of the magnetic field produced by currents $I_1$ and $I_2$. | $B_{\infty \text{ wire}} = \dfrac{\mu_0 I}{2\pi r}$ <br><br> $B_1 = \dfrac{\mu_0 I_1}{2\pi r_1}$  ;  $B_2 = \dfrac{\mu_0 I_2}{2\pi r_2}$ |
|---|---|
| 3. The directions of $B_1$ and $B_2$ can be determined by the right-hand-rule, and are shown in the figure. The total magnetic field at $P$ will be the vector sum of the two magnetic fields. By symmetry, the $z$ components of the magnetic field will cancel, leaving only the $x$ components. <br><br> Because the current in each wire is the same, and each wire is the same distance from point $P$, the magnitudes of each magnetic field are the same. In addition, because of the symmetry $\theta_1$ and $\theta_2$ are also the same. | $\vec{B} = \vec{B}_1 + \vec{B}_2$ <br> $B_{1z} = -B_{2z}$   $\therefore B_z = 0$ <br> $B_x = B_{1x} + B_{2x} = B_1 \sin\theta_1 + B_2 \sin\theta_2$ <br> $\quad = 2B \sin\theta_1 = 2\dfrac{\mu_0 I_1}{2\pi r_1}\sin\theta_1$ <br> $\quad = 2\dfrac{(4\pi \times 10^{-7}\,\text{T·m/A})(20\,\text{A})}{2\pi\,(0.05\,\text{m})}\dfrac{0.03\,\text{m}}{0.05\,\text{m}}$ <br> $\quad = 9.60 \times 10^{-5}\,\text{T}$ <br> $\vec{B}_P = (9.60 \times 10^{-5}\,\text{T})\hat{i}$ |

**Example #4—Interactive.** Three very long, straight, current-carrying wires are at the corners of a square of side $d$, as shown in Figure 27-9. Currents $I_1$ and $I_2$ are directed into the figure, and current $I_3$ is directed out of the figure. $|I_1| = |I_2| = |I_3| = I$. All currents are constant. Find an expression for the magnitude and direction of the magnetic field at point $P$, which is at the fourth corner of the square.

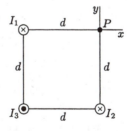

Figure 27-9

**Picture the Problem.** Use the formula for the field due to a long straight wire to find the contribution of each wire to the magnetic field at point $P$. Add the vector components of these magnetic fields. **Try it yourself.** Work the problem on your own, in the spaces provided, to get the final answer.

| 1. Determine the magnitude of the magnetic field at $P$ from $I_1$. | |
|---|---|

| | |
|---|---|
| 2.  Determine the magnitude of the magnetic field at $P$ from $I_2$. | |
| 3.  Determine the magnitude of the magnetic field at $P$ from $I_3$. | |
| 4.  Sketch the direction of the magnetic field vectors from each wire, and determine the $x$ and $y$ components of each magnetic field. | |
| 5.  Add the vector components of each magnetic field to get the total field at point $P$. | $$\vec{B} = \frac{\mu_0}{4\pi}\frac{I}{d}\hat{i} - \frac{\mu_0}{4\pi}\frac{I}{d}\hat{j}$$ |

**Example #5.**  In Figure 27-10, find the magnetic field at point $P$.  The current is 80.0 A.

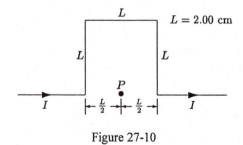

Figure 27-10

**Picture the Problem.**  The total magnetic field at $P$ will be the sum of the contributions of each individual line segment.  Use the result for the magnetic field due to a straight segment of wire to find the field from each segment.

| | |
|---|---|
| 1.  Split the wire into five segments, $a$-$e$, as shown. | 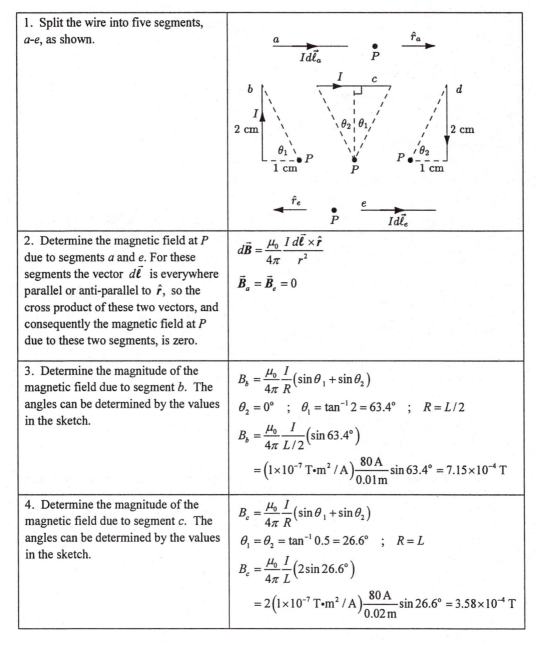 |
| 2.  Determine the magnetic field at $P$ due to segments $a$ and $e$. For these segments the vector $d\vec{\ell}$ is everywhere parallel or anti-parallel to $\hat{r}$, so the cross product of these two vectors, and consequently the magnetic field at $P$ due to these two segments, is zero. | $$d\vec{B} = \frac{\mu_0}{4\pi}\frac{I\,d\vec{\ell}\times\hat{r}}{r^2}$$ $$\vec{B}_a = \vec{B}_e = 0$$ |
| 3.  Determine the magnitude of the magnetic field due to segment $b$.  The angles can be determined by the values in the sketch. | $$B_b = \frac{\mu_0}{4\pi}\frac{I}{R}\left(\sin\theta_1 + \sin\theta_2\right)$$ $$\theta_2 = 0° \quad ; \quad \theta_1 = \tan^{-1}2 = 63.4° \quad ; \quad R = L/2$$ $$B_b = \frac{\mu_0}{4\pi}\frac{I}{L/2}\left(\sin 63.4°\right)$$ $$= \left(1\times10^{-7}\,\text{T·m}^2/\text{A}\right)\frac{80\,\text{A}}{0.01\,\text{m}}\sin 63.4° = 7.15\times10^{-4}\,\text{T}$$ |
| 4.  Determine the magnitude of the magnetic field due to segment $c$.  The angles can be determined by the values in the sketch. | $$B_c = \frac{\mu_0}{4\pi}\frac{I}{R}\left(\sin\theta_1 + \sin\theta_2\right)$$ $$\theta_1 = \theta_2 = \tan^{-1}0.5 = 26.6° \quad ; \quad R = L$$ $$B_c = \frac{\mu_0}{4\pi}\frac{I}{L}\left(2\sin 26.6°\right)$$ $$= 2\left(1\times10^{-7}\,\text{T·m}^2/\text{A}\right)\frac{80\,\text{A}}{0.02\,\text{m}}\sin 26.6° = 3.58\times10^{-4}\,\text{T}$$ |

| 5.  Determine the magnitude of the magnetic field due to segment $d$. The angles can be determined by the values in the sketch. | $B_d = \dfrac{\mu_0}{4\pi}\dfrac{I}{R}\left(\sin\theta_1 + \sin\theta_2\right)$ <br><br> $\theta_1 = 0° \quad ; \quad \theta_2 = \tan^{-1}2 = 63.4° \quad ; \quad R = L/2$ <br><br> $B_d = \dfrac{\mu_0}{4\pi}\dfrac{I}{L/2}\left(\sin 63.4°\right)$ <br><br> $= \left(1\times10^{-7}\ \text{T·m}^2/\text{A}\right)\dfrac{80\,\text{A}}{0.01\,\text{m}}\sin 63.4° = 7.15\times10^{-4}\ \text{T}$ |
|---|---|
| 6.  All non-zero magnetic fields are directed into the page, so the vector addition of the fields in this case reduces to the sum of the magnitudes of the fields. | $B = B_a + B_b + B_c + B_d + B_e$ <br><br> $= 0 + 7.15\times10^{-4}\ \text{T} + 3.58\times10^{-4}\ \text{T} + 7.15\times10^{-4}\ \text{T} + 0$ <br><br> $= 1.79\times10^{-3}\ \text{T}$ <br><br> The field is directed into the page. |

**Example #6—Interactive.**  Find the magnetic field at point $P$ in Figure 27-11.  The curved section is three-fourths of a 15.0-cm-radius circle centered at point $P$.  The current is a constant 40.0 A.

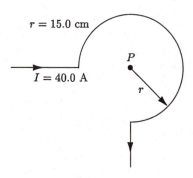

$r = 15.0$ cm

$I = 40.0$ A

$P$

$r$

Figure 27-11

**Picture the Problem.**  The magnetic field is the vector sum of the fields due to each straight section plus the field due to the curved section.  Apply the Biot-Savart law to each section.  **Try it yourself.**  Work the problem on your own, in the spaces provided, to get the final answer.

| 1.  Draw a sketch splitting the wire into three segments:  two straight and one curved. | |
|---|---|
| 2.  Determine the magnetic field at $P$ due to the incoming straight current segment. | |

| | |
|---|---|
| 3. Determine the magnetic field at $P$ due to the outgoing straight current segment. | |
| 4. Determine the magnetic field at $P$ due to the curved segment. Follow the procedure presented on pages 859-860 of the textbook for the magnetic field at the center of a circular current loop. | |
| 5. Determine the total magnetic field at $P$ by adding the fields from the three segments. | $B = 1.26 \times 10^{-4}$ T, directed into the page |

**Example #7.** In the laboratory, two identical coaxial circular coils, called Helmholtz coils, are separated by a distance equal to the radius of one of the coils. These are used to create a region where the magnetic field is fairly uniform. If the radius of the 100-turn coils is 30.0 cm and if the current in each coil is a steady 10.0 A, what is the magnetic field at a point on the axis midway between the coils?

**Picture the Problem.** Use the expression for the magnetic field on the axis of a circular coil. The total field will be the sum of the magnetic fields from each of the two coils.

| | |
|---|---|
| 1. Sketch the physical situation. Since we want the magnetic fields of the two coils to add, the current must be flowing in the same direction in both coils. | |

| | |
|---|---|
| 2.  Determine the magnitude of the magnetic field due to each coil, using the expression for the magnetic field on the axis of a current-carrying coil.  In our case, $x$, the distance from the coil, is simply $R/2$.<br><br>Because of the symmetry, the field from coil 2 is identical to that from coil 1. | $B_{1x} = \dfrac{\mu_0}{4\pi}\dfrac{2\pi R^2 NI}{\left(x^2 + R^2\right)^{3/2}}$<br><br>$= \dfrac{\mu_0}{4\pi}\dfrac{2\pi R^2 NI}{\left(\left(R/2\right)^2 + R^2\right)^{3/2}} = \dfrac{\mu_0}{4\pi}\dfrac{2\pi NI}{\left(1.25\right)^{3/2} R}$<br><br>$= \left(1\times10^{-7}\,\text{T}\right)\dfrac{2\pi\left(100\right)\left(10\,\text{A}\right)}{1.25^{3/2}\left(0.3\,\text{m}\right)} = 1.50\times10^{-3}\,\text{T}$<br><br>$B_{2x} = B_{1x}$ |
| 3.  Find the total magnetic field at the center of the Helmholtz coils. | $B = B_{1x} + B_{2x} = 3.00\times10^{-3}\,\text{T}$ |

**Example #8—Interactive.**  You want to wind a solenoid 3.50 cm in diameter and 16.0 cm long in which the magnetic field will be 250 G when it has a constant current of 3.00 A.  What length of wire do you need?

**Picture the Problem.**  The total length of wire will be equal to the number of turns in the solenoid times the length of wire needed for one turn of the wire.  **Try it yourself.**  Work the problem on your own, in the spaces provided, to get the final answer.

| | |
|---|---|
| 1. Using the expression for the magnetic field inside a long, tightly-wound solenoid, find the required number of turns per unit length of solenoid. | |

| | |
|---|---|
| 2.  From the turn density calculated in step 1, determine the total number of turns in the solenoid. | |
| 3.  Determine the length of wire required for one turn of the solenoid. | |
| 4.  Determine the total length of wire needed by multiplying the results of steps 2 and 3. | $L = 117\,\text{m}$ |

**Example #9.**  A long, straight, thin-walled cylindrical shell of radius $R$ carries a steady current $I$ flowing parallel with the shell's axis.  Find $B$ at all points inside the shell and outside the shell.

**Picture the Problem.**  There is a significant amount of symmetry in this problem, so Ampère's law should be used.  There are two regions in this problem, radii less than $R$ and radii greater than $R$.  Ampère's law will be applied differently for these two regions.

| | |
|---|---|
| 1.  Draw a sketch of the situation, with two ampèrian loops, one with a radius less than $R$ and the other with a radius greater than $R$. | 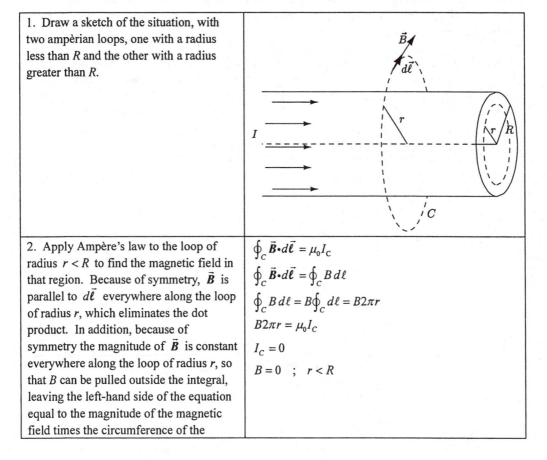 |
| 2.  Apply Ampère's law to the loop of radius $r < R$ to find the magnetic field in that region.  Because of symmetry, $\vec{B}$ is parallel to $d\vec{\ell}$ everywhere along the loop of radius $r$, which eliminates the dot product.  In addition, because of symmetry the magnitude of $\vec{B}$ is constant everywhere along the loop of radius $r$, so that $B$ can be pulled outside the integral, leaving the left-hand side of the equation equal to the magnitude of the magnetic field times the circumference of the | $\oint_C \vec{B} \cdot d\vec{\ell} = \mu_0 I_C$ $\oint_C \vec{B} \cdot d\vec{\ell} = \oint_C B\, d\ell$ $\oint_C B\, d\ell = B\oint_C d\ell = B2\pi r$ $B2\pi r = \mu_0 I_C$ $I_C = 0$ $B = 0 \;\; ; \;\; r < R$ |

| ampèrian loop. | |
|---|---|
| For $r < R$, the loop does not enclose any current, so the magnetic field must be zero inside the conducting shell. | |
| 3.  For the situation with the ampèrian loop outside of the conducting shell, $r > R$, the same symmetry arguments can be made.  However, in this case, the entire current $I$ flows through the loop, so $B \neq 0$. | $\oint_C \vec{B} \cdot d\vec{\ell} = \mu_0 I_C$ <br><br> $\oint_C \vec{B} \cdot d\vec{\ell} = \oint_C B \, d\ell$ <br><br> $\oint_C B \, d\ell = B \oint_C d\ell = B 2\pi r$ <br><br> $B 2\pi r = \mu_0 I_C$ <br><br> $I_C = I$ <br><br> $B = \dfrac{\mu_0 I}{2\pi r} \quad ; \quad r > R$ |

**Example #10—Interactive.**  A long, straight, coaxial cable consists of a solid wire of radius $a$ and a thin-walled conducting coaxial shell of radius $b > a$. The wire carries a steady current $I$ in one direction, and the shell carries the same current in the opposite direction.  Find $B$ at all points inside the wire, between the wire and the shell, and outside the shell.

**Picture the Problem.**  There is a significant amount of symmetry in this problem, so Ampère's law should be used.  There are three regions in this problem, so you will need to apply Ampère's law differently for these three regions, depending on how much current is enclosed.  **Try it yourself.**  Work the problem on your own, in the spaces provided, to get the final answer.

| 1.  Draw a sketch of the situation, with three ampèrian loops, one for each region of interest. | |
|---|---|
| 2.  Determine the amount of current that flows through a loop of radius $r < a$. It will not simply be $I$. | |

| | |
|---|---|
| 3. Now that you know how much current flows through the loop, apply Ampère's law to the loop of radius $r < a$ to find the magnetic field in that region. Remember to make the appropriate symmetry arguments. | $$B = \frac{\mu_0 I r}{2\pi a^2} \quad ; \quad 0 \le r < a$$ |
| 4. Apply Ampère's law to the region $a < r < b$. | $$B = \frac{\mu_0 I}{2\pi r} \quad ; \quad a < r < b$$ |
| 5. Apply Ampère's law to the region $r > b$. | $$B = 0 \quad ; \quad r > b$$ |

**Example #11.**  The 1500-turn winding of a 1.00-cm-diameter solenoid 20.0 cm long carries a current of 1.00 A.  The inside of the solenoid is filled with tungsten at 20.0° C.  What are $B_{app}$, $M$, and $B$ at the center of the solenoid?

**Picture the Problem.**  The applied field is provided by the solenoid.  The susceptibility of tungsten can be found in Table 27-1 on page 876 of the textbook.  The magnetization is related to the susceptibility, and adds a component to the total magnetic field inside the solenoid.

| | |
|---|---|
| 1. Determine the magnetic field produced by the solenoid. | $B_{app} = \mu_0 n I = \mu_0 (N/L) I$  <br> $= (4\pi \times 10^{-7})(1500)/(20\,\text{cm})(1\,\text{A}) = 9.43 \times 10^{-3}\,\text{T}$ |
| 2. Determine the magnetization of the tungsten from its susceptibility and the applied field. | $M = \chi_m \dfrac{B_{app}}{\mu_0} = (6.8 \times 10^{-5}) \dfrac{9.43 \times 10^{-3}\,\text{T}}{4\pi \times 10^{-7}\,\text{N/A}^2} = 0.510\,\text{A/m}$ |
| 3. Determine the total magnetic field inside the solenoid.  Notice that the paramagnetic material had very little effect on the total magnetic field. | $B = B_{app}(1 + \chi_m)$ <br> $= (9.43 \times 10^{-3}\,\text{T})(1 + 6.8 \times 10^{-5}) = 9.43 \times 10^{-3}\,\text{T}$ |

**Example #12—Interactive.** An unmagnetized, cylindrical iron bar has been inserted into the solenoid described in Example #11, and the current in the coil is brought up to 10.0 mA. What are $B_{app}$, $M$, and $B$ at the center of the solenoid?

**Picture the Problem.** Follow the solution to Example #11. The relative permeability of iron can be found in Table 27-2 on page 882 of the textbook. **Try it yourself.** Work the problem on your own, in the spaces provided, to get the final answer.

| | |
|---|---|
| 1. Determine the magnetic field produced by the solenoid. | $B_{app} = 9.42 \times 10^{-5}$ T |
| 2. Determine the magnetization of the tungsten from its susceptibility and the applied field. | $M = 4.12 \times 10^{5}$ A/m |
| 3. Determine the total magnetic field inside the solenoid. | $B = 0.518$ T |

**Example #13.** Aluminum has a density of $2700 \, kg/m^3$ and a molar mass of $0.0270 \, kg/mol$. What is the magnetic dipole moment of an aluminum atom in Bohr magnetons?

**Picture the Problem.** The magnetic susceptibility of aluminum can be found in Table 27-1 of the textbook. Note that it is small and positive, which tells us that aluminum is paramagnetic. Thus, we can use Curie's law, which relates the atomic magnetic dipole moment to other variables.

| | |
|---|---|
| 1. Write down Curie's law. | $M = \dfrac{1}{3} \dfrac{\mu B_{app}}{kT} M_s$ |
| 2. Write the expression used to define magnetic susceptibility. | $M = \chi_m \dfrac{B_{app}}{\mu_0}$ |
| 3. Write the expression for the saturation magnetization. Here $n$ is the number density, $\rho$ is the mass density, $M_{mol}$ is the molar mass, and $N_A$ is Avogadro's number. | $M_s = n\mu = \dfrac{\rho N_A}{M_{mol}} \mu$ |

| | |
|---|---|
| 4. Divide the first expression by the second, and substitute for $M_s$. | $1 = \frac{1}{3}\frac{\mu_0 \mu}{\chi_m kT} M_s = \frac{1}{3}\frac{\mu_0 \mu}{\chi_m kT}\frac{\rho N_A}{M_{mol}}\mu$ |
| 5. Solve the above expression for the magnetic moment $\mu$. | $\mu = \left(\frac{3\chi_m kT M_{mol}}{\mu_0 \rho N_A}\right)^{1/2}$ <br><br> $= \left[\frac{3(2.3\times10^{-5})(1.38\times10^{-23}\ \text{J/K})(298\,\text{K})(0.0270\,\text{kg/mol})}{(4\pi\times10^{-7}\ \text{T·m/A})(2700\,\text{kg/m}^3)(6.02\times10^{23}\ \text{atoms/mol})}\right]^{1/2}$ <br><br> $= 1.94\times10^{-24}\ \text{J/T}$ |
| 6. Convert this magnetic moment into Bohr magnetons. | $\mu = (1.94\times10^{-24}\ \text{J/T})\left(\frac{1\mu_B}{9.27\times10^{-24}\ \text{J/T}}\right) = 0.209\,\mu_B$ |

**Example #14—Interactive.**  Magnesium has a density of $1740\,\text{kg/m}^3$ and a molar mass of $0.0243\,\text{kg/mol}$. What is the magnetic dipole moment of a magnesium atom in Bohr magnetons?

**Picture the Problem.**  This is just like Example #13. **Try it yourself.** Work the problem on your own, in the spaces provided, to get the final answer.

| | |
|---|---|
| 1. Write down Curie's law. | |
| 2. Write the expression used to define magnetic susceptibility. | |
| 3. Write the expression for the saturation magnetization | |
| 4. Divide the first expression by the second, and substitute for $M_s$. | |
| 5. Solve the above expression for the magnetic moment $\mu$. | |

| 6. Convert this magnetic moment into Bohr magnetons. | |
|---|---|
| | $\mu = 0.178\,\mu_B$ |

**Example #15.**  A long iron-core solenoid of $4000\,\text{turns/m}$ carries a current of 10.0 mA. At this current the relative permeability is 1200. (*a*) What is the magnetic field strength within the solenoid? (*b*) Find the current necessary to produce the same field in the solenoid with the iron core removed.

**Picture the Problem.**  The actual magnetic field depends on the relative permeability of the material and the applied magnetic field.

| 1. Find the magnitude of the applied magnetic field, using the expression for the field inside a long solenoid. | $B = \mu_0 n I$ |
|---|---|
| 2. Use the expression relating the actual magnetic field to the applied field and the relative permeability to find the actual magnetic field. | $B = K_m B_{app} = K_m \mu_0 n I$ <br> $= (1200)\left(4\pi \times 10^{-7}\,\text{T·m/A}\right)\left(4000\,\text{m}^{-1}\right)(0.01\,\text{A})$ <br> $= 0.0603\,\text{T}$ |
| 3. Use the expression for the current in the center of a long solenoid to determine the current required to recreate this field without the iron core. | $B = \mu_0 n I$ <br> $I = \dfrac{B}{\mu_0 n} = \dfrac{0.0603\,\text{T}}{\left(4\pi \times 10^{-7}\,\text{T·m/A}\right)\left(4000\,\text{m}^{-1}\right)} = 12.0\,\text{A}$ |

**Example #16—Interactive.**  A toroid is filled with liquid oxygen, which has a magnetic susceptibility of $4 \times 10^{-3}$. The toroid has 4000 turns and carries a current of 10.0 A. Its mean radius is 30.0 cm and the radius of its cross section is 1.00 cm. (*a*) What is the magnetization *M*? (*b*) What is the magnetic field *B*? (*c*) What is the percentage increase in *B* produced by the liquid oxygen?

**Picture the Problem.**  This is very similar to Example #15. **Try it yourself.**  Work the problem on your own, in the spaces provided, to get the final answer.

| 1. Find the magnitude of the applied magnetic field, using the expression for the field inside a tightly wound toroid. | |
|---|---|
| | |

| 2. Find the magnetization from the applied field and the susceptibility of oxygen. | |
|---|---|
| | $M = 84.9\,\text{A/m}$ |
| 3. Determine the magnetic field from the applied magnetic field and the magnetization. | |
| | $B = 0.0268\,\text{T}$ |
| 4. Determine the percentage increase in the magnetic field due to the presence of the oxygen. | |
| | 0.4% |

**Example #17—Interactive.** The magnetic dipole moment of an iron atom is $2.22\,\mu_B$. In an iron bar 5.00 cm long and 0.300 cm in diameter, the magnetization $M$ is 50% of the saturated magnetization $M_s$. The density of iron is $7.9 \times 10^3\,\text{kg/m}^3$, and its molecular mass is $5.58 \times 10^{-2}\,\text{kg/mol}$. (*a*) What is the magnetic dipole moment of the bar? (*b*) What is the magnitude of the ampèrian surface current? (*c*) What is the torque exerted on the bar if it is suspended at right angles to a uniform magnetic field of 0.500 T?

**Picture the Problem.** The magnetic dipole moment of the bar is related to the magnetization of the bar. **Try it yourself.** Work the problem on your own, in the spaces provided, to get the final answer.

| 1. Relate the magnetic dipole moment of the bar to the magnetization of the bar. | |
|---|---|
| 2. We know the magnetization is 50% of the saturation magnetization. Use this to solve for the dipole moment of the bar. | |
| | $\mu = 0.310\,\text{A·m}^2$ |

| | |
|---|---|
| 3. The magnetic field at the center of the bar is $\mu_0 M$, and the magnetic field at the center of a long, tightly wound solenoid with $n$ turns per unit length carrying current $I$ is $\mu_0 nI$. The ampèrian surface current per unit length needed to produce the magnetization $M$ at the center of the bar is $nI$, where $I$ is the current in a long, tightly wound solenoid of the same length and diameter as the bar that is needed to produce the same magnetic field. | $I = 4.39 \times 10^4 \, \text{A}$ |
| 4. The torque on a magnetic dipole equals the cross product of the dipole moment and the applied magnetic field. | $\tau = 0.155 \, \text{m·N}$ |

# Chapter **28**

# Magnetic Induction

## I.   Key Ideas

The emf resulting from a changing (time-varying) magnetic field is called an **induced emf;** any current caused by an induced emf is called an **induced current.** The process by which induced emfs and currents are produced is called **magnetic induction.**

*Section 28-1. Magnetic Flux.* The **magnetic flux** $\phi_m$ through a surface $S$ is equal to the integral of the normal component $B_n$ of the magnetic field on the surface times the area $dA$ of each element of surface area. That is,

$$\phi_m = \int_S B_n \, dA = \int_S \vec{B} \cdot \hat{n} \, dA \qquad \text{Magnetic flux defined}$$

where $\hat{n}$ is the unit normal to the surface. The SI unit of flux is the **weber** (Wb). From the definition of magnetic flux one can see that

$$1 \text{ weber (Wb)} = 1 \text{ T} \cdot \text{m}^2 \qquad \text{Weber defined}$$

*Section 28-2. Induced EMF and Faraday's Law.* Because magnetic forces act at right angles to the velocity of the particles they act on, they never do work. Therefore, we know that an induced emf is not the result of work done by a magnetic field. If the magnetic field is not doing the work, what is? The answer is that a changing magnetic field generates a nonconservative electric field, and it is this nonconservative electric field that circulates and, hence, does work on the charge carriers, which results in the induced emf. (Any vector field $\vec{E}$ for which $\oint_C \vec{E} \cdot d\vec{\ell} \neq 0$, where $C$ is any closed path, is said to circulate.) The equation relating the rate of change of the magnetic field with the induced emf $\mathcal{E}$ is

$$\mathcal{E} = \oint_C \vec{E}_{nc} \cdot d\vec{\ell} = -\frac{d\phi_m}{dt} \qquad \text{Faraday's law}$$

where $\phi_m$ is the magnetic flux through any surface bounded by the curve $C$. This result (that $\mathcal{E} = -d\phi_m / dt$ ) is known as **Faraday's law.**

*Section 28-3. Lenz's Law.* An emf is directional in the sense that if a battery is connected backward in a circuit, the current reverses direction. The directional sense of an induced emf is predicted by a rule called **Lenz's law,** which states

> The induced emf and induced current are in such a direction as to oppose the change that produces them.

That is, the feedback is never positive. The negative sign in Faraday's law is a reminder of this.

When current flows through a long, tightly wound solenoid, a uniform magnetic field $\vec{B}$ is produced inside the solenoid. The magnetic flux per turn of the solenoid is *BA,* where *A* is the area of a plane surface bounded by a turn. The net magnetic flux through the solenoid is the flux per turn times the number of turns. When the current increases, the magnetic field increases, and so does the flux. In accordance with Lenz's law, an induced emf is directed so as to oppose the increasing flux due to the increasing current, and therefore is called a **back emf.**

*Section 28-4. Motional EMF.* Moving a wire through a static magnetic field induces an emf in the wire that is called a **motional emf.** The motional emf $\mathcal{E}$ is related to the magnetic field and the motion of the wire by

$$\mathcal{E} = -\frac{d\phi_m}{dt}$$

where $d\phi_m$ is the magnetic flux through the surface that is swept out by the moving wire during time $dt$. If a wire of length $\ell$, shown in Figure 28-1, moves with velocity $\vec{v}$ through, and perpendicularly to a magnetic field $\vec{B}$, the area $dA$ swept out by the wire during time $dt$ is $\ell v\, dt$. Thus,

$$\mathcal{E} = -\frac{d\phi_m}{dt} = -\frac{B\ell v\, dt}{dt} = -B\ell v \qquad\qquad \text{Motional emf}$$

where the minus sign is a reminder that the feedback is negative. That is, any currents generated by the motion of the wire are in such a direction that the magnetic force acting on the moving wire is directed opposite to its motion. Obviously, this force opposes the motion of the wire. This force will not be long-lived, since a steady current cannot be maintained in an isolated piece of wire.

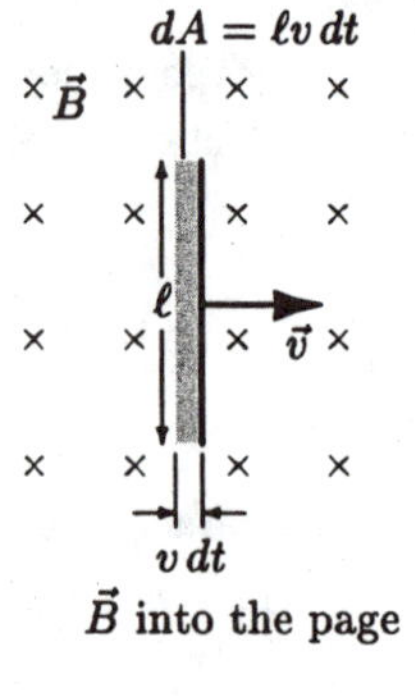

Figure 28-1

*Section 28-5. Eddy Currents.* When induced currents occur in bulk conductors (not thin wires), the induced currents tend to swirl around inside the conductors. These swirling currents are called **eddy currents**. In accordance with Lenz's law, if an eddy current is the result of a motional emf, a magnetic force acts on the moving conductor opposing its motion.

*Section 28-6. Inductance.* At all points in space, the magnetic field due to a current $I$ in a circuit is proportional to the current. Thus, the flux $\phi_m$ of this magnetic field through the circuit, due to the circuit itself, is also proportional to $I$. That is,

$$\phi_m = LI \qquad\qquad \text{Self-inductance } L \text{ defined}$$

where the proportionality constant $L$ is called the **self-inductance** of the circuit. The value of the self-inductance of a circuit depends on the geometrical size and shape of the circuit. The SI unit of inductance is the **henry** (H). From the definition of self-inductance we have

$$1 \text{ henry (H)} = 1 \text{ Wb/A} = 1 \text{ T·m}^2 / \text{A} \qquad\qquad \text{Henry defined}$$

In principle, the self-inductance $L$ of any circuit can be calculated by assuming a current $I$, solving for the resulting magnetic field $\vec{B}$, computing the flux $\phi_m$ due to that current, and then using the expression $L = \phi_m / I$ to determine the inductance. For most circuits this calculation is very difficult. However, it is straightforward for a long, tightly wound solenoid:

$$L = \frac{\phi_m}{I} = \mu_0 n^2 A\ell \qquad\qquad \text{Self-inductance of a solenoid}$$

where $\ell$ is the length of the solenoid, $n$ is the number of turns per unit length, and $A$ is the cross-sectional area. A dimensional analysis of this equation shows us that the permeability of free space $\mu_0$ can be alternatively expressed in henrys per meter:

$$\mu_0 = 4\pi \times 10^{-7} \text{ H/m}$$

Applying Faraday's law to the changing current in a circuit gives us

$$\mathcal{E} = -\frac{d\phi_m}{dt} = -L\frac{dI}{dt}$$

That is, the magnitude of the self-induced emf is proportional to the rate of change of the current in the circuit. The negative sign tells us that when the current is increasing, the self-induced emf opposes the increase, and when the current is decreasing, it opposes the decrease. That is, the feedback is always negative.

Two (or more) circuits that are near each other influence each other by means of their magnetic fields. Consider circuit 1 with current $I_1$, which produces magnetic field $\vec{B}_1$, and circuit 2 with current $I_2$, which produces magnetic field $\vec{B}_2$. The field $\vec{B}_1$ is proportional to $I_1$ and the field $\vec{B}_2$ is proportional to $I_2$. The net magnetic flux $\phi_{m2}$ through circuit 2 is the sum of two parts, one proportional to $I_1$ and one proportional to $I_2$. Thus,

$$\phi_{m2} = L_2 I_2 + M_{2,1} I_1 \qquad\qquad \text{Mutual inductance}$$

where $L_2$ is the self-inductance of circuit 2 and $M_{2,1}$ is called the **mutual inductance** of the two circuits. The mutual inductance depends on the individual geometries of the two circuits and their position and orientation relative to each other. The flux through circuit 1 is

$$\phi_{m1} = L_1 I_1 + M_{1,2} I_2$$

It can be shown that the two mutual inductances are always equal, so it is convenient to drop the subscripts. That is, for any two circuits, $M_{2,1} = M_{1,2} = M$.

The word *inductance* is commonly used as an abbreviation for self-inductance, when no ambiguity will arise.

*Section 28-7. Magnetic Energy.* Like a capacitor, an inductor stores energy. The energy $U$ stored in an inductor is given by

$$U = \tfrac{1}{2} L I^2 \qquad\qquad\qquad \text{Energy stored in an inductor}$$

It is useful to consider the stored energy as being stored in the magnetic field. The energy density $u_m$ of the magnetic field is

$$u_m = \frac{B^2}{2\mu_0} \qquad\qquad\qquad \text{Magnetic-field energy density}$$

*Section 28-8. RL Circuits.* Many circuits that contain coils or solenoids have large self-inductances. Coils or solenoids are often referred to as **inductors,** for which the circuit symbol is a wire coil, as shown in Figure 28-2.

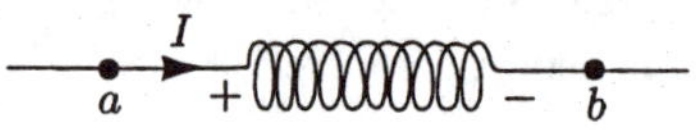

Figure 28-2

In accordance with Lenz's law, an increasing current in an inductor wire like that in Figure 28-2 induces an emf that opposes the current increase. This is because the increasing magnetic field produces a nonconservative electric field opposing the current. No difference in potential is associated with this field because potential functions can be defined only for conservative fields.

A conservative electric field in the direction of the current is also produced in the inductor wire by a distribution of charges on its surface. The changes in potential, which sum to zero in a closed circuit in accordance with Kirchhoff's loop rule, result from this conservative electric field.

If the resistance of the inductor wire is zero, the net electric field in the wire is also zero, in which case the conservative electric field in the wire is equal in magnitude and oppositely directed to the nonconservative electric field. However, if the resistance of the wire is finite, the magnitude of the conservative electric field exceeds that of the nonconservative field, resulting in a net electric field in the direction of the current. This resultant electric field drives the current in the inductor wire.

The change in potential across an inductor results from the presence of the conservative electric field. Because a conservative electric field always points in the direction of decreasing potential, the current through the inductor is in the direction of decreasing potential. Thus, the equation for the change in potential $V_b - V_a$ across a current-carrying inductor is

$$V_b - V_a = \mathcal{E} - Ir = -L\frac{dI}{dt} - Ir \qquad \text{Change in potential across a current-carrying inductor}$$

where $L$ is the self-inductance of the coil and $r$ the resistance of the coil. When solving circuit problems, always use this expression for the change in potential cross an inductor. When the current in an inductor decreases, the nonconservative electric field opposes this decrease; and because the net electric field in the wire drives the current, a surface-charge distribution produces a conservative electric field in the wire directed so as to oppose the current.

As in a battery, the internal resistance of an inductor is not always negligible. This internal resistance is represented in circuit diagrams by placing the symbol for a resistor in series with the inductor.

If a battery with an emf of $\mathcal{E}_0$ is connected across an inductor in series with a resistor, applying Kirchhoff's loop rule results in the equation

$$\mathcal{E}_0 - IR - L\frac{dI}{dt} = 0$$

Integrating this equation we obtain

$$I = \frac{\mathcal{E}_0}{R}\left(1 - e^{-Rt/L}\right) = I_f\left(1 - e^{-t/\tau}\right) \qquad \text{Energizing an inductor in an RL circuit}$$

where $\tau = L/R$ is the time constant and $I_f$ is the current after the circuit has been connected for a long time. Similarly, when an inductor with a current in it is simultaneously disconnected from a battery and connected across a resistor, applying Kirchhoff's loop rule results in the equation

$$-IR - L\frac{dI}{dt} = 0$$

Integrating this equation we obtain

$$I = I_0 e^{-Rt/L} = I_0 e^{-t/\tau} \qquad \text{De-energizing an inductor with a resistor}$$

where $I_0$ is the initial current.

***Section 28-9. Magnetic Properties of Superconductors.*** As a superconductor is cooled below the critical temperature in an applied magnetic field, the magnetic field inside the superconductor becomes zero. Superconducting currents on the surface of the superconductor are induced which produce a magnetic field in the opposite direction of the applied field that exactly cancels it out. That is, a superconductor is a perfect paramagnetic material.

## II.   Physical Quantities and Key Equations

### Physical Quantities

*Weber defined*                    1 weber (Wb) $= 1$ T·m$^2$

*Henry defined*                    1 henry (H) $= 1$ Wb/A $= 1$ T·m$^2$ / A

### Key Equations

*Magnetic flux defined*            $$\phi_m = \int_S B_n \, dA = \int_S \vec{B} \cdot \hat{n} \, dA$$

*Faraday's law*                    $$\mathcal{E} = \oint_C \vec{E} \cdot d\vec{\ell} = -\frac{d\phi_m}{dt}$$

*Motional emf*                     $$\mathcal{E} = -\frac{d\phi_m}{dt} = -\frac{B\ell v \, dt}{dt} = -B\ell v$$

*Self-inductance defined*          $$\phi_m = LI$$

*Self-inductance of a solenoid*    $$L = \frac{\phi_m}{I} = \mu_0 n^2 A\ell$$

*Mutual inductance*                $$\phi_{m2} = L_2 I_2 + M_{2,1} I_1$$

*Energy stored in an inductor*     $$U = \tfrac{1}{2} LI^2$$

*Magnetic-field energy density*    $$u_m = \frac{B^2}{2\mu_0}$$

*Change in potential across a current-carrying inductor*   $$V_b - V_a = \mathcal{E} - Ir = -L\frac{dI}{dt} - Ir$$

*Energizing an inductor in an RL circuit*   $$I = \frac{\mathcal{E}_0}{R}\left(1 - e^{-Rt/L}\right) = I_f\left(1 - e^{-t/\tau}\right)$$

*De-energizing an inductor with a resistor*   $$I = I_0 e^{-Rt/L} = I_0 e^{-t/\tau}$$

*Circuit symbol for inductor*

## III. Potential Pitfalls

Use Faraday's law to determine the magnitude of the induced emf. The minus sign in the equation is a useful reminder of Lenz's law, from which you can deduce the sense of the induced emf.  It is always directed so that any induced current would oppose the change in flux.

A large magnetic flux does not necessarily cause a large induced emf. It is the rate at which the flux changes, not the magnitude of the flux, that determines the induced emf. Thus, a large induced emf is caused by a rapidly changing magnetic flux.

Steady magnetic fields do not induce an emf by themselves. If a coil rotates, there will be a changing flux through it, which will induce an emf.

Inductance (either self or mutual) is a geometric property and depends only on the sizes, shapes, and relative positions of circuits, not on the currents in them.

All circuits have some self-inductance—it's a matter of degree. Further, circuits that contain coils of wire with ferromagnetic cores have a great deal of inductance compared to those that do not. Due to inductance, any change in current is accompanied by negative feedback. Consequently, the current in circuits with significant inductance cannot "instantaneously" change from one value to another, as it can in circuits with negligible inductance. Instead, it must change incrementally.

Lenz's law says that the induced emf is in a direction that opposes not the flux itself, but the change in the flux that produced it. The induced emf may be in a direction that tends to increase or decrease the existing flux, depending on the circumstances.

The fact that the emf from a generator varies sinusoidally with time has nothing to do with the shape of the rotating coil—only with the coil being rotated at constant angular speed in a static, uniform magnetic field.

## IV. True or False Questions and Responses

**True or False**

_____ 1. Any change in the magnetic flux through a circuit results in an induced emf.

_____ 2. The emf induced in a circuit is proportional to the magnetic flux through the circuit.

_____ 3. The electric field induced by a changing magnetic field is conservative.

_____ 4. Lenz's law states that the direction of an induced emf is always opposite to the magnetic field that induced it.

_____ 5. The emf induced in a circuit is always directed so as to reduce the magnetic flux through the circuit.

_____ 6. Two circuits are close to one another but not in physical contact; if the current in one changes, an emf is induced in the other.

_____ 7. Inductance is a geometric property.

_____ 8. The phrase "the flux through a circuit" is short for the phrase "the flux through any surface bounded by a closed circuit."

____ 9. The self-inductance of an isolated circuit is the ratio of the flux through the circuit to the magnetic field at the circuit.

____ 10. If an ideal battery is connected across an inductor, the current rises to its steady-state value in a time of the order of $RL$, where $R$ and $L$ are the resistance and inductance of the inductor.

____ 11. If an ideal battery is connected across an inductor that has negligible resistance, the current rises to its steady-state value with a very short time constant.

____ 12. A current cannot change abruptly in a circuit that has an inductance.

____ 13. No energy is required to sustain a constant current in an inductor with negligible resistance.

____ 14. The energy in a magnetic field is proportional to the current that creates the field.

**Responses to True or False**

1. True.

2. False. The induced emf is proportional to the *rate of change* of the magnetic flux through the circuit.

3. False. It is nonconservative. Otherwise, the electric field would be associated with a potential difference, not an emf. In a conservative electric field, the loop integral $\mathcal{E} = \oint_C \vec{E} \cdot d\vec{\ell}$ is zero. However, if the magnetic flux through $C$ is changing in time, then $\mathcal{E} = -d\phi_m / dt$ is nonzero.

4. False. Lenz's law refers, not to the induced emf, but to the current that would be generated by the induced emf. This current would oppose, not the magnetic field, but the *change* in magnetic flux that induced the current.

5. False. An induced emf due to a change in flux produces a current which produces a magnetic flux that opposes the initial change in flux that induced the emf. If the initial change in flux is an increase, then the flux resulting from induced emf will oppose the increase, but if the initial change in flux is a decrease, then the flux resulting from the induced emf will oppose the decrease.

6. True in most cases. Even though the circuits are not in physical contact, each is affected by the change in flux of the magnetic field due to the current in the other circuit. (For certain very specific geometries this change in flux is zero.)

7. True.

8. True.

9. False. It is the ratio of the flux through the isolated circuit to the current in the circuit.

10. False. The time constant of an $RL$ circuit is $L/R$, not $RL$.

11. False. If the resistance is negligible, then the time constant ($L/R$) is very large for a finite inductance $L$.

12. True. Any abrupt change in current produces a large back emf opposing the change.

13. True.

14. False. The energy is proportional to the square of the current.

## V.  Questions and Answers

**Questions**

1. Two conducting loops with a common axis are placed near each other, as shown in Figure 28-3, and initially the currents in both loops are zero. If a current $I_a$ is suddenly set up in loop *a*, as shown, is there also a current in loop *b*? If so, in which direction? What is the direction of the force that loop *a* exerts on loop *b*?

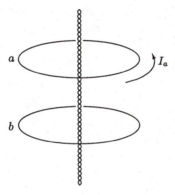

Figure 28-3

2. A common physics demonstration is to drop a bar magnet down a long, vertical aluminum pipe. Describe its motion.

3. An electric field is set up in the conducting bar of Figure 28-4 as the bar moves to the right. What causes this field and in what direction does it point? Does an external force have to act on the bar to keep it moving at a constant speed?

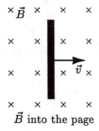

$\vec{B}$ into the page

Figure 28-4

4.  When the switch $S$ is opened in the $RL$ circuit shown in Figure 28-5, a spark jumps between the switch contacts. Why?

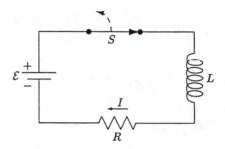

Figure 28-5

**Answers**

1.  A current is induced in loop $b$ while the current in loop $a$ is changing. If the current $I_a$ is in the direction shown and is increasing, the flux of its magnetic field $\vec{B}_a$ through loop $b$ is upward and increasing. In accordance with Lenz's law, the direction of the induced current $I_b$ in loop $b$ is such that the flux of its magnetic field $\vec{B}_b$ through loop $b$ is downward, opposing the change in flux that produced it. This means that $I_b$ is antiparallel to $I_a$. Because the currents are antiparallel, the two loops repel. After $I_a$ stops changing, $I_b = 0$, and there is no force between the loops.

2.  The falling bar magnet induces a current in the wall of the pipe. In accordance with Lenz's law, the direction of this induced current is such that it produces a magnetic field that exerts forces on the magnet that oppose its motion. The speed of the magnet therefore increases less rapidly than in free fall, until it reaches a terminal speed at which the magnitudes of the gravitational and magnetic forces acting on the magnet are equal.

3.  As the conducting bar is moved to the right, a downward magnetic force $-e\vec{v} \times \vec{B}$ acts on each and every free electron in the rod. This causes the electrons to drift downward (down the page), making the lower end of the rod negative and the upper end positive. This charge distribution produces a downward electric field in the rod. After the free charges have stabilized, no current (moving charge) exists in the bar, and so there is no net magnetic force on the bar opposing its motion. Therefore, no external force is needed to keep the bar moving at constant velocity.

4.  The current drops suddenly toward zero when the switch is opened. This current drop induces a very large emf in the inductor, which produces a large potential difference across the switch gap. This potential is large enough to cause dielectric breakdown of the air in the gap; this causes the spark, which allows current to continue to flow for a brief period.

## VI.  Problems, Solutions, and Answers

**Example #1.**    A 30-turn coil with a diameter of 6.00 cm is placed in a constant, uniform magnetic field of 1.00 T directed perpendicular to the plane of the coil. Beginning at time $t = 0$, the field is increased at a uniform rate until it reaches 1.30 T at $t = 10.0$ s. The field remains constant thereafter.   What is the magnitude of the induced emf in the coil at $(a)\, t < 0$,   $(b)$ $t = 5.00\,\text{s}$,  and $(c)\ t > 10.0\,\text{s}$?

**Picture the Problem.**  Use Faraday's law to find the induced emf.  The flux through a multi-turn coil is equal to the flux per turn multiplied by the number of turns.

| | |
|---|---|
| 1.  Write down Faraday's law as a guiding principle for the problem.  In this problem neither the area nor the relative direction of the unit normal to the surface and the magnetic field are changing. | $\mathcal{E} = \left\| -\dfrac{d\phi_{m}}{dt} \right\| = \left\| -\dfrac{d}{dt} N \oint_{C} \vec{B} \cdot \hat{n}\, dA \right\| = \left\| -\dfrac{d}{dt} NBA \right\|$ $= \left\| -NA \dfrac{dB}{dt} \right\|$ |
| 2.  For $t < 0$ the magnetic field, and hence the magnetic flux through the coil is not changing, so the induced emf in this timeframe is zero. | $\mathcal{E} = 0 \quad;\quad t < 0$ |
| 3.  At $t = 5\,\text{s}$ the magnetic field, and hence the flux through the coil is changing.  Find the time rate of change of the magnetic field. | $dB/dt = \left(B_{f} - B_{i}\right)/\Delta t$ |
| 4. Now calculate the induced emf in the coil at $t = 5\,\text{s}$. | $\mathcal{E} = -NA\left(B_{f} - B_{i}\right)/\Delta t$ $= -(30)\pi(0.03\,\text{m})^{2}(1.3\,\text{T} - 1.0\,\text{T})/(10\,\text{s})$ $= 2.55\,\text{mV} \quad;\quad t = 5\,\text{s}$ |
| 5.  For $t > 10\,\text{s}$ the rate of change of the magnetic field is again zero, so the induced emf is also zero. | $\mathcal{E} = 0 \quad;\quad t > 10\,\text{s}$ |

**Example #2—Interactive.**  A 100-turn coil with a 10.0-cm radius is placed in a spatially uniform magnetic field directed perpendicular to the plane of the coil.  The magnetic field changes sinusoidally according to the expression $B = B_{0} \sin(2\pi ft)$, where $B_{0} = 0.100\,\text{T}$ and $f = 100\,\text{Hz}$. $(a)$ Find the maximum emf induced in the coil. $(b)$ What is the magnitude of the magnetic field when the induced emf is maximum?

**Picture the Problem.**  The maximum emf will occur when the change in the magnetic flux, and hence the magnetic field is largest. **Try it yourself.**  Work the problem on your own, in the spaces provided, to get the final answer.

| | |
|---|---|
| 1.  Write down Faraday's law as a guide. | |
| 2.  Find the derivative of the magnetic field. | |

| | |
|---|---|
| 3.  Substitute your result from step 2 into Faraday's law to find the induced emf as a function of time. | |
| 4.  Find the maximum value of the induced emf as a function of time, and the time at which that maximum occurs. | $\mathcal{E}_{max} = 197\,\text{V}$ |
| 5.  Determine the magnitude of the magnetic field when the induced emf is a maximum. | $B = 0$ |

**Example #3.**  A 50-turn square coil with a cross-sectional area of $5.00\,\text{cm}^2$ has a resistance of $20.0\,\Omega$. The plane of the coil is perpendicular to a uniform magnetic field of 1.00 T. The coil is suddenly rotated about the axis shown in Figure 28-6 through an angle of $60°$ over a period of 0.200 s. What charge flows past a point in the coil during this time?

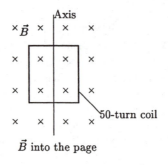

$\vec{B}$ into the page

Figure 28-6

**Picture the Problem.**  Because the direction of the area vector changes relative to the magnetic field, the flux through the coil changes. This change in flux induces an emf in the coil which causes current to flow in the coil. The total charge passing through one point is the integral of current over time.

| 1. Obtain an expression for the total charge passing a single point. We know that current is voltage divided by resistance, and that the voltage is really the induced emf in the loop. Note that the total charge that flows is ultimately independent of the speed at which the loop is rotated. | $Q = \int I\,dt = \int \dfrac{\mathcal{E}}{R}\,dt = -\dfrac{1}{R}\int \dfrac{d\phi_m}{dt}\,dt$ $= -\dfrac{1}{R}\int d\phi_m = -\dfrac{\Delta\phi_m}{R}$ |
|---|---|
| 2. Determine the initial flux through the coil. | $\phi_{m1} = N\oint_C \vec{B}\cdot\hat{n}_1\,dA = NBA\cos 0° = NBA$ |
| 3. Determine the final flux through the coil. After the coil has rotated $60°$, the magnetic field makes this same angle with the unit normal to the surface. | $\phi_{m2} = N\oint_C \vec{B}\cdot\hat{n}_1\,dA = NBA\cos 60°$ |
| 4. Substitute the results of step 2 and 3 into the expression for the charge to solve. | $Q = -\dfrac{\phi_{m2}-\phi_{m1}}{R} = \dfrac{NBA(1-\cos 60°)}{R}$ $= \dfrac{50(1.00\,\text{T})\pi\left(5\times10^{-4}\,\text{m}^2\right)(1-\cos 60°)}{20\,\Omega}$ $= 6.25\times10^{-4}\,\text{C}$ |

**Example #4—Interactive.** As shown in Figure 28-7, a square, 30-turn, 0.82-Ω coil 10.0 cm on a side is between the poles of a large electromagnet that produces a constant, uniform magnetic field of 6.00 kG. As suggested by the figure, the field drops sharply to zero at the edges of the magnet and the coil moves to the right at a constant velocity of 2.00 cm/s. What is the current through the coil wire (a) before the coil reaches the edge of the field, (b) while the coil is leaving the field, and (c) after the coil leaves the field? (d) What is the total charge that flows past a given point in the coil as it leaves the field?

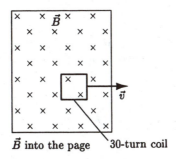

$\vec{B}$ into the page        30-turn coil

Figure 28-7

**Picture the Problem.** Faraday's law gives the induced emf in the coil, which will produce a flow of current in the coil. The flux through the coil changes as it moves out of the region of magnetic field. **Try it yourself.** Work the problem on your own, in the spaces provided, to get the final answer.

| | |
|---|---|
| 1. Determine the change in flux, and hence the induced emf and current through the loop before it reaches the edge of the field. | $I = 0\,\text{A}$ |
| 2. Determine the change in flux, and hence the induced emf and current through the loop after the coil leaves the field. | $I = 0\,\text{A}$ |
| 3. Calculate the rate of change of flux when the coil is partially out of the magnetic field by finding an expression for the flux when the right edge of the coil is a distance $x$ from the edge of the field. A sketch will help. | |
| 4. Relate the velocity of the coil to the rate of change of $x$, and solve for the induced emf. | |
| 5. Determine the current that flows through the coil while it is partially out of the magnetic field. | $I = 0.0439\,\text{A}$ |
| 6. Determine the time require for the coil to leave the magnetic field. | |
| 7. Integrate the current over this time interval to find the total charge that flows past a given point. | $Q = 0.220\,\text{C}$ |

**Example #5.** A tightly wound solenoid of 1600 turns, cross-sectional area of $6.00\,\text{cm}^2$, and length of 20.0 cm carries a current of 2.80 A. (*a*) What is its inductance? (*b*) How much energy is stored in the solenoid?

| | |
|---|---|
| 1. Find the inductance using the expression for the inductance of a long, tightly wound solenoid. | $L = \mu_0 n^2 A\ell = \dfrac{\mu_0 N^2 A}{\ell}$ $= \dfrac{\left(4\pi\times10^{-7}\,\text{H/m}\right)\left(1600\right)^2\left(6\times10^{-4}\,\text{m}^2\right)}{0.20\,\text{m}} = 9.65\times10^{-3}\,\text{H}$ |
| 2. Find the energy stored in the inductor by using the appropriate expression. | $U = \tfrac{1}{2}LI^2 = 0.5\left(9.65\times10^{-3}\,\text{H}\right)\left(2.8\,\text{A}\right)^2 = 0.0378\,\text{J}$ |

**Example #6—Interactive.** A tightly wound solenoid 18.0 cm long with a 2.00-cm diameter is made of 1500 turns of #22 copper wire. Find the (*a*) inductance, (*b*) resistance, and (*c*) time constant of the solenoid.

**Picture the Problem.** The inductance can be found using the same method as in Example #5. You will need the resistivity of copper and the cross-sectional area of #22 copper wire found in Tables 25-1 and 25-2 on page 792 of the textbook. **Try it yourself.** Work the problem on your own, in the spaces provided, to get the final answer.

| | |
|---|---|
| 1. Find the inductance using the expression for the inductance of a long, tightly wound solenoid. | $L = 4.94 \times 10^{-3}\,\text{H}$ |
| 2. Determine the length of wire needed to build this solenoid. | |
| 3. From the length of wire and the information from Tables 25-1 and 25-2 in the textbook, determine the resistance of the wire. | $R = 4.92\,\Omega$ |
| 4. The time constant can be determined from the inductance and the resistance. | $\tau = 1.00\,\text{ms}$ |

**Example #7.** The solenoid in Example #6 is surrounded by a 20-turn circular coil, 3.00 cm in diameter, that is coaxial with the solenoid, as in Figure 28-8. The circular coil is connected across a resistor of very high resistance. What is the magnitude of the induced emf in the coil when the current in the solenoid is changing at a rate of 100 A/s?

Figure 28-8

**Picture the Problem.** The emf induced in the coil is the emf induced by the change of the solenoid current plus the emf induced by the change of the circular-coil current.

| 1. Write an expression for the induced emf in the coil. | $\mathcal{E} = -\dfrac{d\phi_m}{dt} = -\dfrac{d}{dt}\left(L_{coil}I_{coil} + MI_{solenoid}\right)$ |
|---|---|
| 2. The coil has a very high resistance, so it has very little current, so the self-inductance term is essentially zero. As a result, the induced emf in the coil is caused only by the changing magnetic flux through the coil from the solenoid. | $L_{coil}I_{coil} \approx 0$ <br><br> $\mathcal{E} = -\dfrac{d}{dt}MI_{solenoid} = -\dfrac{d\phi_m}{dt}$ |
| 3. Determine the flux through the coil from the solenoid. Because the solenoid only produces a field within its radius, the flux through the coil at radii larger than the radius of the solenoid is zero. | $\phi_{coil} = N_{coil}\displaystyle\oint_C \vec{B}_{solenoid}\cdot\hat{n}_{coil}\,dA_{coil}$ <br><br> $= N_{coil}\left(\mu_0 n_{solenoid}I_{solenoid}\right)A_{solenoid}$ <br><br> $= N_{coil}\left(\mu_0 n_{solenoid}I_{solenoid}\right)\pi r_{solenoid}^2$ |
| 4. The induced emf can now be found. | $\left\|\mathcal{E}\right\| = \left\|-\dfrac{d}{dt}\left[N_{coil}\left(\mu_0 n_{solenoid}I_{solenoid}\right)\pi r_{solenoid}^2\right]\right\|$ <br><br> $= \left\|-N_{coil}\mu_0 n_{solenoid}\pi r_{solenoid}^2 \dfrac{dI_{solenoid}}{dt}\right\|$ <br><br> $= \left\|-(20)\left(4\pi\times10^{-7}\right)\left[1500/(0.18\,\text{m})\right]\pi(0.01\,\text{m})^2\left(100\,\text{A/s}\right)\right\|$ <br><br> $= \left\|6.58\,\text{mV}\right\|$ |

**Example #8—Interactive.** A 20-turn circular coil, 1.60 cm in diameter, is inside and coaxial with the solenoid in Example #6 as shown in Figure 28-9. The circular coil is connected across a resistor of very high resistance.  What is the magnitude of the emf induced in the circular coil when the current in the solenoid changes at a rate of 100 A/s?

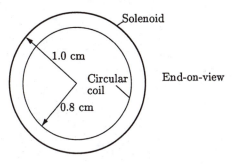

Figure 28-9

**Picture the Problem.**  Follow the procedure used in Example #7.  Be careful when determining the flux through the coil. **Try it yourself.**  Work the problem on your own, in the spaces provided, to get the final answer.

| | |
|---|---|
| 1. Write an expression for the induced emf in the coil. | |
| 2. Determine the emf due to the self-inductance of the coil. | |
| 3. Determine the flux through the coil from the solenoid. | |
| 4. The induced emf can now be found. | $\mathcal{E} = 4.21\,\text{mV}$ |

**Example #9.** After the switch is closed in Figure 28-10, the current begins to flow. At the instant at which half the power supplied by the battery is being dissipated in the resistor, what is the current and how fast is it changing?

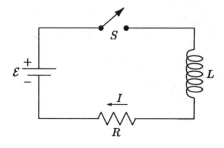

Figure 28-10

**Picture the Problem.** If half the power delivered by the battery reaches the resistor, the other half goes to the inductor.

| | |
|---|---|
| 1. Solve for the current by equating half the power delivered by the battery to the power dissipated in the resistor. | $0.5P_{battery} = P_{resistor}$ <br> $0.5\mathcal{E}I = I^2 R$ <br> $I = \mathcal{E}/(2R)$ |
| 2. Solve for the rate of change of the current by equating half of the battery power to the power delivered to the inductor. | $0.5P_{battery} = P_{inductor}$ <br> $0.5\mathcal{E}I = -\mathcal{E}_{back}I$ <br> $0.5\mathcal{E}I = LI(dI/dt)$ <br> $dI/dt = \mathcal{E}/(2L)$ |

**Example #10—Interactive.** The switch in Figure 28-11 is closed and the current through the resistor at some instant is 6.00 A to the left. What is the rate of change of the current at this instant? Is the current increasing or decreasing? What is the potential difference $V_b - V_a$?

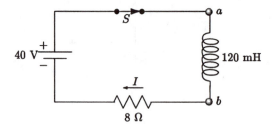

Figure 28-11

**Picture the Problem.** Apply Kirchhoff's loop rule to the circuit and solve for the potential difference across the inductor. Use this to find the rate of change of the current. **Try it yourself.** Work the problem on your own, in the spaces provided, to get the final answer.

| 1. Apply Kirchhoff's loop rule to find the potential difference across the inductor. | |
|---|---|
| | $V_b - V_a = +8.00$ V |
| 2. Because $V_b > V_a$, the inductor is encouraging the clockwise flow of current, so the current must be decreasing. | |
| 3. Determine the rate of change of the current using the expression for the voltage gain or drop across an inductor. | |
| | $dI / dt = 66.7$ A/s |

**Example #11—Interactive.** A "Rowland ring," shown in Figure 28-12, can be used to study the properties of ferromagnetic materials. A toroidal coil is wound around a ring-shaped sample of the material, and an electrically separate pickup coil is wound around the toroid. In this case, the inner and outer radii of the toroid are both approximately 10.0 cm, and it is wound with 600 turns. The unmagnetized sample inside the toroid has a cross-sectional diameter of 0.800 cm, and the pickup coil has 50 turns with a resistance of $8.00\,\Omega$. When the current in the toroid is increased from zero to 1.40 A, a total charge of $3.30 \times 10^{-6}$ C flows in the pickup coil. Find $B_{app}$, $M$, and $B$ in the coil.

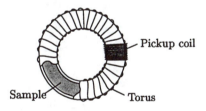

Figure 28-12

**Picture the Problem.** The total charge that flows through the pickup coil is due to the changing flux through the pickup coil. **Try it yourself.** Work the problem on your own, in the spaces provided, to get the final answer.

| 1. Relate the total flow of charge to the change in magnetic flux through the pickup coil. | |
|---|---|
| | |

| | |
|---|---|
| 2.  Find an expression for the total magnetic flux through the pickup coil, in terms of $B$. | |
| 3.  Use your expressions from steps 1 and 2 to solve for $B$. | $B = 0.0105\,\text{T}$ |
| 4.  Find $B_{\text{app}}$ from the expression for the magnetic field inside a tightly would toroid derived in Chapter 27. | $B_{\text{app}} = 1.60 \times 10^{-3}\,\text{T}$ |
| 5.  Use your values from steps 3 and 4 to find $M$. | $M = 7080\,\text{A/m}$ |

# Chapter **29**

# **Alternating Current Circuits**

## I.   Key Ideas

Almost all of our electrical energy is produced by electromagnetic induction in ac generators and is delivered as alternating-current (ac) electrical potential energy. An ac generator produces an emf that varies sinusoidally with time. The study of sinusoidal emf is important, not only because most electrical power is produced in that form but also because voltages and currents that are not sinusoidal can be analyzed in terms of sinusoidal components using Fourier analysis.

*Section 29-1. AC Generators.* One type of ac generator consists of a coil of wire rotating in a static magnetic field $B$. An expression for the emf $\mathcal{E}$ of such a generator, obtained via Faraday's law, $\mathcal{E} = d\phi_m / dt$, is

$$\mathcal{E} = NBA\omega \sin\left(\omega t + \delta\right) = \mathcal{E}_{\text{peak}} \sin\left(\omega t + \delta\right) \qquad \text{Terminal voltage of an ideal ac generator}$$

where $N$ is the number of turns, $A$ is the area bounded by a single turn, $\omega$ is the coil's angular speed, and $\mathcal{E}_{\text{peak}} = NBA\omega$.

   The same device (a coil in a static magnetic field) can be used as an ac **motor.** Instead of rotating the coil mechanically, another ac generator is connected across the coil. As a result there is a current through the coil. The torque due to the magnetic forces causes the coil to start rotating. As it gains speed there is a back emf resulting from its motion through the magnetic field. In accord with Lenz's law, this back emf tends to oppose the emf of the other ac generator. When the motor is first turned on, the back emf is negligible and the current is very large, being limited only by the coil's resistance.

*Section 29-2. Alternating Current in a Resistor.* Consider the circuit shown in Figure 29-1, with a resistor connected across the terminals of an ideal ac generator, whose terminal voltage is given by $\mathcal{E} = \mathcal{E}_{\text{peak}} \cos \omega t$. [For convenience, we have chosen $\delta = \pi / 2$ so that $\mathcal{E}_{\text{peak}} \sin\left(\omega t + \delta\right)$ equals $\mathcal{E}_{\text{peak}} \cos \omega t$.] The voltage drop $V_R$ across the resistor is the terminal voltage of the generator. In accordance with Ohm's law, the current through the resistor is related to this voltage drop by the equation $V_R = IR$, so

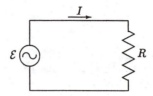

Figure 29-1

$$I = \frac{V_R}{R} = \frac{\mathcal{E}_{peak}}{R} \cos \omega t = I_{peak} \cos \omega t$$

where

$$I_{peak} = \frac{\mathcal{E}_{peak}}{R} \qquad\qquad \text{Maximum current through a resistor}$$

Over a complete cycle, both the cosine function and the sine function are negative to the same degree that they are positive, thus their average values are both zero. The average values of $\cos^2 \theta$ and $\sin^2 \theta$ are both $\frac{1}{2}$. To understand why this is so, recall that $\cos^2 \theta + \sin^2 \theta = 1$; so the average value of ($\cos^2 \theta + \sin^2 \theta$) equals 1. The average values of $\cos^2 \theta$ and $\sin^2 \theta$ over a complete cycle are equal, so $\left(\cos^2 \theta\right)_{av} = \left(\sin^2 \theta\right)_{av} = \frac{1}{2}$.

The instantaneous power delivered to a resistor is the product of the current and the voltage drop in the direction of the current. Thus,

$$\begin{aligned} P = IV_R &= \left(I_{peak} \cos \omega t\right)\left(\mathcal{E}_{peak} \cos \omega t\right) \\ &= I_{peak}^2 R \cos^2 \omega t \end{aligned} \qquad \text{Power delivered to a resistor}$$

where we have substituted $I_{peak} R$ for $\mathcal{E}_{peak}$. The average value of $\cos^2 \theta$ over one cycle ($2\pi$ rad) is $\frac{1}{2}$, so the average power $P_{av}$ is

$$P_{av} = \frac{1}{2} I_{peak}^2 R$$

**RMS Values.** The root-mean-square (rms) value of the current $I_{rms}$ is defined as

$$I_{rms} = \sqrt{\left(I^2\right)_{av}} \qquad\qquad \text{root-mean-square (rms) value}$$

where the average is taken over a complete cycle. Because the average value of $\cos^2 \theta$ is $\frac{1}{2}$, the rms value of $\cos \theta$ is

$$\left(\cos \theta\right)_{rms} = \sqrt{\left(\cos^2 \theta\right)_{av}} = \frac{1}{\sqrt{2}}$$

Thus,

$$I_{rms} = \left( I_{max} \cos \omega t \right)_{rms}$$
$$= \sqrt{\left( I_{max}^2 \cos^2 \omega t \right)_{av}} = I_{max} \sqrt{\left( \cos^2 \omega t \right)_{av}}$$
$$= \frac{1}{\sqrt{2}} I_{max}$$

Therefore, the average power being delivered to a resistor is

$$P_{av} = \frac{1}{2} I_{max}^2 R = I_{rms}^2 R \qquad\qquad \text{Average power dissipated by a resistor}$$

***Section 29-3. Alternating Current Circuits.*** The voltage drop across an ideal inductor in the direction of the current is $V_L = L(dI / dt)$. If an ideal inductor is connected across the terminals of a generator, as shown in Figure 29-2, the terminal voltage of the generator equals the voltage drop across the inductor. Thus,

$$\frac{dI}{dt} = \frac{\mathcal{E}_{peak}}{L} \cos \omega t$$

which integrates to

$$I = \frac{\mathcal{E}_{peak}}{\omega L} \sin \omega t = I_{peak} \sin \omega t$$

where

$$I_{peak} = \frac{\mathcal{E}_{peak}}{\omega L} = \frac{\mathcal{E}_{peak}}{X_L}$$

and the **inductive reactance** is

$$X_L = \omega L \qquad\qquad \text{Inductive reactance}$$

Dividing both sides of this equation by $\sqrt{2}$ gives

$$I_{rms} = \frac{\mathcal{E}_{rms}}{\omega L} = \frac{\mathcal{E}_{rms}}{X_L} \qquad\qquad \text{rms current through an inductor}$$

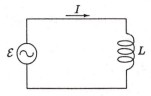

Figure 29-2

The instantaneous power delivered to an inductor is the product of the current and the voltage drop in the direction of the current. That is,

$$P = IV_L = \left( I_{peak} \sin \omega t \right)\left( \mathcal{E}_{peak} \cos \omega t \right)$$
$$= \tfrac{1}{2} I_{peak}\, \mathcal{E}_{peak} \sin 2\omega t$$

where we have used the identity $\sin 2\theta = 2\sin\theta\cos\theta$. Because the average of $\sin 2\omega t$ over one cycle is zero, the average power delivered to an ideal inductor is zero.

The voltage drop across a capacitor in the direction of the current is $V_C = Q/C$. If a capacitor is connected across the terminals of a generator, as shown in Figure 29-3, the terminal voltage of the generator equals the voltage drop across the capacitor. Thus,

$$Q = CV_c = C\mathcal{E}_{peak} \cos \omega t$$

which can be differentiated to give

$$I = -\omega C \mathcal{E}_{peak} \sin \omega t = -I_{peak} \sin \omega t$$

where

$$I_{max} = \omega C \mathcal{E}_{peak} = \frac{\mathcal{E}_{peak}}{X_C}$$

and the **capacitive reactance** is

$$X_C = 1/\omega C$$                                                       Capacitive reactance

Dividing both sides of this equation by $\sqrt{2}$ gives

$$I_{rms} = \omega C \mathcal{E}_{rms} = \frac{\mathcal{E}_{rms}}{X_C}$$               rms current "through" a capacitor

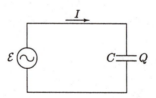

Figure 29-3

The instantaneous power delivered to a capacitor is the product of the current and the voltage drop in the direction of the current. That is,

$$P = IV_C = \left( -I_{peak} \sin \omega t \right)\left( \mathcal{E}_{peak} \cos \omega t \right)$$
$$= \tfrac{1}{2} I_{peak}\, \mathcal{E}_{peak} \sin 2\omega t$$

where we have again used the identity $\sin 2\theta = 2\sin\theta\cos\theta$. Because the average of $\sin 2\omega t$ over one cycle is zero, the average power delivered to a capacitor is also zero.

Most circuits do not consist of a single resistor, capacitor, or inductor connected across the terminals of a generator, so we need to express the rms current through each device in terms of the voltage drop across that device rather than in terms of the terminal voltage of a generator. Therefore, we have

$$I_{rms} = \frac{V_{R,rms}}{R}$$

$$I_{rms} = \frac{V_{L,rms}}{\omega L} = \frac{V_{L,rms}}{X_L}$$

$$I_{rms} = \omega C V_{C,rms} = \frac{V_{C,rms}}{X_C}$$

***Section 29-4. Phasors.*** For a resistor the voltage drop is in phase with the current; for an inductor the voltage drop leads the current by a phase angle of 90°; and for a capacitor the voltage drop lags behind the current by a phase angle of 90°. If the three components are connected in series, then the current through each of them is $I = I_{peak} \cos\theta$, where $\theta = \omega t + \delta$ is the phase of the current. The voltages are

$$V_R = V_{R,peak} \cos\theta$$

$$V_L = V_{L,peak} \sin\theta$$

$$= V_{L,peak} \cos\left(\theta + (\pi/2)\right)$$

and

$$V_C = -V_{C,peak} \sin\theta$$

$$= V_{C,peak} \cos\left(\theta - (\pi/2)\right)$$

and the voltage drop $V_{tot}$ across the series combination is given by

$$V_{tot} = V_{R,peak} \cos\theta + V_{L,peak} \cos\left(\theta + (\pi/2)\right) + V_{C,peak} \cos\left(\theta - (\pi/2)\right)$$

With each voltage drop proportional to the cosine of an angle, you can find the *x* component of a resultant vector by adding the *x* components of the vectors. The two-dimensional vectors, shown in Figure 29-4, corresponding to the voltage drops are called **phasors** and have magnitudes equal to the maximum voltages $V_{R,peak}$, $V_{L,peak}$, and $V_{C,peak}$. The three phasors rotate with the same angular velocity $\omega$ This means that relative to each other their phases are fixed. To find the voltage drop across the series combination, you add the phasors (using vector addition of course) and then take the *x* component of the sum.

***Section 29-5. LC and RLC Circuits Without a Generator.*** Applying Kirchhoff's loop rule to a circuit consisting of an initially charged capacitor and an inductor, shown in Figure 29-5, gives us an equation that is identical to the equation for simple harmonic motion. The solution to this equation is that the charge on the capacitor and the current vary as

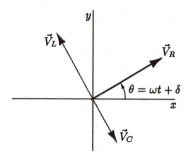

Figure 29-4

$$Q = Q_{peak} \cos\left(\omega t + \delta\right)$$

and

$$I = \frac{dQ}{dt} = -I_{peak} \sin\left(\omega t + \delta\right)$$

where $\omega = 1/\sqrt{LC}$, and $\delta$ is the initial phase of the charge. When the magnitude of the charge is at a maximum, the current is zero, and all the energy is stored in the capacitor as electrical potential energy. When the magnitude of the current is at a maximum, the charge is zero, and all the energy is stored in the inductor as magnetic potential energy.

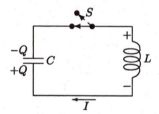

Figure 29-5

If a resistor is included in our circuit, we obtain an equation identical to that of a linearly damped harmonic oscillator.

***Section 29-6. Driven RLC Circuits.*** Applying Kirchhoff's loop rule to a circuit consisting of a generator connected to a **series** combination of an inductor, a capacitor, and a resistor, shown in Figure 29-6, gives us the equation

$$L\frac{d^2Q}{dt^2} + R\frac{dQ}{dt} + \frac{1}{C}Q = \mathcal{E}_{peak} \cos \omega t$$

for which the steady-state solution is

$$I = I_{peak} \cos\left(wt - \delta\right)$$

where the **phase angle** $\delta$ is given by

$$\tan \delta = \frac{X_L - X_C}{R}$$    Phase angle

and the peak current is

$$I_{peak} = \frac{\mathcal{E}_{peak}}{Z}$$

where

$$Z = \sqrt{R^2 + (X_L - X_C)^2}$$    Impedance for a series $RLC$ combination

Here $X_L - X_C$ is called the **total reactance** of the circuit, and $Z$ the **impedance** of the circuit.

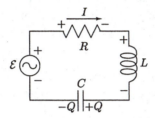

Figure 29-6

The peak current is largest when the total reactance, $X_L - X_C$, is zero. The expression for the frequency $\omega_{res}$ at which this occurs, called the **natural** or **resonance frequency,** can be computed by equating the expressions for the inductive and capacitive reactances and solving for the frequency. That is,

$$X_L = X_C \quad \text{or} \quad \omega_{res} L = \frac{1}{\omega_{res} C}$$

Thus,

$$\omega_{res} = \frac{1}{\sqrt{LC}}$$    Natural or resonance frequency

When $X_L - X_C = 0$, the phase angle $\delta$ is zero; so, at resonance, the current is in phase with the generator voltage.

The average power delivered to a series $RLC$ combination is all delivered to the resistor. This power is

$$P_{av} = V_{rms} I_{rms} \cos \delta$$    Average power for a series $RLC$ combination

where $V_{rms}$ and $I_{rms}$ are, respectively, the rms voltage drop across and the rms current through the series-connected three-element combination. The quantity $\cos \delta$ is called the **power factor.**

$$\cos \delta$$    Power factor

The "power meters" found on most homes do not measure the integrated power or energy but rather the integrated product $V_{rms}I_{rms}$, which just happens to equal to the energy when the power factor is one.

Plots of $P_{av}$ versus frequency are called **resonance curves.** The measure of the sharpness of the resonance is called the **Q factor,** which is

$$Q = \frac{2\pi E}{|\Delta E|} = \frac{\omega_{res}L}{R} \approx \frac{\omega_{res}}{\Delta \omega} \qquad \text{Q factor}$$

where $\Delta \omega$, the full width of the resonance curve at half the peak power, is called the **resonance width.**

When an inductor, a capacitor, and a resistor are connected in parallel across the terminals of a generator, it is the currents that add; but because they are not in phase, it is convenient to represent them as phasors and add them vectorally. The impedance is related to the resistance and the reactances by the equation

$$\frac{1}{Z} = \sqrt{\left(\frac{1}{R}\right)^2 + \left(\frac{1}{X_L} - \frac{1}{X_C}\right)^2} \qquad \text{Impedance for a parallel } RLC \text{ combination}$$

The peak current is related to the amplitude of the voltage supplied by the generator by

$$I_{peak} = \frac{\mathcal{E}_{peak}}{Z}$$

***Section 29-7. The Transformer.*** A step-up transformer is a device that increases voltage at the expense of current. A step-down transformer does the opposite. An ideal transformer consumes no energy, so that the power delivered to the primary circuit equals the power delivered by the secondary circuit. Because the two coils have the same magnetic flux per turn through them, the induced emf in each coil is proportional to the number of its turns. In an ideal transformer,

$$\frac{V_{1,rms}}{V_{2,rms}} = \frac{I_{2,rms}}{I_{1,rms}} = \frac{N_1}{N_2} \qquad \text{Relations for an ideal transformer}$$

where $N_1$ and $N_2$ are the numbers of turns in the primary and secondary coils.

Our ability to step voltages up and down makes long-distance power transmission, at high voltage and low current, with acceptable energy losses possible.

# II.  Physical Quantities and Key Equations

**Physical Quantities**

*Reactance* $\qquad\qquad\qquad X(\Omega)$

## Key Equations

*Terminal voltage of an ideal ac generator*  $\mathcal{E} = \mathcal{E}_{peak} \sin(\omega t + \delta) = NBA\omega \sin(\omega t + \delta)$

*Definition of root-mean-square (rms)*  $I_{rms} = \sqrt{\left(I^2\right)_{av}}$

*rms current through a resistor*  $I_{rms} = \dfrac{V_{R,rms}}{R}$ , voltage and current in phase

*Average power dissipated by a resistor*  $P_{av} = \frac{1}{2}I_{max}^2 R = I_{rms}^2 R$

*Inductive reactance*  $X_L = \omega L$

*rms current through an inductor*  $I_{rms} = \dfrac{V_{L,rms}}{\omega L} = \dfrac{V_{L,rms}}{X_L}$ , voltage leads current by 90°

*Capacitive reactance*  $X_C = \dfrac{1}{\omega C}$

*rms current "through" a capacitor*  $I_{rms} = \dfrac{V_{C,rms}}{1/\omega C} = \dfrac{V_{C,rms}}{X_c}$ , voltage lags current by 90°

*Total reactance for a series LC combination*  $X = X_L - X_C$

*Impedance defined*  $Z = \dfrac{V_{rms}}{I_{rms}}$

*Impedance for a series RLC combination*  $Z = \sqrt{R^2 + \left(X_L - X_C\right)^2}$

*Phase angle*  $\tan \delta = \dfrac{X_L - X_C}{R}$

*Power factor*  $\cos \delta$

*Natural or resonance frequency*  $\omega_{res} = \dfrac{1}{\sqrt{LC}}$

*Average power for a series LCR combination*  $P_{av} = V_{rms} I_{rms} \cos \delta$

*Impedance for a parallel RLC combination*  $\dfrac{1}{Z} = \sqrt{\left(\dfrac{1}{R}\right)^2 + \left(\dfrac{1}{X_L} - \dfrac{1}{X_C}\right)^2}$

*Q factor*  $Q = \dfrac{2\pi E}{|\Delta E|} = \dfrac{\omega_{res} L}{R} \approx \dfrac{\omega_{res}}{\Delta \omega}$

*Relations for an ideal transformer*

$$\frac{V_{1,\text{rms}}}{V_{2,\text{rms}}} = \frac{I_{2,\text{rms}}}{I_{1,\text{rms}}} = \frac{N_1}{N_2}$$

## III. Potential Pitfalls

The average value of the square of any sinusoidal function over one or more complete cycles is $\frac{1}{2}$. Also, for any sinusoidally varying function, the maximum value, or amplitude, equals $\sqrt{2}$ times the rms value.

In an ac circuit the instantaneous current through a device is not necessarily in phase with the instantaneous voltage drop across it. Impedance is the ratio of either the maximum or the rms values of voltage and current, not the instantaneous values.

A reactance is not a resistance, although both are measured in ohms. There is no power dissipation in a purely reactive device for which the voltage drop and current are 90° out of phase.

Capacitive and inductive reactance have opposite frequency dependences. A capacitor has a very large reactance at low frequencies and a very small reactance at high frequencies; the reverse is true for an inductor.

Because of phase differences, maximum or rms voltage drops in a series ac circuit cannot simply be added. One way to add them is vectorially, using phasors.

The behavior of a series $RLC$ circuit near resonance depends on all three quantities, but the resonance frequency is determined by $L$ and $C$ independent of $R$.

## IV. True or False Questions and Responses

**True or False**

_____ 1.   The emf of an ac generator varies sinusoidally with time.

_____ 2.   For a voltage that varies sinusoidally with time, the maximum voltage equals $\sqrt{2}$ times the average voltage.

_____ 3.   A capacitor acts like a short circuit at very high frequencies.

_____ 4.   The power dissipated by an inductor with negligible internal resistance is also negligible.

_____ 5.   The voltage drop across a resistor is in phase with the current.

_____ 6.   The voltage drop across a capacitor lags behind the current by 90°.

_____ 7.   Relative to each other, the vectors on phasor diagrams are fixed.

_____ 8.   The current through the resistor in a series $RLC$ combination is in phase with the current "through" the capacitor.

_____ 9. The current through the inductor in a series *RLC* combination is 180° out of phase with the current "through" the capacitor.

_____ 10. The voltage drop across the resistor in a series *RLC* combination is in phase with the voltage drop across the capacitor.

_____ 11. The voltage drop across the inductor in a series *RLC* combination is 180° out of phase with the voltage drop across the capacitor.

_____ 12. The voltage drop across the resistor in a parallel *RLC* circuit is in phase with the voltage drop across the capacitor.

_____ 13. The current through the resistor in a parallel *RLC* circuit is in phase with the current "through" the capacitor.

_____ 14. The angle between two vectors in a phasor diagram is the phase difference between the quantities represented by the vectors.

_____ 15. At resonance, the impedance of a series *RLC* circuit is a minimum.

_____ 16. At resonance, the impedance of a series *RLC* circuit is equal to the resistance.

_____ 17. A transformer whose primary coil has 10 times as many turns as its secondary coil normally delivers about 10 times more power than it receives.

_____ 18. Transformers are designed to maximize the mutual inductance of the primary and secondary coils.

**Responses to True or False**

1. True.

2. False. The average voltage is zero. The maximum voltage equals $\sqrt{2}$ times the rms voltage.

3. True. The impedance of a capacitor equals its reactance, which equals $1/\omega C$. For high frequencies the reactance is very small.

4. True.

5. True.

6. True.

7. True.

8. True. The instantaneous currents are the same because they are in series.

9. False. The instantaneous currents are the same because they are in series.

10. False. The voltage drop across the resistor is in phase with the current, whereas the voltage

drop across the capacitor lags the current by 90°. Thus, the voltage drop across the resistor leads the voltage drop across the capacitor by 90°.

11. True.

12. True. The instantaneous voltage drops are the same because they are in parallel.

13. False. It is the voltage drops that are in phase. The current in a resistor is in phase with the voltage drop, but the current in a capacitor leads the voltage drop by 90°. Thus, the current through the resistor lags the current through the capacitor by 90°.

14. True.

15. True.

16. True.

17. False. The average power delivered by an ideal transformer equals the average power it receives. Transformers are not power sources, so they cannot deliver more average power than they receive. In fact, they deliver less average power than they receive due to internal dissipative processes.

18. True.

## V.  Questions and Answers

### Questions

1. Why is the total or equivalent reactance of a series $LC$ combination the difference between the inductive and capacitive reactances rather than the sum?

2. Why is electric power for domestic use transmitted at very high voltages and stepped down to 110 V by a transformer near the point of consumption?

3. The reactance of an inductor increases with increasing frequency. Why?

4. What is meant by the statement "The voltage drop across an inductor leads the current by 90°"?

5. In a coil that has both resistance and inductance, does the phase angle between the current through the coil and the voltage drop across it vary with the frequency?

### Answers

1. The instantaneous current in each device is the same because they are in series. The voltage drop across the inductor leads the current by 90°, whereas the voltage drop across the capacitor lags the current by 90°; thus, the two voltage drops are 180° out of phase. The equivalent reactance of the combination is the ratio of the rms voltage drop across the combination to the

rms current through it. Because the two voltage drops are 180° out of phase, the rms voltage drop across the combination equals the difference of the individual rms voltage drops; so the equivalent reactance equals the difference between the individual reactances. That is,

$$X_{eq} = \frac{V_{rms}}{I_{rms}} = \frac{V_{L,rms} - V_{C,rms}}{I_{rms}}$$

$$= \frac{V_{L,rms}}{I_{rms}} - \frac{V_{C,rms}}{I_{rms}} = X_L - X_C$$

Represented as a phasor, the voltage across the combination is $\vec{V} = \vec{V}_L + \vec{V}_C$. Because $\vec{V}_L$ and $\vec{V}_C$ are oppositely directed, the magnitude of $\vec{V}$ equals the difference between their magnitudes.

2. It saves copper. The primary power loss in transmission equals $I_{rms}^2 R$, where $R$ is the resistance of the transmission lines, and the rate at which energy is transported equals $V_{rms} I_{rms}$. Thus, at high voltage, power can be transmitted with a lower current, and the lower the current, the less the transmission loss. The other way to reduce transmission loss would be to decrease the resistance of the transmission lines. The only practical way to do that is to use thicker wire, which requires more of the metal the wires are made of. The voltages are stepped down for consumer safety.

3. The reactance of an inductor increases with increasing frequency because, in accordance with Faraday's law, the greater the rate of change of current the greater the back emf.

4. It means that the current peaks one-fourth of a period after the voltage drop peaks. One-fourth of a period corresponds to a phase difference of 90°.

5. An inductor with resistance is equivalent to a series $LR$ combination. The tangent of the phase angle equals the ratio of the inductive reactance to the resistance. The phase angle varies with frequency because the inductive reactance does.

## VI. Problems, Solutions and Answers

**Example #1.** A sinusoidal voltage of 40.0 V rms and a frequency of 100 Hz is applied to (*a*) a 100-Ω resistor, (*b*) a 0.200-H inductor, and (*c*) a 50.0-$\mu$F capacitor. Find the peak value of the current in each case.

**Picture the Problem.** Use the impedance for the appropriate circuit element. Convert the rms current to peak current.

| | |
|---|---|
| 1. For a resistor, the impedance is simply the resistance. | $I_{rms} = \dfrac{V_{rms}}{R}$ |
| 2. Convert the rms current into a peak current, and solve. | $I_{peak} = \sqrt{2}\dfrac{V_{rms}}{R} = \sqrt{2}\dfrac{40\,\text{V}}{100\,\Omega} = 0.566\,\text{A}$ |
| 3. Find the reactance of the inductor. | $X_L = \omega L = 2\pi fL = 2\pi(100\,\text{Hz})(0.2\,\text{H}) = 126\,\Omega$ |

| 4. Using the reactance, calculate the peak current through the inductor. | $I_{peak} = \sqrt{2}\dfrac{V_{rms}}{X_L} = \sqrt{2}\dfrac{40\,V}{126\,\Omega} = 0.449\,A$ |
|---|---|
| 5. Find the reactance of the capacitor. | $X_C = \dfrac{1}{\omega C} = \dfrac{1}{2\pi fC} = \dfrac{1}{2\pi(100\,Hz)(50\times10^{-6}\,F)} = 31.8\,\Omega$ |
| 6. Using the reactance, calculate the peak current through the capacitor. | $I_{peak} = \sqrt{2}\dfrac{V_{rms}}{X_C} = \sqrt{2}\dfrac{40\,V}{31.8\,\Omega} = 1.78\,A$ |

**Example #2—Interactive.** A sinusoidal voltage of 50.0 V (peak) at a frequency of 400 Hz is applied to a capacitor of unknown capacitance. The current in the circuit is 400 mA rms. What is the capacitance?

**Picture the Problem.** Determine the reactance of the circuit, and use that to solve for the capacitance. **Try it yourself.** Work the problem on your own, in the spaces provided, to get the final answer.

| 1. Convert the peak voltage to rms voltage. | |
|---|---|
| 2. Use the rms current and voltage to compute the reactance of the circuit. | |
| 3. Use the reactance to calculate the capacitance. | $C = 4.50\,\mu F$ |

**Example #3.** The circuit in Figure 29-7 consists of a 1.50-$\mu$F capacitor and a 100-$\Omega$ resistor connected in series with a generator supplying a 40.0-V rms voltage at a frequency of 300 Hz. (*a*) Find the phase difference between the voltage supplied by the generator and the current. (*b*) Find the rms current.

**Picture the Problem.** Use a phasor diagram with vectors for the voltage across all three components. The rms current is the ratio of $V_{R,rms}$ to R.

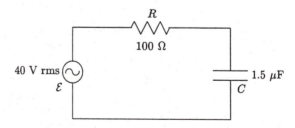

Figure 29-7

| 1. Find the capacitive reactance. | $X_C = \dfrac{1}{\omega C} = \dfrac{1}{2\pi f C} = \dfrac{1}{2\pi(300\text{ Hz})(1.5\times 10^{-6}\text{ F})} = 354\,\Omega$ |
|---|---|
| 2. Find the phase angle. The negative result indicates that the voltage lags the current. | $\tan\delta = \dfrac{X_L - X_C}{R} = \dfrac{-X_C}{R} = -\dfrac{354\,\Omega}{100\,\Omega}$ <br><br> $\delta = \tan^{-1}\left(-\dfrac{354}{100}\right) = -74.2°$ |
| 3. Use these results to draw a phasor diagram. Note that the angular position of the voltage source, $\vec{V}$ is arbitrary, but the positions of the voltage drop across the resistor and capacitor are fixed relative to it. The direction of $\vec{V}_R$ relative to $\vec{V}$ is given by $\delta$. Also note that $\vec{V}_R + \vec{V}_C = \vec{V}$ in order to satisfy Kirchhoff's loop rule. | |
| 4. The current is always in phase with the voltage drop across the resistor, so we can find the rms current once we know the rms voltage across the resistor. | $V_{R,\text{rms}} = V_{\text{rms}}\cos\delta = (40\text{ V})\cos 74.2° = 10.9\text{ V}$ <br><br> $I_{\text{rms}} = \dfrac{V_{\text{rms}}}{R} = \dfrac{10.9\text{ V}}{100\,\Omega} = 0.109\text{ A}$ |

**Example #4—Interactive.** The resistance of an induction coil is $50.0\,\Omega$ and its inductance is 0.10 H. The coil is connected to a 120-V rms, 60-Hz power line. Find (*a*) the rms current, (*b*) the power drawn from the line, and (*c*) the power factor.

**Picture the Problem.** You will need to find the impedance of the circuit. All the power drawn from the line goes into joule heating of the resistor. **Try it yourself.** Work the problem on your own, in the spaces provided, to get the final answer.

| 1. Find the impedance of this circuit. | |
|---|---|
| 2. Use the impedance to find the rms current. | $I = 1.92\text{ A}$ |
| 3. Find the phase angle. | |

| 4. Calculate the power factor. | |
|---|---|
| | 0.799 |
| 5. Calculate the power delivered to the device. | |
| | $P_{av} = 184\,\text{W}$ |

**Example #5.** Figure 29-8 shows a series *RLC* circuit with $L = 0.0100\,\text{H}$, $R = 220\,\Omega$, and $C = 0.100\,\mu F$. The amplitude of the driving voltage is $\mathcal{E}_{max} = 150\,\text{V}$. At a frequency of 1000 Hz, what power is being supplied by the generator?

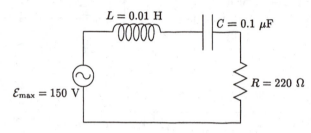

Figure 29-8

**Picture the Problem.** All the power supplied by the generator is dissipated by the resistor. The impedance of the circuit can be used to determine the current, and hence the power supplied.

| 1. Find the inductive reactance. | $X_L = \omega L = 2\pi f L = 2\pi (1000\,\text{Hz})(0.01\,\text{H}) = 62.8\,\Omega$ |
|---|---|
| 2. Find the capacitive reactance. | $X_C = \dfrac{1}{\omega C} = \dfrac{1}{2\pi f C} = \dfrac{1}{2\pi (1000\,\text{Hz})(10^{-7}\,\text{F})} = 1590\,\Omega$ |
| 3. Calculate the impedance of the circuit. | $Z = \sqrt{R^2 + (X_L - X_C)^2}$ $= \sqrt{(220\,\Omega)^2 + (62.8\,\Omega - 1592\,\Omega)^2} = 1540\,\Omega$ |
| 4. Calculate the maximum current through the circuit. | $I_{max} = \dfrac{\mathcal{E}_{max}}{Z} = \dfrac{150\,\text{V}}{(1540\,\Omega)} = 0.0974\,\text{A}$ |
| 5. Calculate the power delivered by the generator. All this power is dissipated by the resistor, because the capacitor and inductor are nondissipative. | $P = \tfrac{1}{2} I_{max}^2 R = \tfrac{1}{2}(0.0974\,\text{A})^2 (220\,\Omega) = 1.04\,\text{W}$ |

**Example #6—Interactive.** A series *RLC* circuit is driven by 115 V rms from an ac power line of unknown frequency. The power drawn is 65.0 W and the current in the circuit is 1.00 A rms. Find (*a*) the resistance and (*b*) the net reactance of the circuit.

**Picture the Problem.** All the power is being dissipated in the resistor, which allows you to calculate the resistance. The reactance then follows. **Try it yourself.** Work the problem on your own, in the spaces provided, to get the final answer.

| | |
|---|---|
| 1. Determine the resistance from the current and power. | $R = 65.0\,\Omega$ |
| 2. Determine the net impedance from the voltage and current. | |
| 3. Use the impedance to find the net reactance of the circuit. | $X_{net} = 94.9\,\Omega$ |

**Example #7.** In the series *RLC* circuit in Figure 29-9, find the maximum voltages $V_R$, $V_C$, and $V_L$ across the resistor, capacitor, and inductor, respectively, (*a*) at a frequency of 100 Hz and (*b*) at resonance.

**Picture the Problem.** The impedance of the circuit at each frequency can be used to determine the current, and then the voltage across each element.

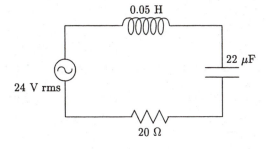

Figure 29-9

| | |
|---|---|
| 1. Calculate the impedance of the circuit at each frequency.  At resonance, the impedance is the resistance. | $Z_{resonance} = 20\,\Omega$ <br><br> $X_L = \omega L = 2\pi f L = 2\pi(100\,\text{Hz})(0.05\,\text{H}) = 31.4\,\Omega$ <br><br> $X_C = \dfrac{1}{\omega C} = \dfrac{1}{2\pi f C} = \dfrac{1}{2\pi(100\,\text{Hz})(2.2\times10^{-5}\,\text{F})} = 72.3\,\Omega$ <br><br> $Z_{100\,\text{Hz}} = \sqrt{R^2 + (X_L - X_C)^2}$ <br><br> $\qquad = \sqrt{(20\,\Omega)^2 + (31.4\,\Omega - 72.3\,\Omega)^2} = 45.5\,\Omega$ |
| 2. Determine the maximum current at 100 Hz from the impedance. | $I_{max,\,100\,\text{Hz}} = \dfrac{\mathcal{E}_{max}}{Z} = \dfrac{\sqrt{2}\,\mathcal{E}_{rms}}{Z} = \dfrac{\sqrt{2}\,24\,\text{V}}{(45.5\,\Omega)} = 0.746\,\text{A}$ |
| 3. Determine the maximum voltage across all three components at 100 Hz.  The three voltages don't add to the maximum voltage provided by the generator because the maximum voltages are out of phase. | $V_{R,max} = I_{max}R = (0.746\,\text{A})(20\,\Omega) = 14.9\,\text{V}$ <br><br> $V_{L,max} = I_{max}X_L = (0.746\,\text{A})(31.4\,\Omega) = 23.4\,\text{V}$ <br><br> $V_{C,max} = I_{max}X_C = (0.746\,\text{A})(73.2\,\Omega) = 54.6\,\text{V}$ |
| 4. Determine the maximum current at resonance. | $I_{max,\,100\,\text{Hz}} = \dfrac{\mathcal{E}_{max}}{R} = \dfrac{\sqrt{2}\,\mathcal{E}_{rms}}{R} = \dfrac{\sqrt{2}\,24\,\text{V}}{(20\,\Omega)} = 1.70\,\text{A}$ |
| 5. Find the maximum voltage across the resistor at resonance. | $V_{R,max} = I_{max}R = (1.70\,\text{A})(20\,\Omega) = 34.0\,\text{V}$ |
| 6. Find the resonance frequency of the circuit. | $\omega_0 = \dfrac{1}{\sqrt{LC}} = \dfrac{1}{\sqrt{(0.05\,\text{H})(2.2\times10^{-5}\,\text{F})}} = 954\,\text{rad/s}$ |
| 7. Determine the maximum voltage across the inductor and resistor.  At resonance, the voltage across these elements must cancel each other, so their maximum voltages must be the same. | $V_{L,max} = V_{C,max} = I_{max}X_L = I_{max}\omega L$ <br><br> $\qquad = (1.70\,\text{A})(954\,\text{rad/s})(0.05\,\text{H}) = 81.1\,\text{V}$ |

**Example #8—Interactive.**    A series $RLC$ circuit consists of an $8.00\text{-}\Omega$ resistor, a 100-mH inductor, and a $5.00\text{-}\mu\text{F}$ capacitor.  A signal generator applies 5.00 V rms to this circuit.  (*a*) At what frequency is the maximum power delivered to this circuit? (*b*) What is this maximum power?

**Picture the Problem.**    The maximum power is delivered when the generator frequency is the resonance frequency of the circuit. **Try it yourself.**  Work the problem on your own, in the spaces provided, to get the final answer.

| 1. Calculate the resonance frequency of the circuit, which is the frequency at which the greatest power is delivered. | $f_{res} = 225\,\text{Hz}$ |
|---|---|
| 2. At the resonance frequency, the impedance is equal to the resistance. Calculate the power dissipated by the resistor. The capacitor and inductor are not dissipative elements. | $P_{av} = 3.13\,\text{W}$ |

**Example #9.** When a certain *RLC* series combination is connected to the power line (110 V rms, 60 Hz), it draws 700 W of power, and the current leads the line voltage by 50°. Find (*a*) the rms current through the combination, and (*b*) its resistance. (*c*) Is the resonance frequency greater or less than 60 Hz?

**Picture the Problem.** Use the phase angle, power, and voltage to find the rms current. The power drawn is dissipated in the resistance. Because the current leads the voltage, the net reactance is more capacitive than inductive.

| 1. Determine the rms current drawn by the device. | $P_{av} = V_{rms} I_{rms} \cos\delta$ <br><br> $I_{rms} = \dfrac{P_{av}}{V_{rms}\cos\delta} = \dfrac{700\,\text{W}}{(110\,\text{V})\cos 50°} = 9.90\,\text{A}$ |
|---|---|
| 2. All the power is dissipated in the resistor, so we can find the resistance. | $P_{av} = I_{rms}^2 R$ <br><br> $R = \dfrac{P_{av}}{I_{rms}^2} = \dfrac{700\,\text{W}}{(9.90\,\text{A})^2} = 7.14\,\Omega$ |
| 3. Because the voltage lags behind the current, the reactance is dominated by the capacitive reactance. It is clear that the frequency of 60 Hz is less than the resonance frequency of the circuit. | $X_C > X_L$ <br><br> $\dfrac{1}{\omega C} > \omega L$ <br><br> $\omega^2 < \dfrac{1}{LC}$ <br><br> $\omega < \dfrac{1}{\sqrt{LC}} = \omega_{res}$ |

**Example #10—Interactive.** A $8.00 \times 10^{-2}\text{-}\Omega$ power cord is used to deliver 1500 W of power. (*a*) If the power is delivered at 12.0 V rms, how much power is dissipated in the power cord (assuming the current and voltage are in phase)? (*b*) If the power is delivered at 120 V rms, how much power is dissipated in the power cord (again assuming the current and voltage are in phase)?

**Picture the Problem.**  Not all the power will be dissipated in the power cord.  Some will be dissipated by the device it is powering.  **Try it yourself.**  Work the problem on your own, in the spaces provided, to get the final answer.

| | |
|---|---|
| 1. Find the current required to deliver 1500 W of power at 12 V. | |
| 2. Determine the average power dissipated by the cord when this much current flows through it. | $P_{av} = 1250\,\text{W}$ |
| 3. Find the current required to deliver 1500 W of power at 12 V. | |
| 4. Determine the average power dissipated by the cord when this much current flows through it. | $P_{av} = 12.5\,\text{W}$ |

**Example #11.**  The 400-turn primary coil of a step-down transformer is connected to a 110-V rms ac line.  The secondary coil is to supply 15.0 A at 6.30 V rms.  Assuming no power loss in the transformer, find (*a*) the number of turns in the secondary coil and (*b*) the current in the primary coil.

**Picture the Problem.**  The voltage ratio equals the ratio of the number of turns of the coils.  The input power must be the same as the output power.

| | |
|---|---|
| 1. Determine the number of turns in the secondary coil.  A fractional number of turns is not practical. | $\dfrac{V_{2,\text{rms}}}{V_{1,\text{rms}}} = \dfrac{N_2}{N_1}$ <br><br> $N_2 = \dfrac{V_{2,\text{rms}}N_1}{V_{1,\text{rms}}} = \dfrac{(6.3\,\text{V})(400)}{110\,\text{V}} = 22.9 = 23\,\text{turns}$ |
| 2. Find the required input current by using the fact that the output power must equal the input power since there are no power losses. | $P_{\text{out}} = P_{\text{in}}$ <br><br> $V_{\text{rms, out}}I_{\text{rms, out}} = V_{\text{rms, in}}I_{\text{rms, in}}$ <br><br> $I_{\text{rms, in}} = \dfrac{V_{\text{rms, out}}I_{\text{rms, out}}}{V_{\text{rms, in}}} = \dfrac{(6.3\,\text{V})(15\,\text{A})}{(110\,\text{V})} = 0.859\,\text{A}$ |

# Chapter **30**

# Maxwell's Equations and Electromagnetic Waves

## I.   Key Ideas

In the 1860s, about a century after Coulomb established the first quantitative electrical relation, a Scotsman named James Clerk Maxwell worked out the details of a complete classical theory of electricity and magnetism. This theory is described by a group of equations called Maxwell's equations. Maxwell's contribution was to realize that a time-varying electric field can generate a magnetic field in the same way a current does.

***Section 30-1. Maxwell's Displacement Current.*** Ampère's law ( $\oint_C \vec{B} \cdot d\vec{\ell} = \mu_0 I$ ) is valid only for steady-state situations, that is, situations in which all currents and all charge distributions are constant in time. The current $I$ that appears in Ampère's law is the rate of flow of charge through any surface $S$ that is bounded by the closed curve $C$. Maxwell modified this relation to include situations in which the currents and charges are not constant. The resulting relation is

$$\oint_C \vec{B} \cdot d\vec{\ell} = \mu_0 \left( I_d + I \right) \qquad\qquad \text{Generalized form of Ampère's law}$$

where $I_d$ is the displacement current. The displacement current through a surface $S$ is defined as the product of the permittivity of free space and the rate of change of the flux of the electric field through $S$, whereas $I$ is the rate of flow of charge through the same surface. That is,

$$I_d = \varepsilon_0 \frac{d\phi_e}{dt} = \varepsilon_0 \frac{d}{dt} \oint_s E_n \, dA \qquad\qquad \text{Displacement current defined}$$

The displacement current is not actually a current; that is, it is not a flow of charge through a surface. However, it contributes to the net magnetic field in the same way that a current does.

***Section 30-2. Maxwell's Equations.*** By modifying Ampère's law to include time-varying situations, Maxwell fitted the final piece of the electromagnetic puzzle into place. His equations specifying the relation between charges, currents, and electric and magnetic fields in free space are

$$\oint_S \vec{E} \cdot \hat{n} \, dA = \frac{1}{\varepsilon_0} Q_{\text{inside}} \qquad \text{Gauss's law for electric fields}$$

$$\oint_S \vec{B} \cdot \hat{n} \, dA = 0 \qquad \text{Gauss's law for magnetic fields}$$

Maxwell's equations

$$\oint_C \vec{E} \cdot d\vec{\ell} = -\frac{d}{dt} \int_S \vec{B} \cdot \hat{n} \, dA \qquad \text{Faraday's law}$$

$$\oint_C \vec{B} \cdot d\vec{\ell} = \mu_0 I + \mu_0 \varepsilon_0 \frac{d}{dt} \int_S \vec{E} \cdot \hat{n} \, dA \quad \text{Ampère-Maxwell's law}$$

The first of these equation, Gauss's law, is a mathematical statement that electric field lines may begin and end only at electric charges; the second is a mathematical statement that magnetic field lines never begin or end, and so magnetic monopoles do not exist; the third, Faraday's law of inductance, states that a time-varying magnetic field generates a nonconservative electric field; and the fourth, Ampère's law with the Maxwell correction, states that a magnetic field can be generated either by currents or by time-varying electric fields.

*Section 30-3. Electromagnetic Waves.* Maxwell's equations predict the existence of waves that consist of oscillating electric and magnetic fields. In empty space the electric and magnetic fields are perpendicular to each other ($\vec{E} \cdot \vec{B} = 0$) and both fields are perpendicular to the direction of propagation in such a way that $\vec{E} \times \vec{B}$ is in the direction of propagation. In SI units the magnitudes of these fields are such that

$$E = cB \qquad \text{Field magnitudes for electromagnetic waves}$$

These waves propagate with a speed, to three significant figures, of

$$c = \frac{1}{\sqrt{\mu_0 \varepsilon_0}} = 3.00 \times 10^8 \text{ m/s} \qquad \text{Speed of electromagnetic waves in free space}$$

Because the electric and magnetic fields are perpendicular to the direction of propagation, electromagnetic waves are transverse and can be polarized. Electromagnetic waves traveling parallel to the $x$ axis are said to be **linearly polarized** if, at any fixed point, the tip of the $\vec{E}$ vector moves back and forth in a line perpendicular to the $x$ axis; they are **circularly polarized** if, at a fixed point, the tip of the $\vec{E}$ vector moves in a circle whose plane is perpendicular to the $x$ axis.

*The Electromagnetic Spectrum.* Electromagnetic waves exist over a broad range of frequencies, from radio waves with frequencies of $10^6$ Hz or less to gamma rays with frequencies of more than $10^{20}$ Hz. Visible light is a narrow band of frequencies in this broad spectrum, in the range from $4.3 \times 10^{14}$ Hz to $7.9 \times 10^{14}$ Hz. Before the development of Maxwell's equations, the connection between optics (the study of light) and electricity and magnetism was unknown. Maxwell's equations resulted in the unification of these two previously distinct branches of physics.

*Energy and Momentum in an Electromagnetic Wave.* The electromagnetic energy density $u$ is the sum of the electric energy density $u_e$ and the magnetic energy density $u_m$. This energy density is

$$u = u_e + u_m$$

$$= \frac{1}{2}\varepsilon_0 E^2 + \frac{B^2}{2\mu_0} = \varepsilon_0 E^2 = \frac{B^2}{\mu_0} = \frac{EB}{\mu_0 c} \qquad \text{Energy density of electromagnetic waves}$$

The intensity of a wave—the power delivered per unit area—is the product of the energy density and the velocity. An associated vector quantity is the Poynting vector $\vec{S}$

$$\vec{S} = \frac{\vec{E} \times \vec{B}}{\mu_0} \qquad \text{Poynting vector}$$

The Poynting vector's magnitude is the instantaneous intensity of the wave, and its direction is the direction of the wave's propagation. Thus, the average intensity $I$ is

$$I = u_{av} c = \frac{E_{rms} B_{rms}}{\mu_0} = \frac{E_0 B_0}{2\mu_0} = \left| \vec{S} \right|_{av} \qquad \text{Intensity of electromagnetic waves}$$

The magnitude of the momentum $p$ carried by an electromagnetic wave is $1/c$ times the energy $U$ carried by the wave, that is

$$p = \frac{U}{c} \qquad \text{Momentum and energy carried by an electromagnetic wave}$$

Consider an electromagnetic wave incident on some surface. If the surface is nonreflective, the wave's energy and momentum are both transferred to the surface. Momentum transferred per unit time is force. Thus, the intensity divided by $c$ is a force per unit area, which is a pressure. This pressure is called **radiation pressure** $P_r$.

$$P_r = \frac{I}{c} = \frac{E_0 B_0}{2\mu_0 c} = \frac{\varepsilon_0 E_0^2}{2\mu_0 c^2} = \frac{B_0^2}{2\mu_0} \qquad \text{Radiation pressure}$$

If the wave is reflected, then the radiation pressure equals $1/c$ times the sum of the incident intensity and the reflected intensity.

*Dipole Radiation.* A common broadcast antenna is an **electric dipole antenna,** which consists of two conducting rods, one pointing upward and one pointing downward as in Figure 30-1. The charges on the upper and lower rods are always opposite and are periodically exchanged. Such an antenna produces a wave that is linearly polarized with the polarization parallel with the antenna. (The polarization direction of an electromagnetic wave is defined to be in the direction of the electric field vector.) The angular distribution of the intensity of the radiation broadcast by an electric dipole antenna is given by the relation

$$I = I_0 \sin^2 \theta \qquad \text{Electric dipole antenna angular-intensity distribution}$$

where $\theta$ is the angle between the propagation direction and the antenna rods, shown in Figure 30-2. In this figure the angular dependence of the intensity distribution is illustrated by the oval curve, which is a plot of the relative intensity on the surface of a large sphere centered on the antenna. The length of the arrow is proportional to the intensity broadcast in the direction indicated by the arrow.

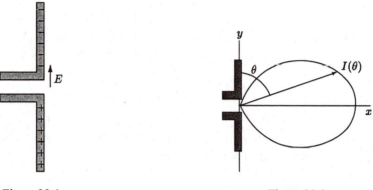

Figure 30-1                                Figure 30-2

An identical dipole antenna can be used to detect the radiation. Alternatively, a loop antenna, which is commonly used to detect UHF television waves, detects radiation in accordance with Faraday's law. This means that changes in the magnetic flux through the loop induce a current in it.

***Section 30-4. The Wave Equation for Electromagnetic Waves.*** For an electromagnetic wave traveling in free space and traveling parallel to the x axis, the electric and the magnetic fields both obey the wave equations

$$\frac{\partial^2 \vec{E}}{\partial x^2} = \frac{1}{c^2}\frac{\partial^2 \vec{E}}{\partial t^2} \text{ and } \frac{\partial^2 \vec{B}}{\partial x^2} = \frac{1}{c^2}\frac{\partial^2 \vec{B}}{\partial t^2}$$    Wave equations for $\vec{E}$ and $\vec{B}$

## II.  Physical Quantities and Key Equations

### Physical Quantities

*Speed of electromagnetic waves in free space*        $c = \dfrac{1}{\sqrt{\mu_0 \varepsilon_0}} = 3.00 \times 10^8$ m/s

### Key Equations

*Ampère's law with the Maxwell correction*        $\oint_C \vec{B} \cdot d\vec{\ell} = \mu_0 \left( I_d + I \right)$

*Displacement current defined*        $I_d = \varepsilon_0 \dfrac{d\phi_e}{dt}$

*Field magnitudes for electromagnetic waves*        $E = cB$

| | | |
|---|---|---|
| *Energy density of electromagnetic waves* | | $u = u_e + u_m = \varepsilon_0 E^2 = \dfrac{B^2}{\mu_0} = \dfrac{EB}{\mu_0 c}$ |

*Maxwell's equations*

$$\oint_S \vec{E} \cdot \hat{n} \, dA = \frac{1}{\varepsilon_0} Q_{\text{inside}} \qquad \text{Gauss's law for electric fields}$$

$$\oint_S \vec{B} \cdot \hat{n} \, dA = 0 \qquad \text{Gauss's law for magnetic fields}$$

$$\oint_C \vec{E} \cdot d\vec{\ell} = -\frac{d}{dt} \int_S \vec{B} \cdot \hat{n} \, dA \qquad \text{Faraday's law}$$

$$\oint_C \vec{B} \cdot d\vec{\ell} = \mu_0 I + \mu_0 \varepsilon_0 \frac{d}{dt} \int_S \vec{E} \cdot \hat{n} \, dA \qquad \text{Ampère-Maxwell's law}$$

*Poynting vector*

$$\vec{S} = \frac{\vec{E} \times \vec{B}}{\mu_0}$$

*Intensity of electromagnetic waves*

$$I = \frac{E_{\text{rms}} B_{\text{rms}}}{\mu_0} = \frac{E_0 B_0}{2\mu_0} = \left| \vec{S} \right|_{\text{av}}$$

*Momentum carried by an electromagnetic wave*

$$p = \frac{U}{c}$$

*Radiation pressure*

$$P_r = u_{\text{av}} c = \frac{I}{c} = \frac{E_0 B_0}{2\mu_0 c} = \frac{\varepsilon_0 E_0^2}{2\mu_0 c^2} = \frac{B_0^2}{2\mu_0}$$

*Electric dipole antenna angular-intensity distribution* $I = I_0 \sin^2 \theta$

*Wave equations for* $\vec{E}$ *and* $\vec{B}$

$$\frac{\partial^2 \vec{E}}{\partial x^2} = \frac{1}{c^2} \frac{\partial^2 \vec{E}}{\partial t^2} \qquad \frac{\partial^2 \vec{B}}{\partial x^2} = \frac{1}{c^2} \frac{\partial^2 \vec{B}}{\partial t^2}$$

## III. Potential Pitfalls

Remember: The original form of Ampère's law holds only when there are no time-varying electric fields. This occurs only when all currents and all charge distributions are constant. However, the modified Ampère's law holds even when electric fields vary with time.

Don't forget: Radiation pressure is exerted on a surface by a beam leaving the surface as well as by one incident on the surface.

The magnitude of the Poynting vector is the instantaneous intensity, not the average intensity.

## IV. True or False Questions and Responses

**True or False**

_____ 1.   The displacement current is the rate of flow of free charges.

_____ 2.  The magnetic field produced by a displacement current has identical properties to the magnetic field produced by an ordinary current.

_____ 3.  Maxwell's equations apply only to $\vec{E}$ and $\vec{B}$ fields that vary sinusoidally with time.

_____ 4.  The $\vec{E}$ and $\vec{B}$ fields of electromagnetic waves in free space (free space refers to regions where no matter is present) are in phase.

_____ 5.  Both the $\vec{E}$ and the $\vec{B}$ fields of electromagnetic waves in free space are perpendicular to the direction of propagation.

_____ 6.  If the $\vec{E}$ and $\vec{B}$ fields of an electromagnetic wave in free space are expressed in SI units, then $\vec{E} \times \vec{B} = c\hat{k}$, where $c$ is the speed of light and $\hat{k}$ is the unit vector in the direction of propagation of the wave.

_____ 7.  The Poynting vector is the instantaneous radiation pressure.

_____ 8.  The total energy density in an electromagnetic wave in free space is equally divided between that of the electric field and that of the magnetic field.

_____ 9.  The Poynting vector of an electromagnetic wave in free space is in the direction of wave propagation.

**Responses to True or False**

1.  False. The displacement current through a surface is the electric permittivity of free space times the rate of change of the flux of the electric field through the surface. It does not represent the rate of flow of free charges.

2.  True.

3.  False. Maxwell's equations apply to all $\vec{E}$ and $\vec{B}$ fields.

4.  True.

5.  True.

6.  False. The direction of propagation of electromagnetic waves is in the direction of $\vec{E} \times \vec{B}$, but $\vec{E} \times \vec{B} = \mu_0 \vec{S}$, not $c\hat{k}$. Also, the product $EB$ does not have the same dimensions as speed.

7.  False. The magnitude of the Poynting vector is the instantaneous intensity, not the instantaneous radiation pressure.

8.  True.

9.  True.

# V.  Questions and Answers

**Questions**

1.  It is easier to verify that a time-varying magnetic field produces an electric field (electromagnetic induction) than to verify that a time-varying electric field produces a magnetic field. Why?

2.  Which of Maxwell's equations denies the existence of magnetic monopoles?

3.  The radiation pressure exerted on a perfectly reflecting surface is $2I/c$. This is twice the value given by the equation for radiation pressure. Why is there a factor of 2 in this relation?

4.  Many television sets come with a loop antenna for the UHF channels. How should it be oriented relative to the direction of the transmitting antenna, assuming that the wave is transmitted by an electric dipole antenna?

5.  What role does the displacement current play in the theory of electromagnetic waves?

**Answers**

1.  The electric field produced by a time-varying magnetic field induces currents that are strong enough to detect easily. For example, a transformer can be used to light a light bulb. The magnetic fields produced by time-varying electric fields are weak and, therefore, detection requires more sensitive equipment.

2.  The equation $\oint_S \vec{B}\cdot\hat{n}\, dA = 0$, or Gauss's law for magnetics, denies the existence of magnetic monopoles. The net flux through a closed surface would be nonzero if a magnetic monopole existed inside the surface.

3.  The radiation pressure $P_r = I/c$ is the momentum per unit time per unit area carried by a wave. On a perfectly reflecting surface both the incident wave and the reflected wave carry momentum, and both exert pressure on the surface. Thus, the net pressure is twice the pressure exerted by a wave incident on a perfectly absorbing surface.

4.  The loop antenna works by induction. The changing magnetic field induces an emf and, therefore, a current in the loop. Thus, the loop must be oriented so that the magnetic field of the wave produces a changing magnetic flux through the loop, preferably with the magnetic field perpendicular to the plane of the loop. This means that the plane of the loop should contain the electric dipole moment of the broadcast antenna.

5.  In free space the displacement current is the only source of the magnetic field. Without it there could be no electromagnetic waves.

# VI. Problems, Solutions, and Answers

**Example #1.**  A parallel-plate capacitor, shown in Figure 30-3, has circular plates of radius 1.00 m separated by a 0.500-mm thick sheet of mica. (Mica's dielectric constant $\kappa$ is 5.4.) Charge is flowing onto one plate and off the other plate at a rate of 5.00 A. (*a*) Find the displacement current through a 1.500-m-radius circular surface *S*, coaxial with the capacitor, that is in the gap between the plates. (*b*) Find the current through *S*. (*c*) Find the magnetic field on the perimeter of *S*.

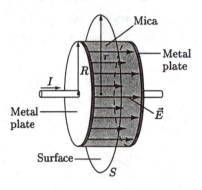

Figure 30-3

**Picture the Problem.** This example requires the use of Ampère's law with the Maxwell correction. The changing electric flux through the surface is the key element of this problem.

| | |
|---|---|
| 1. Find the displacement current. The electric field is everywhere parallel to the normal of the surface vector. The electric field is also uniform. Because the area is constant, we only need to worry about changes in the electric field causing a change in the electric flux. | $I_\mathrm{d} = \varepsilon \dfrac{d\phi_e}{dt} = \varepsilon_0 \dfrac{d}{dt} \int_S \vec{E} \cdot \hat{n}\, dA = \varepsilon_0 \dfrac{d}{dt} \int_S E\, dA$ $= \varepsilon_0 \dfrac{d}{dt} E \int_S dA = \varepsilon_0 \dfrac{d}{dt} EA = \varepsilon_0 A \dfrac{dE}{dt}$ |
| 2. Sketch the physical situation to guide your determination of the electric field. | 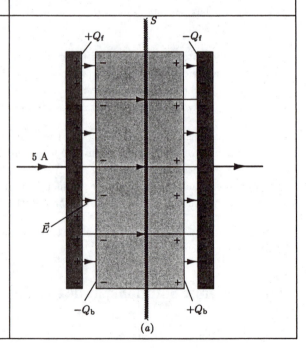 |

| | |
|---|---|
| | <br>(b) |
| 3. Determine the electric field that gives rise to the flux through the surface. The electric field produced by the free charge is reduced by the dielectric. | $E = \dfrac{E_0}{\kappa} = \dfrac{Q_f}{\kappa \varepsilon_0 A}$ |
| 4. Substitute this electric field into the expression for the displacement current and solve. | $I_d = \varepsilon_0 A \dfrac{d}{dt}\left(\dfrac{Q_f}{\kappa \varepsilon_0 A}\right) = \dfrac{1}{\kappa}\dfrac{dQ_f}{dt} = \dfrac{1}{5.4}(5\,\text{A}) = 0.926\,\text{A}$ |
| 5. Solve for the actual physical current that flows through the surface. This current is due to the polarization of the bound charges in the mica. According to page 775 in Chapter 24 of the textbook, the bound charge is related to the free charge. | $I = I_p = \dfrac{dQ_b}{dt} = \dfrac{d}{dt}\left(\dfrac{\kappa-1}{\kappa}Q_f\right) = \dfrac{\kappa-1}{\kappa}\dfrac{dQ_f}{dt}$<br><br>$= \dfrac{5.4-1}{5.4}(5\,\text{A}) = 4.07\,\text{A}$ |
| 6. Apply Ampère's law with the Maxwell correction to find the magnetic field along surface $S$. Because of the symmetry in the problem, the magnetic field is everywhere parallel to $d\vec{\ell}$, and is also uniform. | $\displaystyle\int_C \vec{B}\boldsymbol{\cdot} d\vec{\ell} = \mu_0\left(I + I_d\right)$<br><br>$B 2\pi r = \mu_0\left(I + I_d\right)$<br><br>$B = \dfrac{\mu_0\left(I + I_d\right)}{2\pi r}$ |
| 7. Relate the polarization and displacement currents to the free charge. | $I_p + I_d = \dfrac{d}{dt}\left[\left(\dfrac{\kappa-1}{\kappa}\right)Q\right]_f + \dfrac{d}{dt}\left(\dfrac{1}{\kappa}Q_f\right) = \dfrac{dQ_f}{dt}$ |
| 8. Solve for the magnetic field. | $B = \dfrac{\left(4\pi \times 10^{-7}\,\text{T}\boldsymbol{\cdot}\text{m/A}\right)(5\,\text{A})}{2\pi(1.5\,\text{m})} = 6.67 \times 10^{-7}\,\text{T}$ |

**Example #2—Interactive.** For the situation described in Example #1, find the magnetic field in the mica at a distance of 30.0 cm from the axis of the capacitor. Mica's magnetic susceptibility $\chi_m$ is negligible.

**Picture the Problem.**  This problem will be solved in the same way we solved Example #1. However, not all of the polarization current will contribute to the flux through the surface. **Try it yourself.**  Work the problem on your own, in the spaces provided, to get the final answer.

| | |
|---|---|
| 1.  Sketch the physical situation to guide your problem-solving. | |
| 2.  Find the displacement current. The electric field is everywhere parallel to the normal of the surface vector. The electric field is also uniform. Because the area is constant, we only need to worry about changes in the electric field causing a change in the electric flux. | |
| 3.  Determine the electric field that gives rise to the flux through the surface. The electric field produced by the free charge is reduced by the dielectric. | |
| 4.  Substitute this electric field into the expression for the displacement current and solve. | |
| 5.  Solve for the actual physical current that flows through the surface. This current is due to the polarization of the bound charges in the mica. According to page 775 in Chapter 24 of the textbook, the bound charge is related to the free charge. | |
| 6.  Apply Ampère's law with the Maxwell correction to find the magnetic field along surface $S$. Because of the symmetry in the problem, the magnetic field is everywhere parallel to $d\vec{\ell}$ and is also uniform. | |
| 7.  Relate the polarization and displacement currents to the free charge. | |

| 8. Solve for the magnetic field. | $B = 3.00 \times 10^{-7}$ T |
|---|---|

**Example #3.** Show by direct substitution that the wave function $\vec{B} = \left[ B_0 \cos\left( kx - \omega t \right) \right] \hat{j}$ satisfies the wave equation

$$\frac{\partial^2 \vec{B}}{\partial x^2} = \frac{1}{c^2} \frac{\partial^2 \vec{B}}{\partial t^2}$$

given that $\omega = kc$.

| 1. Find the second derivative of $\vec{B}$ with respect to $x$. You really only need to worry about the magnitude, since the vector direction is constant in both $x$ and $t$. | $\dfrac{\partial^2 B}{\partial x^2} \left[ B_0 \cos\left( kx - \omega t \right) \right] = -k^2 B_0 \cos\left( kx - \omega t \right)$ |
|---|---|
| 2. Find the second derivative of $\vec{B}$ with respect to $t$. | $\dfrac{\partial^2 B}{\partial t^2} \left[ B_0 \cos\left( kx - \omega t \right) \right] = -\omega^2 B_0 \cos\left( kx - \omega t \right)$ |
| 3. Substitute into the wave equation and demonstrate you have an equality. | $\dfrac{\partial^2 \vec{B}}{\partial x^2} = \dfrac{1}{c^2} \dfrac{\partial^2 \vec{B}}{\partial t^2}$ <br><br> $-k^2 B_0 \cos\left( kx - \omega t \right) = \dfrac{-\omega^2 B_0 \cos\left( kx - \omega t \right)}{c^2}$ <br><br> $-k^2 B_0 \cos\left( kx - \omega t \right) = -k^2 B_0 \cos\left( kx - \omega t \right)$ |

**Example #4—Interactive.** Show by direct substitution that the wave function $\vec{E} = \left[ E_0 \cos\left( kx - \omega t \right) + E_0 \sin\left( kx - \omega t \right) \right] \hat{k}$ satisfies the wave equation

$$\frac{\partial^2 \vec{E}}{\partial x^2} = \frac{1}{c^2} \frac{\partial^2 \vec{E}}{\partial t^2}$$

given that $\omega = kc$. **Try it yourself.** Work the problem on your own, in the spaces provided, to get the final answer.

| 1. Find the second derivative of $\vec{E}$ with respect to $x$. | |
|---|---|

| 2. Find the second derivative of $\vec{E}$ with respect to $t$. | |
|---|---|
| 3. Substitute into the wave equation and demonstrate you have an equality. | |

**Example #5.**  A long, straight wire of radius $a$ has a resistance per unit length $b$ and is carrying a steady current $I$. Calculate the Poynting vector at the surface of the wire, and from it calculate the energy flow per unit time through the surface into a unit length of the wire.  How does this compare with the power being dissipated in the wire?

**Picture the Problem.**  You can use the equation for the magnetic field of a long straight wire for the magnetic field, and use Ohm's law to determine the electric field.

| 1. Draw a sketch of the situation. The electric field will be in the direction of current flow, and the direction of the magnetic field and Poynting vector can be determined by applying right-hand-rules. | 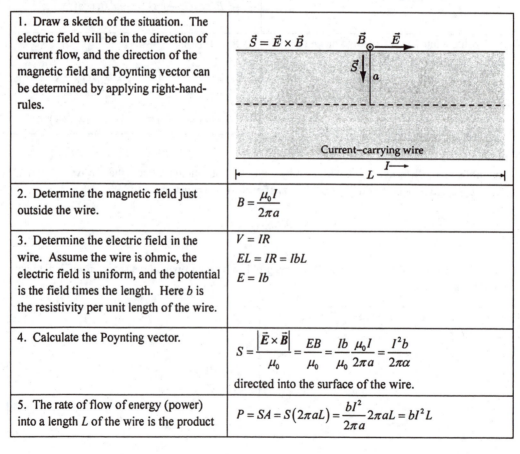 |
|---|---|
| 2. Determine the magnetic field just outside the wire. | $B = \dfrac{\mu_0 I}{2\pi a}$ |
| 3. Determine the electric field in the wire.  Assume the wire is ohmic, the electric field is uniform, and the potential is the field times the length.  Here $b$ is the resistivity per unit length of the wire. | $V = IR$ <br> $EL = IR = IbL$ <br> $E = Ib$ |
| 4. Calculate the Poynting vector. | $S = \dfrac{\left\lvert \vec{E} \times \vec{B} \right\rvert}{\mu_0} = \dfrac{EB}{\mu_0} = \dfrac{Ib}{\mu_0}\dfrac{\mu_0 I}{2\pi a} = \dfrac{I^2 b}{2\pi a}$ <br> directed into the surface of the wire. |
| 5. The rate of flow of energy (power) into a length $L$ of the wire is the product | $P = SA = S(2\pi a L) = \dfrac{bI^2}{2\pi a}\,2\pi a L = bI^2 L$ |

| of the Poynting vector and the area of the surface. | |
|---|---|
| 6. Determine the power dissipated in a length $L$ of the wire. This rate of power dissipation is the same as the rate at which electromagnetic energy is delivered to the wire. | $P = I^2 R = I^2 bL$ |

**Example #6—Interactive.** At the top of the earth's atmosphere the intensity of sunlight is $1.35\,kW/m^2$. Find the area of a 100% reflecting surface, upon which the force of sunlight is $10^{-5}\,N$ (about the weight of a mosquito). Speculate as to whether you can feel the radiation pressure from the sunlight.

**Picture the Problem.** The net radiation pressure is the pressure exerted by the incident wave plus the pressure exerted by the reflected wave. **Try it yourself.** Work the problem on your own, in the spaces provided, to get the final answer.

| 1. Find the radiation pressure on a perfectly absorbing surface from the intensity. | |
|---|---|
| 2. Find the radiation pressure resulting from the reflection radiation. | |
| 3. Determine the surface area required to feel the given force from the total radiation pressure you calculated above. This area is approximately your entire surface area. It is unlikely you can feel this pressure, since it would be the force of a mosquito landing, spread out over this entire area. I can hardly feel a mosquito land on my hand, when the contact area is much smaller. | $A = 1.11\,m^2$ |

**Example #7.** A local radio station broadcasts with 10.0 kW of power at a frequency of 1340 kHz. If you are detecting this signal with a 30.0 cm straight-wire antenna, a distance of 5.00 km from the antenna, what is the maximum voltage that goes to the amplifier and tuner? Assume that the antenna radiates uniformly in all directions (it is not a dipole antenna). How must your antenna be oriented relative to the incoming signal?

**Picture the Problem.** The electric field strength can be determined from the intensity of the signal. Assume the electric field strength is uniform along the length of the wire.

| 1. Determine the intensity of the signal at your location. The antenna must radiate into a full sphere, 5 km away. | $I = \dfrac{P}{A} = \dfrac{P}{4\pi r^2} = \dfrac{10{,}000\,\text{W}}{4\pi(5000\,\text{m})^2} = 3.18 \times 10^{-5}\,\text{W/m}^2$ |
|---|---|
| 2. Find the electric field strength by relating it to the intensity. | $I = \dfrac{E_0^2}{2\mu_0 c}$ <br><br> $E_0 = \sqrt{2\mu_0 c I}$ <br><br> $\quad = \sqrt{2\left(4\pi \times 10^{-7}\,\text{N} \cdot \text{m/A}\right)\left(3 \times 10^8\,\text{m/s}\right)\left(3.18 \times 10^{-5}\,\text{W/m}^2\right)}$ <br><br> $\quad = 0.155\,\text{V/m}$ |
| 3. For a uniform electric field, $\Delta V = \vec{E} \cdot \vec{L}$. The maximum voltage will occur for the maximum value of the electric field, and when the electric field vector is parallel to the wire. | $V_{max} = E_0 L = (0.155\,\text{V/m})(0.30\,\text{m}) = 46.5\,\text{mV}$ |

**Example #8—Interactive.** For the broadcast antenna in the previous example, what is the smallest radius circular loop antenna you would need to produce the same size signal? How must this antenna be oriented relative to the field to achieve maximum sensitivity?

**Picture the Problem.** A loop antenna is sensitive to changes in magnetic flux, via Faraday's law, rather than the electric field. **Try it yourself.** Work the problem on your own, in the spaces provided, to get the final answer.

| 1. Determine the intensity of the signal at your location. The antenna must radiate into a full sphere, 5 km away. | |
|---|---|
| 2. Find the magnetic field strength by relating it to the intensity. | |

| | |
|---|---|
| 3. Find an expression for the changing magnetic flux through a circular antenna of area $A$. Remember that the magnetic field varies sinusoidally. | |
| 4. The maximum voltage signal will occur when the rate of change of the flux is at a maximum. Find an expression for the maximum value of the rate of change of flux. | |
| 5. The maximum rate of change of flux is supposed to produce a signal of 46.5 mV. Use this to solve for the radius of the circular loop. | $r = 1.85\,\text{m}$ |

# Chapter **31**

# Properties of Light

## I.  Key Ideas

***Section 31-1.  Wave–Particle Duality.***[1] The wave nature of light was contemplated by Isaac Newton and his contemporaries, but Newton favored a theory in which the particle nature of light dominated. The wave nature of light was first demonstrated persuasively by Thomas Young and his contemporaries. Young observed the interference pattern of two coherent light sources produced by illuminating a pair of narrow, parallel slits with a single source.[2] The wave theory of light culminated in 1860 with Maxwell's prediction of electromagnetic waves. The particle nature of light was first proposed by Albert Einstein in 1905 in his explanation of the photoelectric effect. A particle of light called a **photon** has energy $E$ that is related to the frequency $f$ and wavelength $\lambda$ of the light wave by the Einstein equation

$$E = hf = \frac{hc}{\lambda}$$
Einstein equation for photon energy

where $c$ is the speed of light and $h$ is Planck's constant:

$$h = 6.626 \times 10^{-34} \text{ J} \cdot \text{s} = 4.136 \times 10^{-15} \text{ eV} \cdot \text{s}$$
Planck's constant

The propagation of light can be described both by its wave properties, via classical electrodynamics, and by its particle properties, via quantum electrodynamics. However, energy and momentum exchanges between light and matter are always both localized and quantized. This duality between light's particle nature and wave nature is a general property of nature. For example, electrons and other so-called particles can exhibit wave properties under certain conditions.

---

[1] The wave-particle duality of light and electrons is discussed in detail in Chapter 34.

[2] General wave properties such as propagation, reflection, refraction, interference, and coherence are discussed in Chapters 15 and 16.

**Section 31-2. Light Spectra.** Using a prism, Newton demonstrated that the spectrum of sunlight was composed of all colors, with the intensity of the different colors being approximately equal. Such a combination of colors produces the sensation of white light. Because the spectrum of sunlight contains a continuous range of wavelengths, it is called a **continuous spectrum.** The light emitted by the atoms in low-pressure gases contains only a discrete set of wavelengths which, upon refraction (or diffraction through an array of parallel slits), produce a **line spectrum.**

*Section 31-3. Sources of Light.   Line Spectra.* Many common sources of visible light are transitions of the outer electrons in atoms. Normally an atom is in its ground state with its electrons at their lowest allowed energy levels. The lowest energy electrons are closest to the nucleus and are tightly bound, forming a stable inner core. The one or two electrons in the highest energy states are much farther from the nucleus and are relatively easily excited to vacant higher energy states. These outer electrons are responsible for the energy changes in the atom that result in the emission or absorption of visible light.

When an atom collides with another atom or with a free electron, or absorbs electromagnetic energy, the outer electrons can be excited to higher energy states. After about $10^{-8}$ s, these outer electrons spontaneously make transitions to lower energy states with the emission of a photon. This process, called **spontaneous emission,** is random; the photons emitted from two different atoms are not correlated, even though their frequencies may be the same. The frequency of the light wave is related to the energy by the Einstein equation, $\Delta E = hf$. The wavelength of the emitted light is then

$$\lambda = \frac{c}{f} = \frac{hc}{hf} = \frac{hc}{|\Delta E|}$$

where $|\Delta E|$ is both the energy difference between the emitting atom's initial and final states and the energy of the emitted photon. Because electrons reside only in specific energy levels in atoms, they can make only a particular set of transitions from high to low energy states. Thus, the emission spectrum of light from single atoms or atoms in low-pressure gases consists of a set of sharp discrete lines, that is specific wavelengths or photon energies, that are characteristic of the element.

*Continuous Spectra.* When atoms are close together and interact strongly, as in liquids and solids, the energy levels of the individual atoms are spread out into energy bands, resulting in an essentially continuous range of energy band levels. Thus, when the outer electrons of these atoms are excited, there is a continuous spectrum of possible transition energies, which results in a continuous emission spectrum instead of a discrete line spectrum.

In incandescent objects such as the filament of a light bulb or the surface of the sun, the electrons are randomly excited by frequent atom-atom collisions. This results in a broad spectrum of thermal radiation, and the light emitted by such incandescent objects is a continuous spectrum. The thermal radiation emitted by an object below about 1000 K is concentrated in the infrared region and is therefore not visible. The surface of the sun, which has a temperature of 6000 K, emits a continuous spectrum. Over the visible range, the intensity of the sun's radiation is nearly uniformly distributed.

*Absorption, Scattering, and Stimulated Emission.* Radiation is emitted when an atom makes a transition from an excited state to a state of lower energy; radiation is absorbed when an atom makes a transition from a lower state to a higher state. When atoms are irradiated with a continuous spectrum of radiation, the transmitted spectrum shows dark lines corresponding to absorption of light at discrete wavelengths. The absorption spectra of atoms were the first line spectra observed. Since atoms and molecules at temperatures typical to our environment are in either their ground states or low-lying excited states, absorption spectra usually have fewer lines over a given range of wavelengths than do emission spectra.

Several interesting phenomena can occur when a photon is incident on an atom. If the energy of the incoming photon is too small to excite the atom to an excited state, the atom remains in its ground state and the photon is said to be scattered. Since the incoming and outgoing, or scattered, photons have the same energy, the scattering is said to be elastic. If the wavelength of the incident light is large compared with the size of the atom, the scattering can be described in terms of classical electromagnetic theory and is called **Rayleigh scattering** after Lord Rayleigh, who worked out the theory in 1871. The probability of Rayleigh scattering varies as $1/\lambda^4$. This means that when sunlight is scattered by air molecules, the blue light is scattered much more readily than red light, which accounts for the bluish color of the sky. When you look directly at the sun during sunrise or sunset, the blue light has been removed by Rayleigh scattering. Only the longer wavelengths remain to enter your eye, which accounts for the reddish, orange color that you see.

**Inelastic scattering** occurs when the incident photon has more than enough energy to cause the atom to make a transition to an excited state. A photon with energy $hf'$ is scattered, but its energy is less than that of the incident photon, $hf$, by $\Delta E$, the energy absorbed by the atom as it transitions to its excited state. Inelastic scattering of light from molecules was first observed by the Indian physicist C. V. Raman and is often referred to as **Raman scattering.**

If the energy of the incident photon is just equal to the difference in energy between the ground state and the first excited state of the atom, the atom makes a transition to its first excited state. After a short delay, it decays by spontaneous emission back to the ground state, and a photon is emitted whose energy is equal to that of the incident photon. This multistep process is called **resonance absorption.**

If the energy of the incident photon is great enough to excite the atom past its first excited state to one of its higher excited states, the atom then loses its energy by spontaneous emission as it makes one or more transitions to lower energy states. A common example occurs in **fluorescence,** where the atom is excited by ultraviolet light and emits visible light as it returns to its ground state. Since the lifetime of a typical excited atomic energy state is of the order of $10^{-8}$ s, fluorescence is often a short-lived process. However, some excited states have much longer lifetimes—of the order of milliseconds or occasionally seconds or even minutes. Such a state is called a **metastable state. Phosphorescent materials** have very long-lived metastable states, and so emit light long after the original excitation.

In the above cases of spontaneous emission, the phase of the emitted photon is not correlated with the phase of the incident photon. However, **stimulated emission** results in photons travelling in phase with one another. In stimulated emission an atom or molecule is initially in an excited state of energy $E_2$, and the energy of the incident photon is equal to $E_2 - E_1$, where $E_1$ is the energy of a lower state or the ground state. The oscillating electromagnetic field associated with the incident photon stimulates the excited atom or molecule, which then emits a photon in the same direction as the incident photon and in phase with it. In stimulated emission, the phase of the light emitted from one atom is related to that emitted by every other atom, so the resulting light is coherent.

*Lasers.* The laser (*l*ight *a*mplification by *s*timulated *e*mission of *r*adiation) is a device that produces a strong beam of coherent photons by stimulated emission. For a laser to work, we must first have a population inversion. That is, more of the atoms have to be in an excited, but metastable, state of energy $E_2$ than in the ground state $E_1$ so that stimulated emission will dominate over absorption. (Normally, the atoms in a system at temperatures typical to our environment will be in the ground state.) A population inversion can be achieved by a variety of methods in which atoms are "pumped" up to energy levels of higher energy than $E_2$ by absorbing energy supplied by an auxiliary source. The atoms then decay down to the metastable state $E_2$ by either spontaneous emission or by nonradiative transitions such as those due to collisions. When some of the atoms at $E_2$ decay to the ground state by spontaneous emission, they emit photons. These photons stimulate other atoms in the same metastable state to emit photons with the same energy in the same phase and direction. The emitted photons and incident photons combine to form a more intense light beam.

If this laser medium is placed between two mirrors, each successive bounce of the light stimulates additional coherent emission of photons, which increases the intensity of the light. This system is called a laser cavity. If one of the mirrors is not 100% reflecting, so that a small fraction of the light can be transmitted through it, a laser beam is produced.

*Section 31-4. The Speed of Light.* The speed of light can be deduced by measuring the time light requires to traverse a known distance. This speed is very high, so unless the distance traveled is large the time interval is very small. The first evidence that the speed of light is finite came from Römer's seventeenth-century observations of the occultations of the moons of Jupiter. In the nineteenth century the first laboratory measurements of the speed of light were made by using a toothed wheel, a rotating mirror, or a similar device to periodically interrupt the light reflected from a distant mirror. Current experimental measurements have excellent agreement with Maxwell's theoretical expression for the speed $c$ of electromagnetic waves in empty space

$$c = \frac{1}{\sqrt{\varepsilon_0 \mu_0}} = 3.00 \times 10^8 \text{ m/s} \qquad \text{Speed of light}$$

*Section 31-5. The Propagation of Light. Huygens' Principle.* The propagation of light waves in a given physical situation can be modeled by **Huygens' principle:**

Each point on a wavefront can be considered as a point source of secondary hemispherical wavelets that propagate with the same speed and wavelength as the primary wave. At some later time, the primary wavefront is the envelope of all the secondary wavelets. (An envelope is a surface tangent to each of a family of surfaces.)

Huygens' principle depicts the nature of wave propagation.

*Fermat's Principle.* An alternative description of the propagation of light is given by **Fermat's principle:**

> The path taken by light from one point to another in a given situation is that for which the travel time is a minimum.

Fermat's principle is a statement of the conditions necessary for the constructive interference of a propagating wavefront with itself. Fermat's principle can be stated as

$$\frac{dt}{dx} = 0 \qquad\qquad \text{Fermat's principle}$$

where $t$ is the time for light to travel along a path, and $x$ is a parameter that specifies the path. That is, the travel time $t$ is a minimum if $dt / dx = 0$.

The laws of reflection and refraction can both be deduced from Fermat's principle.

*Section 31-6. Reflection and Refraction.  Reflection.* When electromagnetic waves encounter a boundary between two transparent materials or between a transparent material and an opaque material, reflection occurs. Reflection from a smooth surface is **specular** (mirrorlike), whereas reflection from a rough surface is **diffuse.** In the case of specular reflection, the directions of propagation of the incident and the reflected wavefronts make equal angles, called the **angle of incidence** and **angle of reflection,** with the normal to the reflecting boundary. This **law of reflection** follows from Huygens' principle.

$$\theta_i = \theta_r \qquad\qquad \text{Law of reflection}$$

*Refraction.* The major optical characteristics of a transparent material are summarized by its **index of refraction** $n$, which is the ratio of the speed of light in vacuum to that in the material.

$$n = \frac{c}{v} \qquad\qquad \text{Index of refraction}$$

where $v$ is the speed of light in the material. The speed of light in most materials is less than that in a vacuum, so indices of refraction are generally greater than one. The fraction of the light intensity that is reflected at a boundary depends on the indices of refraction in the two media, the angle at which light strikes the boundary, and the state of polarization of the incident light.  In the special case of normal incidence of light with an intensity of $I_0$, the reflected intensity is

$$I = \left(\frac{n_1 - n_2}{n_1 + n_2}\right)^2 I_0 \qquad\qquad \text{Intensity of reflected light at normal incidence}$$

When electromagnetic waves are incident on a boundary between transparent media, some of the energy is transmitted across the boundary. The change in the direction of propagation of the waves as they cross the boundary is called refraction. Refraction at a boundary depends on the wave speeds (or alternatively, the refractive indices) of both media. The relation between the directions of propagation of the incident and transmitted waves, called the **law of refraction** (or **Snell's law**) is

$$n_1 \sin \theta_1 = n_2 \sin \theta_2$$                                    Law of refraction (Snell's law)

where $n_1$ and $n_2$ are the indices of refraction, and $\theta_1$ and $\theta_2$ are the angles between the directions of propagation and the normal to the boundary. The angle between the direction of propagation of the refracted (transmitted) wave and the normal is called the **angle of refraction.**

When light propagating in a medium is incident on a boundary between that medium and a medium of lower refractive index, for sufficiently large angles of incidence, the law of refraction predicts the impossible—that the sine of the angle of refraction is greater than one. For these large angles of incidence no refraction occurs, and the light is totally reflected back into the first medium. This **total internal reflection** explains why light traveling along optical fibers remains in the fibers even when they bend, and can also explain mirages. The angle of incidence $\theta_c$ for which the law of refraction predicts that the sine of the angle of refraction equals one (that the angle of refraction equals 90°) is called the **critical angle.**

$$\sin \theta_c = \frac{n_2}{n_1}$$                                    Critical angle for total internal reflection

*Dispersion.* The speed at which light propagates in a material varies with the frequency of the light. As a result, when white light is refracted at a boundary, each frequency (color) is refracted through a different angle. Thus, the light is separated into its component colors. The dispersion of white light in water droplets in the atmosphere gives rise to the rainbow.

*Section 31-7. Polarization.* Light is a transverse wave and thus can be **polarized;** that is, different transverse directions of the oscillation of the electric field can be distinguished. An electromagnetic wave produced by a source that is small compared with the wavelength is usually polarized. A wave is polarized if the plane in which its electric field oscillates is fixed or rotates in a simple way; it is unpolarized if the electric field direction varies randomly. The polarization state of light is affected by its interactions with matter in four distinct ways: by absorption, reflection, scattering, and by passing through a birefringent (doubly refracting) material.

A sheet of Polaroid absorbs any component of the electric field that is perpendicular to the transmission axis of the sheet. Thus, any light transmitted through such a sheet is polarized in a plane parallel with the sheet's transmission axis. If unpolarized light is incident on an ideal polarizing sheet, the intensity of the transmitted light is half of the intensity of the incident light. If linearly polarized light is incident on an ideal polarizing sheet, the intensity of the transmitted light $I$ is related to the intensity of the incident light $I_0$ according to Malus' law

$$I = I_0 \cos^2 \theta$$                                    Malus' law

where $\theta$ is the angle between the transmission axis and the direction of polarization of the incident light. Polarizing sheets are frequently used in pairs. The first sheet, called a **polarizer,** polarizes light and the second sheet, called an **analyzer,** determines the plane of polarization of polarized light.

When light shines on a small object such as a molecule or a dust particle, the electric field of the light causes the electrons in the object to oscillate. These oscillating charges, which constitute an electric dipole antenna, emit light as described in Chapter 30 on pages 978-980 of the text. The light emitted by an electric dipole antenna is linearly polarized.

When unpolarized light is reflected from the boundary between two transparent media such as air and water, the reflected light is partially polarized. When the angle between the reflected and refracted ray is 90°, the reflected light is completely polarized parallel to the plane of the surface, and the angle of incidence is called the polarizing angle $\theta_p$. This condition, along with the laws of reflection and refraction, give us the relation (Brewster's law) between the **polarizing angle** and the indexes of refraction:

$$\tan\theta_p = \frac{n_2}{n_1}$$  Polarizing (Brewster's) angle

where the incident wave is in the medium with refractive index $n_1$.

## II. Physical Quantities and Key Equations

### Physical Quantities

Speed of light in a vacuum  $c = \dfrac{1}{\sqrt{\varepsilon_0\mu_0}} = 3.00\times10^8$ m/s

Planck's constant  $h = 6.626\times10^{-34}$ J•s $= 4.136\times10^{-15}$ eV•s

### Key Equations

Energy of a photon  $E = hf = \dfrac{hc}{\lambda}$

Law of reflection  $\theta_i = \theta_r$

Intensity of reflected light at normal incidence  $I = \left(\dfrac{n_1-n_2}{n_1+n_2}\right)^2 I_0$

Fermat's Principle  $\dfrac{dt}{dx} = 0$

Index of refraction  $n = \dfrac{c}{v}$

Law of refraction (Snell's law)  $n_1\sin\theta_1 = n_2\sin\theta_2$

*Critical angle for total internal reflection*     $\sin \theta_c = \dfrac{n_2}{n_1}$

(Use the law of refraction, letting $\theta_1 = \theta_c$, $\theta_2 = 90°$, and $n_1 > n_2$.)

*Malus' law*     $I = I_0 \cos^2 \theta$

*Polarizing (Brewster's) angle*     $\tan \theta_p = \dfrac{n_2}{n_1}$

(Use the law of refraction, letting $\theta_1 = \theta_p$ and $\theta_2 = 90° - \theta_p$.)

## III. Potential Pitfalls

The laws of reflection and refraction can be explained by *either* a wave or a particle theory of light. It is the implications of one model or the other that enables us to experimentally distinguish between the two models. For example, when light travels from air to water the light bends toward the normal. Newton's particle theory explains this by asserting light consists of particles that travel faster in water than in air. Huygens' wave theory explains it by asserting light consists of waves that travel slower in water than in air. By measuring the speed of light in both air and water, Foucault demonstrated that light travels more slowly in water than in air. This observation discredited Newton's particle theory of light.

Light is polarized (or more precisely, its state of polarization is affected) by reflection at *any* angle between 0 and 90°. The polarizing angle (Brewster's angle) is the angle at which all of the reflected light is linearly polarized, and its direction of polarization is parallel with the surface.

The wave–particle debate about the nature of light was resolved in the early part of the 20th century. In the current model, light has both wave and particle attributes. Light exhibits primarily wave characteristics in propagation and particle characteristics in interactions with atoms, molecules, and nuclei, such as absorption and emission.

Unlike mechanical waves such as sound, light requires no medium to propagate; the electromagnetic field travels in matter-free space. Thus, some wave phenomena, such as the Doppler effect, that depend on the medium of propagation may behave differently for light than for mechanical waves.

The index of refraction is the speed of light in a vacuum divided by its speed in a material; you are likely to invert this division if you aren't careful! Remember that the index of refraction typically is a dimensionless number greater than one.

Light passes from one medium into another with its frequency unchanged. When the speed of propagation changes, the wavelength changes in such a way that the ratio of speed to wavelength remains the same.

Total internal reflection occurs only for light propagation *from* the medium with the *lower* speed of light into that with the higher speed. That is, from the medium of higher refractive index into that with the lower refractive index.

The laws of reflection and refraction always use angles measured with respect to the normal of the surface, not angles measured from the surface.

## IV. True or False Questions and Responses

**True or False**

_____ 1. The first successful measurements of the speed of light were those of Galileo in the eighteenth century.

_____ 2. Both the laws of reflection and refraction can be derived from Huygens' principle.

_____ 3. Both the laws of reflection and refraction can be derived from Fermat's principle.

_____ 4. Between any two points, light follows a path for which its travel time is a minimum.

_____ 5. At a boundary, light is reflected at an angle equal to the angle of incidence.

_____ 6. If light is incident normally on an air–glass boundary, the intensity of the reflected light is greater than the intensity of the transmitted light.

_____ 7. A wavefront is a surface of constant phase, like a crest or a trough.

_____ 8. Longitudinal waves cannot be polarized.

_____ 9. If light falls on a boundary at the polarizating angle, both the refracted and the reflected light are completely polarized.

_____ 10. Unpolarized light reflected from a boundary between transparent media is completely polarized only if the angle of incidence is the polarizing angle.

_____ 11. Light from the sky is polarized because of the polarization associated with the scattering of sunlight by the air molecules.

_____ 12. A ray is a directed line that is perpendicular to the wavefronts of a wave.

**Responses to True or False**

1. False. The first were Römer's observations of the moons of Jupiter. Galileo made attempts at the measurement of light speed, but his methods were completely inadequate. Besides, he was long dead by the eighteenth century.

2. True.

3. True.

4. True.

5. True for specular reflection.

6. False. Only about 4% of the incident intensity is reflected.

7. True.

8. True. There is only one possible direction for the disturbance in a longitudinal wave: along the direction of propagation.

9. False. Only the reflected light is completely polarized.

10. True. The state of polarization is affected by reflection at any angle. Only at the polarizing angle is the reflected light completely polarized.

11. True.

12. True.

# V.  Questions and Answers

## Questions

1. Arguments can be invented to explain diffraction phenomena—perhaps not very convincingly—on the basis of a particle theory of light. Can interference phenomena be similarly explained?

2. In some situations light propagates through a medium with a continuously varying index of refraction. The refractive index of a gas, for example, is proportional to its density, so the index of refraction of the atmosphere decreases with increasing altitude. What would the path of a light ray from the sun look like in such a situation?

3. Galileo's attempt to measure the speed of light is described on pages 1005-1006 of the textbook. Estimate the time it took light to travel the round trip. I know you haven't been given enough information; improvise.

4. A timely measurement of the speed of light in a transparent medium such as glass or water would have settled the debate between the wave and particle theories of light. How?

5. When light propagates across a boundary into a second transparent medium with a higher index of refraction, its wavelength decreases but its frequency remains the same. Why?

6. Which way should the transmission axis be oriented in antiglare polarizing sunglasses?

7. Why is the highway so much harder to see when you are driving on a rainy night than on a dry one?

8.  Blue light is bent more by a glass prism than red light. Which color of light travels faster in glass?

9.  Why does the oar you are rowing with appear to be bent at the water surface?

**Answers**

1.  I don't think so. At some places in an interference pattern, light from two sources cancels; at others, it adds. That is, for particles $1+1=2$ always. For waves, sometimes $1+1=0$ or $1+1=4$. If this isn't beyond the scope of classical particles, then nothing is.

2.  The sketch shows a beam of light in a medium like the atmosphere in which the refractive index decreases as you go upward. The light in the lower portion of the wavefronts travels less rapidly than the light in the upper portion. In this situation the wavefronts in the beam turn downward as shown in Figure 31-1. When the sun is just above the horizon, the light from the bottom edge of the sun is bent downward to a greater degree than is the light from the upper edge of the sun. Thus, the sun appears flattened when it is near the horizon.

Figure 31-1

3.  If the hills were a kilometer apart, it would have taken only about seven millionths of a second for the light to travel the round-trip distance of two kilometers. What Galileo was measuring was his assistant's reaction time, not the travel time of the light.

4.  Both theories can be made to predict the law of refraction. However, to explain the observed direction of refraction, the particle model requires that the speed of light in, say, glass be greater than that in air. On the other hand, to explain the same observation, the wave model requires the speed of light in glass to be less than that in air. Thus, a measurement of the speed of light in glass would confirm one or the other. As it happened, this measurement was not technically feasible until after the observation of interference had already confirmed the wave theory.

5.  The frequency of a light wave in any medium is the number of crests that pass any given point per unit time. When this wave is incident on a boundary, the rate at which the crests arrive at any point, including points on the boundary, is equal to the frequency in the first medium. The rate at which these crests leave the boundary and enter the second medium is the frequency in the second medium. Because the crests do not pile up at the boundary, the two frequencies must be equal. The index of refraction is greater in the second medium, so the crests travel more slowly than they did in the first medium. Because they leave the boundary at the same rate (frequency) at which they arrive, the crests are closer together in the second medium; they have to be because they have a shorter distance to travel in the time between the arrivals of successive crests at the boundary. The distance separating adjacent crests is the wavelength,

which is the product of the speed of the wave and the time between the arrival of the crests. Hence, the wavelength is shorter in the medium in which the wave travels more slowly.

6. Most of the glare you want to reduce is reflection off horizontal surfaces—the sun off a highway or the surface of a swimming pool, for instance. This reflected light is partially polarized in a horizontal plane. To block this glare you want the transmission axis of the polarizing lenses to be vertical.

7. Reflection from a dry highway is diffuse; thus, some of the light from your headlamps is reflected back at you, which enables you to see the road surface. When it rains, the water surface reflects the light specularly; so most of the headlamp beam is reflected forward, and little or none of the light gets back to your eye.

8. Refraction occurs because the light changes speeds as it travels from air to glass. The smaller the difference in speeds, the less the refraction. Thus, the red light, which bends less, travels faster in the glass than the other colors.

9. Because of the refraction at the air–water boundary, you see the underwater portion of the oar at a different position from where it actually is, causing the shaft of the oar to appear bent at the water's surface.

# VI. Problems

**Example #1.** In Foucault's experiment to measure the speed of light (see Figure 31-13 on page 1008 of the textbook), suppose that the distance between the fixed mirror and the rotating mirror is $L = 10.0$ km. What is the lowest rate of rotation of the octagonal mirror, in rev/min, at which you could see light in the viewing telescope?

**Picture the Problem.** The time it takes the light to travel from the octagonal mirror to the fixed mirror and back must be the time required for one-eighth of a revolution of the octagonal mirror.

| 1. Determine the time required for light to travel to the fixed mirror and back. | $t = 2L/c$ |
|---|---|
| 2. Determine the angle through which the mirror rotates during this time. | $\theta = \omega t = \omega \dfrac{2L}{c}$ |
| 3. Solve for the angular velocity $\omega$. | $\omega = \dfrac{c\theta}{2L} = \dfrac{\left(3 \times 10^8 \text{ m/s}\right)\left(1/8 \text{ rev}\right)}{2\left(10{,}000 \text{ m}\right)}$ $= \left(1880 \text{ rev/s}\right)\left(60 \text{ s/min}\right) = 113{,}000 \text{ rev/min}$ |

**Example #2—Interactive.** Light traveling in air is incident normally on a slab of flint glass of refractive index 1.63. What fraction of the incident light energy passes through the slab?

**Picture the Problem.** How much light energy is reflected? The rest must be transmitted. **Try it yourself.** Work the problem on your own, in the spaces provided, to get the final answer.

| | |
|---|---|
| 1. Calculate the intensity of the light reflected at the air-glass interface when the light enters the glass. | |
| 2. Determine the intensity of light transmitted through the air-glass interface. | |
| 3. Calculate the intensity of the light reflected at the glass-air interface as the light leaves the glass. | |
| 4. Determine the intensity of the transmitted light through the entire slab. | $I_{transmitted} = 0.889 I_0$ |

**Example #3.** Light in air is incident on a piece of plastic, making an angle of 48° with the normal to the surface. The light refracts into the material in a direction 32° from the normal. What is the speed of light in the plastic?

**Picture the Problem.** The speed of light in the plastic can be found from the index of refraction.

| | |
|---|---|
| 1. Use the law of refraction to determine the index of refraction of the plastic. The light starts out in air, which is subscript 1, and ends up in the plastic, subscript 2. | $n_1 \sin\theta_1 = n_2 \sin\theta_2$<br><br>$n_2 = \dfrac{n_1 \sin\theta_1}{\sin\theta_2} = \dfrac{(1)\sin 48°}{\sin 32°} = 1.40$ |
| 2. Determine the speed of light in the plastic from the index of refraction. | $n_2 = \dfrac{c}{v_2}$<br><br>$v_2 = \dfrac{c}{n_2} = \dfrac{3\times10^8 \text{ m/s}}{1.40} = 2.14\times10^8 \text{ m/s}$ |

**Example #4—Interactive.** The index of refraction of a certain piece of glass is 1.47. Red light of wavelength 635 nm in a vacuum is incident on the surface of the glass. What is the wavelength of the light in the glass?

**Picture the Problem.**  Wavelength is related to the frequency and speed of the wave.  The frequency of light is constant, independent of the material through which it propagates.  **Try it yourself.**  Work the problem on your own, in the spaces provided, to get the final answer.

| | |
|---|---|
| 1.  Determine the frequency of the light in air. | |
| 2.  Determine the speed of light in the glass from its index of refraction. | |
| 3.  Because the frequency remains constant, you can calculate the wavelength of the light in the glass from the speed calculated in step 2 and the frequency calculated in step 1. | $\lambda = 432\,\text{nm}$ |

**Example #5.**  The object in Figure 31-2 is a distance  $d = 0.850\,\text{m}$  below the surface of the water.  How far from the end of the dock, distance $D$ in the figure, must the object be if it cannot be seen from any point above the surface?  The index of refraction of water is 1.33.

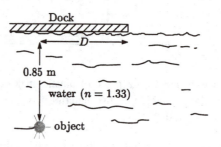

Figure 31-2

**Picture the Problem.**  If the object cannot be seen from a point above the surface, then no light from the object can be transmitted through the surface into the air.  Therefore, total internal reflection must be occurring at the water-air interface.  Assume the surface of the water is perfectly flat (no waves) to make the calculations easier.

| | |
|---|---|
| 1.  Sketch the situation, showing the distances and angles involved.  Remember the angles are always measured with respect to the normal of the surface. | |

| 2. Use the law of refraction to solve for the critical angle, which is the smallest angle for which total internal reflection can occur. | $n_1 \sin \theta_c = n_2 \sin 90°$ $\sin \theta_c = \dfrac{n_2}{n_1} = \dfrac{1}{1.33} = 0.752$ $\theta_c = 48.8°$ |
|---|---|
| 3. Use this angle to solve for $D$. | $\tan \theta_c = \dfrac{D}{d}$ $D = d \tan \theta_c = (0.85\,\text{m}) \tan 48.8° = 0.971\,\text{m}$ |

**Example #6—Interactive.** A flat glass surface with $n = 1.54$ has a layer of water, $n = 1.33$, of uniform thickness directly above the glass. At what minimum angle of incidence must light in the glass strike the glass-water interface for it to be totally internally reflected at the water-air interface?

**Picture the Problem.** Work this problem in reverse. Start with the fact that there is total internal reflection at the water-air interface. This gives you the incident angle at that interface, which must be equal to the refracted angle at the glass-water interface, from which you can find the incident angle for the glass-water interface. **Try it yourself.** Work the problem on your own, in the spaces provided, to get the final answer.

| 1. Make a sketch of the situation so you can easily see all the angles involved and their relationships to each other. | |
|---|---|
| 2. Use the law of refraction to determine the critical angle at the water-air interface. | |
| 3. The critical angle will be equal to the refracted angle at the glass-water interface. Use this fact to find the incident angle on the glass-water interface. (Note that the angle would still be 40.5° even if no water was present.) | $\theta_{inc} = 40.5°$ |

**Example #7.** A ray of light is incident at an angle of $45°$ on a slab of glass $1.00$ cm thick as shown in Figure 31-3. The index of refraction of the glass is $1.51$. (*a*) Show that the emergent ray is parallel to the incident ray. (*b*) Find the lateral displacement *d* between the incident and emergent ray.

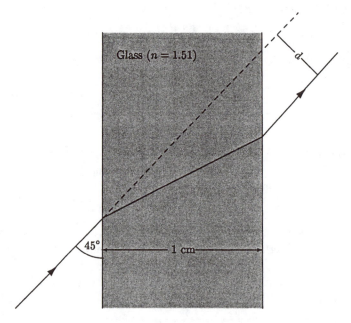

Figure 31-3

**Picture the Problem.** Use the law of refraction to show that the rays are parallel. The displacement can be determined from geometrical considerations.

| | |
|---|---|
| 1. Sketch the situation, showing all the angles involved. | 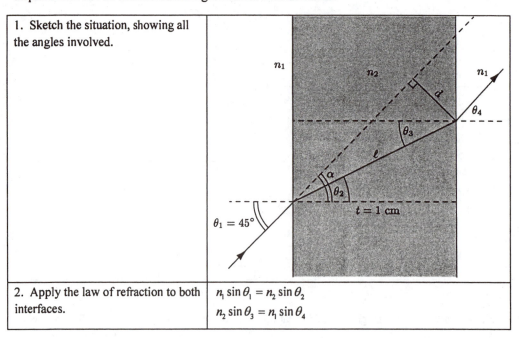 |
| 2. Apply the law of refraction to both interfaces. | $n_1 \sin \theta_1 = n_2 \sin \theta_2$<br>$n_2 \sin \theta_3 = n_1 \sin \theta_4$ |

| 3. Because $\theta_2$ and $\theta_3$ are alternate interior angles, they are equal. This allows us to reduce the two law of refraction equations to one, and solve for $\theta_4$. Since $\theta_1$ and $\theta_4$ are equal, the incoming and outgoing rays are parallel. | $n_1 \sin\theta_1 = n_1 \sin\theta_4$ <br> $\theta_1 = \theta_4$ |
|---|---|
| 4. Find $d$ in terms of angles and distances given. Note that we cannot find $d$ unless we know $\theta_2$. | $d = \ell \sin\alpha$ <br> $\alpha = \theta_1 - \theta_2$ <br> $\cos\theta_2 = \dfrac{t}{\ell}$ <br> $\ell = \dfrac{t}{\cos\theta_2}$ <br> $d = \dfrac{t}{\cos\theta_2}\sin\left(\theta_1 - \theta_2\right)$ |
| 5. Find $\theta_2$ from the law of refraction. | $n_1 \sin\theta_1 = n_2 \sin\theta_2$ <br> $\theta_2 = \sin^{-1}\left(\dfrac{n_1}{n_2}\sin\theta_1\right) = \sin^{-1}\left(\dfrac{1}{1.51}\sin 45°\right) = 27.9°$ |
| 6. Substitute in to find $d$. | $d = \dfrac{(1\,\text{cm})\sin\left(45° - 27.9°\right)}{\cos 27.9°} = 0.333\,\text{cm}$ |

**Example #8—Interactive.** A block of transparent plastic shown in Figure 31-4 has an index of refraction of 1.31. Light is incident on the upper surface with an angle of incidence $\theta$. For what values of $\theta$ will total internal reflection occur at the vertical face of the block?

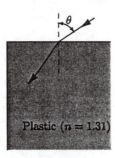

$\theta$

Plastic ($n = 1.31$)

Figure 31-4

**Picture the Problem.** Assume total internal reflection on the side. The critical angle is the smallest angle for total internal reflection to occur. Relate this angle to the refracted angle at the top of the block, and then to the incident angle. **Try it yourself.** Work the problem on your own, in the spaces provided, to get the final answer.

| | |
|---|---|
| 1.  Sketch the situation, making sure to label all important angles. | |
| 2.  Determine the critical angle for total internal reflection. | |
| 3.  Relate the critical angle to the refracted angle at the top of the block. | |
| 4.  Use the law of refraction to relate the incident angle to the critical angle. | |
| 5.  Determine whether increasing or decreasing the incident angle will still cause total internal reflection to occur. | $\theta < 57.8°$ |

**Example #9.**  The polarizing angle for light that passes from water, $n = 1.33,$ into a certain plastic is $61.4°$.  What is the critical angle for total internal reflection of the light passing from this plastic into air?

**Picture the Problem.**  Use the polarizing angle to find the index of refraction of the plastic.  Use that to find the critical angle.

| | |
|---|---|
| 1.  Draw a sketch showing the polarization angle. | 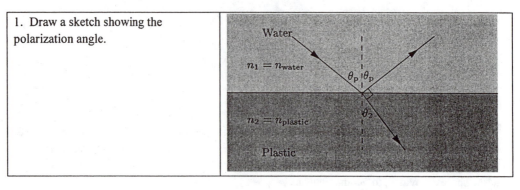 |

| 2. Use the polarizing angle to find the index of refraction of the plastic. | $\tan\theta_p = \dfrac{n_2}{n_1}$ <br><br> $n_2 = n_{plastic} = n_1 \tan\theta_p = n_{water}\, \tan\theta_p$ |
|---|---|
| 3. Draw a sketch showing the total internal reflection. | |
| 4. Determine the critical angle for light traveling from plastic to air. | $n_1 \sin\theta_c = n_2 \sin 90°$ <br><br> $n_{plastic}\sin\theta_c = 1$ <br><br> $\theta_c = \sin^{-1}\left(\dfrac{1}{n_{plastic}}\right)$ <br><br> $\theta_c = \sin^{-1}\left(\dfrac{1}{n_{water}\tan\theta_p}\right) = \sin^{-1}\left(\dfrac{1}{1.33\tan 61.4°}\right) = 24.2°$ |

**Example #10—Interactive.** The radius of the orbit of Mars around the sun is $2.28\times10^{11}$ m and that of the Earth's orbit is $1.50\times10^{11}$ m. Space travelers on Mars use radio waves to communicate with the Earth. What is the maximum value of the time delay for their signal to reach the Earth? What is the minimum value of the delay?

**Picture the Problem.** The time delay is simply distance divided by speed. **Try it yourself.** Work the problem on your own, in the spaces provided, to get the final answer.

| 1. Calculate the maximum distance, which will be the sum of the two radii. | |
|---|---|
| 2. Determine the maximum time. | $t = 1260\,\text{s} = 21.0\,\text{min}$ |
| 3. Calculate the minimum distance. | |
| 4. Calculate the minimum time. | $t = 260\,\text{s} = 4.33\,\text{min}$ |

**Example #11.** Unpolarized light of intensity $I_0$ is incident on two ideal polarizing sheets that are placed with their transmission axes perpendicular to one another. An additional polarizing sheet is then placed between these two, with its transmission axis oriented at 30° to that of the first. What is the intensity of the light passing through the stack of polarizing sheets? What orientation of the middle sheet enables the three-sheet combination to transmit the greatest amount of light?

**Picture the Problem.** The first polarizing sheet will reduce the total intensity in half, because it will block out half of the unpolarized light. After passing through the first sheet, the light is linearly polarized, so Malus' law can be use to calculate the fraction of the incident intensity transmitted by the second and third sheets.

| | |
|---|---|
| 1. Determine the amount of light incident on the second polarizer. It is half of the incident, unpolarized light. | $I = 0.5I_0$ |
| 2. Determine the amount of light that gets through the second polarizing sheet, to the third sheet, using Malus' law. Because of the orientation of the second sheet, the light leaving the second sheet will have a polarization rotated 30° with respect to the incoming polarization. | $I = 0.5I_0 \cos^2 30°$ |
| 3. Use Malus' law again to find the amount of light that passes through the third polarizing sheet, which is now at 60° with respect to the incoming polarized light. | $I = 0.5I_0 \cos^2 30° \cos^2 60° = 0.0938I_0$ |
| 4. To determine the angle for maximum transmission, let the angle of the first two polarizers be $\theta$, so the angle between the second and third polarizers is $90-\theta$. Trigonometric identities help simplify this expression significantly. | $I = 0.5I_0 \cos^2 \theta \cos^2 (90-\theta) = 0.5I_0 \cos^2 \theta \sin^2 \theta$ $= \frac{1}{2}I_0 \frac{1}{4}\sin^2 2\theta = \frac{1}{8}I_0 \sin^2 2\theta$ |
| 5. Set the derivative of the intensity to zero to find transmission maxima and minima. | $\frac{dI}{d\theta} = \frac{1}{2}I_0 \sin 2\theta \cos 2\theta = 0$ $\theta = 0°, 45°, 90°$ |
| 6. Substitute these values back into the original expression for the intensity and see which one gives a maximum value. | $\theta = 45°$ |

# Chapter **32**

# Optical Images

## I.   Key Ideas

Light propagates along straight lines through a uniform medium as long as the sizes of the obstacles or apertures it encounters are large compared with the wavelength(s) of the light. This is the domain of geometric (ray) optics. The direction of the light may change at the interfaces between different transparent media.

*Section 32-1. Mirrors.*

*Plane Mirrors.* The statement that "although you see with light, you cannot see light" often startles people. Light cannot be seen in the sense that if you look across the table at your dining companion, you cannot see the light from the candle that passes upward through the space between you and illuminates the ceiling. Anything that you do see distinctly is either an object, a virtual image, or a real image.

A source of light is called an **object** if light radiates or is scattered from each point on its surface. Some objects, like the filament of an incandescent light bulb, generate their own light whereas others, like the face of your dining companion, merely reflect light diffusely. (Things that reflect light specularly, like mirrors, cannot be seen and are not, in this sense of the word, objects.)

You see an object because the light from a point on the object that enters your eye is refracted by your cornea and lens to a point on your retina. If light from a point source is reflected from a smooth surface, such as a **plane mirror,** the light rays enter the eye as shown in Figure 32-1 just as if they came from a point behind the mirror. This apparent source is a **virtual image** of the actual source—virtual in the sense that light does not actually emanate from the image. The image formed by a plane mirror is at the same distance behind the plane of the mirror as the source of light is in front of it. This image can be viewed by an eye located anywhere in the shaded region of Figure 32-1. The image formed by one reflecting surface can be imaged again by another reflection surface.

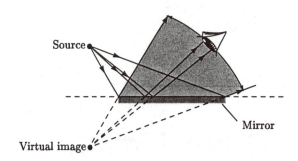

Figure 32-1

***Spherical Mirrors.*** Light from a point source that is reflected by a concave spherical mirror as shown in Figure 32-2 can form a **real image**—real in the sense that, after reflecting from the mirror, the light rays actually converge to the image point and then diverge from it. Viewed by the eye in the figure, the rays diverge from the image point just as if the image point were a point source. All rays emanating from a point source that remain close to the symmetry axis of a spherical mirror, or paraxial rays, will, after reflection, either converge to or diverge from a point image. The nonparaxial rays from the point source that strike the mirror will, after reflection, cross near but not through the image point, thereby producing a blurred image. (How close a ray is to being paraxial determines how close to the image point it passes.)

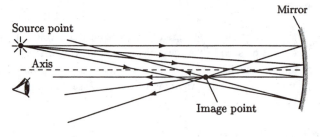

Figure 32-2

Imaging systems are reversible; that is, if the light from an object located at position *A* forms an image of the object at location *B*, then the light from an object located at position *B* forms an image at location *A*.

Parallel paraxial rays shown in Figure 32-3 are reflected by a concave spherical mirror to an image point on the *focal plane,* a plane located halfway between the mirror and its center of curvature. If the parallel incident rays are also parallel with the axis, the location of the image point, called the **focal point** of the mirror, is on the axis. The **focal length** of the mirror is defined as the distance from the focal point to the center of the mirror (the vertex). Thus, the focal length equals one-half the radius of curvature of the mirror. Any point source located closer to the mirror than the focal point forms a virtual image located behind the mirror. That is, paraxial rays from point sources that are located closer to the mirror than the focal length will, on reflection, diverge as if they came from a point behind the mirror.

The formation of an image point by a convex mirror is illustrated in Figure 32-4.

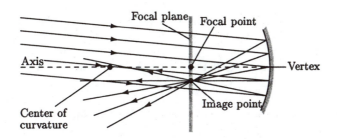

Figure 32-3

We can calculate the location of an image point by using a formula relating it to the location of the source point and the focal length, where $s$, $s'$, $f$, and $r$ are the directed distances from the mirror to the source point, image point, focal point, and center of curvature, respectively. The positive direction is toward the source. The sign conventions, which allow the mirror formula to be valid for both concave and convex mirrors, are as follows:

If the object point is on the same side of the mirror as the incident light, the distance $s$ is positive.

If the image point is on the same side of the mirror as the reflected light, the distance $s'$ from the mirror to the image point is positive, otherwise it is negative.

If the center of curvature of the mirror is on the same side of the mirror as the reflected light, the mirror's radius of curvature $r$ is positive, otherwise it is negative. The mirror equation is established by applying the law of reflection to paraxial rays.

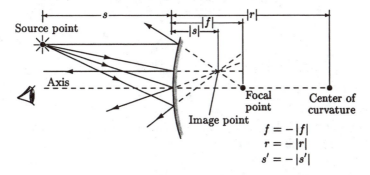

$$f = -|f|$$
$$r = -|r|$$
$$s' = -|s'|$$

Figure 32-4

The mirror equation is

$$\frac{1}{s}+\frac{1}{s'}=\frac{1}{f} \quad \text{where} \quad f=\frac{r}{2}$$     Mirror equation

If all we need is the approximate location of an image, we can locate it quickly by drawing a diagram (called a **ray diagram**) of the system and the principal rays. The *principal rays* are easy to construct. Figures 32-5 and 32-6 are ray diagrams that show image formation, using the principal rays for a concave and a convex mirror, respectively. The principal rays from the source point are:

(1) The *parallel* ray, incident parallel to the axis of the system, is reflected through the focal point.

(2) The *focal* ray, incident on the mirror on a line through the focal point, is reflected parallel to the axis.

(3) The *radial* ray, incident on the mirror on a line through the center of curvature, is reflected back on itself.

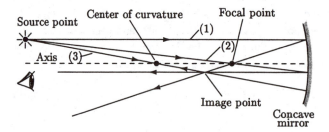

Figure 32-5

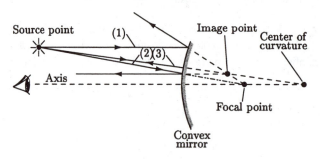

Figure 32-6

In each figure, an eye has been drawn at a position from which you could view the image. To see an object or image, the light entering your eye must be light that is diverging from points on an object or image.

Each point on an extended object is a point source of light. The image of such an object is the set of image points formed by the reflection of light from the object. For an extended object, the size of the image is usually not the same as the size of the object, and the ratio of the image size to the object size is called the **lateral magnification** of the image. This magnification is defined to be positive for an erect image and negative for an inverted image. The formula for the lateral magnification $m$ is

$$m = \frac{y'}{y} = \frac{-s'}{s}$$

Lateral magnification

where $y$ and $y'$ are the distances of the object and the image points above the axis, and $s$ and $s'$ are distances from the mirror to the object (the object distance) and from the mirror to the image (the image distance) as shown in Figure 32-7.

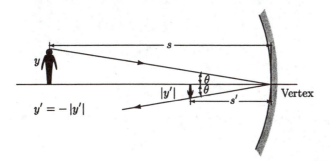

Figure 32-7

### Section 32-2. Lenses.

**Images Formed by Refraction.** Refraction of light at the boundary between transparent media can also form images. At a spherical boundary, paraxial rays from a point source, upon refraction, either converge to or diverge from a point image. The sign convention for the radius of curvature of a spherical boundary between transparent media is that the radius is positive if the center of curvature is on the same side of the boundary as the refracted light, and negative otherwise. The equation relating the location of the object and the location of the image in Figure 32-8 can be derived using the law of refraction, and is

$$\frac{n_1}{s} + \frac{n_2}{s'} = \frac{n_2 - n_1}{r}$$

Spherical boundary equation

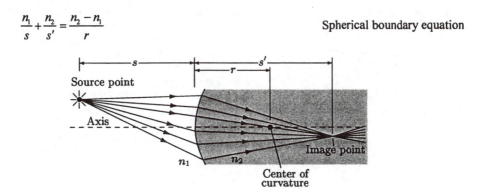

Figure 32-8

If, after refraction, the rays from a source point continue to diverge as in Figure 32-9, there is a virtual image point on the same side of the boundary as the incident light. For such images the spherical boundary equation gives a negative value for $s'$. The sign convention for $s'$ is that if the location of the image point is on the same side of the boundary as the refracted light, $s'$ is positive; otherwise it is negative.

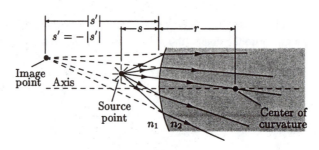

Figure 32-9

The lateral magnification $m$ of a spherical boundary is

$$m = \frac{y'}{y} = \frac{-n_1 s'}{n_2 s}$$    Lateral magnification

***Thin Lenses.*** Many applications for imaging by refraction involve thin lenses. A lens is a piece of transparent material with spherical surfaces on either side. The focal length of a thin lens depends on the two radii of curvature and the refractive index of the lens material. The focal point of a lens is the location at which an incident beam that is parallel with the axis converges to or diverges from. Because light can enter a lens from either side, a lens has two focal points, one on each side. For a thin lens these focal points are equidistant from the lens.

A lens that is thicker at its middle than at its edges converges light ( $f > 0$ ), and a lens that is thicker at its edges than at its middle diverges light ( $f < 0$ ). The location of an image point can be calculated using a formula relating it to the location of the source point and the focal length. This formula, which is identical with the mirror equation, is

$$\frac{1}{s} + \frac{1}{s'} = \frac{1}{f}$$    Thin-lens equation

where the focal length $f$ is related to the index of refraction $n$ of the lens material and to the radii of curvature $r_1$ and $r_2$ of the front and back surfaces of the lens by the equation

$$\frac{1}{f} = (n-1)\left(\frac{1}{r_1} - \frac{1}{r_2}\right)$$    Lens-maker's equation

As for mirrors, images formed by lenses can be located by drawing ray diagrams; but here the principal rays are (1) the *parallel* ray, (2) the *focal* ray, and (3) the undeflected *central* ray through the center of the lens. These are illustrated for thin lenses in Figures 32-10 and 32-11.  Note that although there are two refracting surfaces, for the purposes of ray diagrams, we assume there is only one refracting surface, located in the center of the lens.

Practical optical instruments often consist of a sequence of several optical elements (thin lenses, refracting surfaces, and mirrors). Such a system is analyzed by taking the image formed by each element as the object that is imaged by the next element. In such an analysis, converging light is sometimes incident on an element. The location of the object (a virtual object) associated with this converging light is on the opposite side of the element from the incident light. The object distance *s* associated with the location of a virtual object is negative.

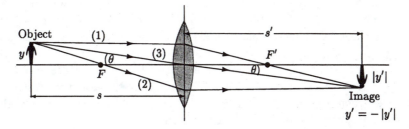

Figure 32-10

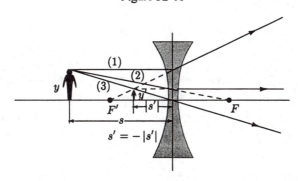

Figure 32-11

The same equation applies to lens or mirror, positive or negative lens, concave or convex mirror, real or virtual image, *provided* we follow consistent sign conventions. The following sign conventions apply to all three cases:

(1) The object distance *s* is positive for a real object (an object from which the incident light rays diverge), and is negative for a virtual object (an object toward which the incident light rays converge).

(2) The image distance *s'* is positive and the image is real if the image is on the same side of the element as the reflected or refracted light.

(3) The focal length is calculated in terms of the radius of curvature of the reflecting or refracting surface(s). The radius of curvature of a reflecting or refracting surface is positive if the center of curvature is on the same side of the surface as the reflected or refracted light, and negative otherwise. The focal length of a mirror is one-half the radius of curvature, whereas the focal length of a thin lens is related to the radii of curvature and the index of refraction by the lens-maker's formula.

The refracting **power of a lens** is the inverse of its focal length; if the focal length is in meters, the power is given in reciprocal meters, called **diopters** (D).

$$P = \frac{1}{f}$$      Power of a lens

***Section 32-3. Aberrations.*** Ideally, all rays from a point source that are incident on an imaging element, such as a mirror, lens, or boundary between transparent media, focus at a single image point. For a spherical imaging element this happens (to a close approximation) for paraxial rays but fails for nonparaxial rays. If all the rays from a point source do not focus at a single point, a blurring of the image, called **aberration,** results. The blurring of the image that results when the object is on the axis is called **spherical aberration.** Additional aberrations called *coma* and *astigmatism* result from imaging the light from off-axis point sources. These aberrations result from the law of refraction and the geometry of spherical surfaces. Because of the dependence of the index of refraction on wavelength, images made from light with more than one wavelength are smeared; this type of distortion is called **chromatic aberration.**

***Section 32-4. Optical Instruments.***

**The Eye.** The eye is our most familiar and important optical instrument. The pupil is a variable aperture that controls the amount of light admitted to the eye. Light coming into the eye is focused primarily by the front surface (the cornea), and to a lesser extent by the lens, onto the photosensitive real surface (the retina), which is about 2.5 cm behind the cornea. The eye has a focusing power of about 60 diopters (D), with the cornea having 43 D and the lens 17 D. Slight changes in focal length (accommodation) are made by muscles that change the shape of the lens, in order to image objects at different distances. In providing accommodation, the focusing power of the lens varies by about 6 D.

**The Simple Magnifier.** The apparent size of an object seen by the eye is determined by the size of the image on the retina; this, in turn, is determined by the *angle* the object subtends at the eye. For the unaided eye, the greatest apparent (angular) size, and thus the most distinct vision, occurs when the object is at the *near point,* the closest distance from the eye for clear vision. The distance from the eye to the near point varies from eye to eye, but the standard value for this distance is 25 cm (about 10 in). When a converging lens is placed closer to the object than the focal length, a virtual image of the object is formed by the refracted light. Looking into the lens, the viewer sees this image, which has a greater angular size than the object when viewed at the near point of the unaided eye. A lens used in this manner is called a simple magnifier or magnifying glass. The angular magnification $M$ of a simple magnifier is the ratio of the angular size $\theta$ of the virtual image to the angular size $\theta_0$ of the object when viewed by the unaided eye at the near point. When the eye is very close to the lens and when the image is at infinity,

$$M = \frac{\theta}{\theta_0} = \frac{x_{np}}{f} \qquad \text{Angular magnification of simple magnifier}$$

where $f$ is the focal length of the lens and $x_{np}$ is the distance of the near point to the eye. In microscopes and telescopes, the lens that you look into (the eyepiece or ocular) is a simple magnifier that is used to provide angular magnification of a real image formed by another lens.

**The Compound Microscope.** There is a limit to the angular magnification that a simple magnifier can provide because a large angular magnification means a short focal length, and (in accordance with the lens-maker's formula) producing a short focal length requires a material with a high index of refraction and surfaces with short radii of curvature. Transparent materials with indexes of refraction higher than two or so are not available, and surfaces with short radii of curvature require physically small lenses. Unfortunately, small lenses have low light-gathering power and thus form dim images. So we use a compound microscope when a simple magnifier is unable to provide sufficient angular magnification.

In a compound microscope, a high-power (short focal length) converging lens (the objective) makes a large real image of a small object. This image is the object for a second converging lens (the eyepiece), which is used as a simple magnifier to provide angular magnification. The net angular magnification (magnifying power) of a compound microscope is

$$M = m_o M_e = -\frac{L}{f_o}\frac{x_{np}}{f_e} \qquad \text{Magnifying power of a compound microscope}$$

where $m_o$ is the lateral magnification associated with the objective lens, $M_e$ is the angular magnification associated with the eyepiece, $f_o$ and $f_e$ are the focal lengths of the objective and eyepiece, respectively, and the tube length $L$ is the distance between the objective and the eyepiece.

*The Telescope.* We use a telescope to examine objects that are too distant to be viewed with a simple magnifier. In a telescope, a converging lens with a long focal length forms a real image of the object near its focal point. This image is the object for the eyepiece, which is used as a simple magnifier to provide angular magnification. The net angular magnification (magnifying power) of a telescope is

$$M = \frac{\theta_e}{\theta_o} = -\frac{f_o}{f_e} \qquad \text{Magnifying power of a telescope}$$

where $\theta_o$ is the angle subtended by the object at the unaided eye and $\theta_e$ is the angle subtended by the virtual image (located at infinity) at the eye when looking through the eyepiece. A major purpose of astronomical telescopes is to gather light. Thus, an objective lens with a large diameter is desirable. Terrestrial telescopes are designed somewhat differently because an upright final image is important.

## II.  Physical Quantities and Key Equations

### Physical Quantities

1 diopter (D) $= 1 \text{ m}^{-1}$

### Key Equations

*Mirror equation*
$$\frac{1}{s} + \frac{1}{s'} = \frac{1}{f} \quad f = \frac{r}{2}$$

*Spherical boundary equation*
$$\frac{n_1}{s} + \frac{n_2}{s'} = \frac{n_2 - n_1}{r}$$

| | |
|---|---|
| *Thin-lens equation* | $\dfrac{1}{s} + \dfrac{1}{s'} = \dfrac{1}{f}$ |
| *Lens-maker's equation* | $\dfrac{1}{f} = (n-1)\left(\dfrac{1}{r_1} - \dfrac{1}{r_2}\right)$ |
| *Lateral magnification of thin lenses and mirrors* | $m = \dfrac{y'}{y} = \dfrac{-s'}{s}$ |
| *Lateral magnification of spherical boundaries* | $m = \dfrac{y'}{y} = \dfrac{-n_1 s'}{n_2 s}$ |
| *Power of a lens* | $P = \dfrac{1}{f}$ |
| *Magnifying power of a simple magnifier* | $M = \dfrac{\theta}{\theta_\circ} = \dfrac{x_{np}}{f}$ |
| *Magnifying power of a compound microscope* | $M = \dfrac{\theta_o}{\theta_e} = m_o M_e = -\dfrac{L}{f_o}\dfrac{x_{np}}{f_e}$ |
| *Magnifying power of a telescope* | $M = \dfrac{\theta_e}{\theta_o} = m_o M_e = -\dfrac{f_o}{f_e}$ |

## III. Potential Pitfalls

The formulas for calculating image positions for mirrors and lenses must be used with proper signs for all quantities, or the answers will be nonsense. Always keep these *sign conventions* in mind:

An object is real, and the object distance is positive, if the object is on the same side of the lens or mirror as the light was traveling *before* it was reflected or refracted; otherwise the object is virtual, and the object distance is negative.

An image is real, and the image distance is positive, if the image forms on the same side of the lens or mirror as the light is traveling *after* it is reflected or refracted; otherwise the image is virtual, and the image distance is negative.

The radius of curvature of a reflecting or refracting surface is positive if the center of curvature is on the same side of the surface as the light is traveling *after* it is reflected or refracted; otherwise it is negative.

By convention, we draw ray diagrams with the light proceeding to the right from a source at the left. This is *not* what determines signs. The image distance from a lens is negative if the image is on the side *not traversed by refracted light* (that is, the same side as the source, and thus is virtual)—not because the image is to the left of the lens.

A negative lateral magnification corresponds to an inverted image.

The radius of curvature of a plane surface is infinite.

The refracting power of a lens is the *inverse* of its focal length; a stronger lens has a smaller, not a larger, focal length.

Don't think that the lens does most of the eye's focusing. The front surface of the eye (the cornea) does most of the focusing. The lens is the apparatus that changes the focus of the cornea-lens system, enabling us to focus on objects over a wide range of distances from the eye.

Don't confuse angular magnification with lateral magnification. Many optical instruments, such as binoculars, microscopes, and simple magnifiers, form a virtual image which is viewed by the eye. For these instruments, the parameter of interest is the angular magnification (magnifying power). The larger the angular magnification of the instrument, the larger the size of the real image formed on the retina. The size of this real image is proportional not to the lateral size of the virtual image but to its angular size.

Don't confuse telescope objectives with microscope objectives. The lateral magnification of a telescope objective is greater for an objective with longer focal length; the longer the focal length of the objective the larger the size of the real image it produces. The lateral magnification of a microscope objective is greater for an objective with shorter focal length; the shorter the focal length of the objective the larger the size of the real image it produces at the focal plane of the eyepiece.

## IV. True or False Questions and Responses

**True or False**

_____ 1.   The image formed by a plane mirror is the same size as the object.

_____ 2.   A real image is formed where light rays actually converge to a point.

_____ 3.   A virtual image cannot be seen but must be projected on a screen.

_____ 4.   The image of a real object formed by a convex mirror is always virtual.

_____ 5.   The image formed by a concave mirror is always real.

_____ 6.   Paraxial rays are always parallel to the axis of a mirror or lens.

_____ 7.   The focal point of an optical element is the point at which light from an infinitely distant source on the symmetry axis is brought to a focus.

_____ 8.   The focal length of a spherical mirror is twice its radius of curvature.

_____ 9.   Only real images can be located by drawing ray diagrams.

_____ 10.   The eye does not distinguish real from virtual images.

____ 11. A negative image distance necessarily corresponds to a virtual image.

____ 12. Refraction at a plane boundary between transparent media does not form images.

____ 13. A thin lens is one whose refractive index is not much different from 1.

____ 14. Light rays from a point source located at a focal point that are incident on a thin lens will pass, after refraction, through the other focal point.

____ 15. When the upper half of a converging lens is missing, a real image formed by the lens will lack its lower half.

____ 16. The lens-maker's formula can be used to calculate the focal length of a thin lens in terms of its physical properties.

____ 17. A converging lens has a negative focal length.

____ 18. A perfectly spherical reflecting surface is free of spherical aberrations.

____ 19. A perfectly spherical reflecting surface is free of chromatic aberrations.

____ 20. Focusing by the eye is accomplished primarily by the lens.

____ 21. Accommodation in the eye is accomplished by the lens.

____ 22. The near point of the nearsighted eye is necessarily closer to the eye than is normal.

____ 23. A simple magnifier is a converging lens of short focal length.

____ 24. The image formed by a lens used as a simple magnifier is virtual.

____ 25. The objective lens of a compound microscope is usually a converging lens of long focal length.

____ 26. The objective lens of an astronomical telescope is usually a diverging lens of long focal length.

**Responses to True or False**

1. True.

2. True.

3. False. Only a real image can be projected on a screen. In the case of a virtual image, the light you see only appears to diverge from the image.

4. True.

5. False. The light from an object that is *outside* the focal point of a concave mirror reflects to form a real image. Light from an object that is inside the focal point diverges upon reflection

off the mirror.

6. False. They are rays that remain near enough to the axis (but not necessarily parallel to it), so we can use small-angle approximations to analyze their reflection and refraction.

7. True.

8. False. It is *half* the radius of curvature.

9. False. Both real and virtual images can be located this way.

10. True. Whether that light actually diverges from the image point, or only appears to do so, cannot be discerned by the viewer.

11. True.

12. False. If it did not form images, you could not "see" fish under water when looking at them from above the surface of the water.

13. False. A thin lens is one whose thickness is small compared with the radii of curvature of its surfaces.

14. False. Any ray incident on a thin lens from a focal point is refracted parallel to the axis.

15. False. An entire image is formed by any portion of the lens. The greater the area of the portion, the brighter the image.

16. True. The lens-maker's equation relates the focal length to the radii of curvature of the lens's two spherical surfaces and to the index of refraction of the material the lens is made of.

17. False. The focal length of a converging lens is positive.

18. False. Spherical aberration is the failure of point-to-point focusing that exists *because* the surface is spherical.

19. True. The angle of reflection equals the angle of incidence for all wavelengths (colors). Thus chromatic aberration does not occur in reflective optical elements (mirrors).

20. False. Most of the focusing takes place at the front surface of the eye (the cornea), and the lens does the fine tuning.

21. True.

22. False. A nearsighted eye is one in which the relaxed lens focuses a nearby object. The eye that is neither nearsighted nor farsighted focuses on very distant objects when relaxed. "Too near" a near point is no defect as long as the eye can accommodate to infinite distance. However, a nearsighted eye is usually associated with a short near-point distance.

23. True. The focal length must be shorter than the near-point distance of the user's eye or the angular magnification will be less than 1.

24. True.

25. False. It usually is a converging lens of very short focal length.

26. False. It usually is a converging lens of long focal length.

# V.  Questions and Answers

### Questions

1.  Objects under water, seen from above the water's surface, appear to be at less than their actual depth. Why?

2.  A plane mirror seems to invert your image left and right but not up and down. Why is this?

3.  What is the cause of chromatic aberration?

4.  Figure 32-12 contains cross-sectional sketches of several thin lenses. Which of these are converging lenses?

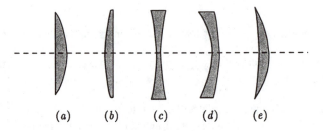

$(a)$      $(b)$      $(c)$      $(d)$      $(e)$

Figure 32-12

5.  What is the focal length of a plane mirror?

6.  Is it possible for a convex mirror to produce a real image?

7.  Is it possible that a positive lens in air can become a negative lens when immersed in a fluid such as water? If so, under what circumstances?

8.  For a certain lens, both radii of curvature are positive. Is it a converging lens or a diverging lens? Or do you need additional information to tell?

9.  What is the minimum height that a plane mirror must have for you to be able to see your whole body reflected in it?

10. Under what circumstances do an object and its image (formed by a spherical mirror) coincide? Is the image real or virtual under these circumstances? Is it erect or inverted?

11. Answer question 10 for a thin lens.

12. A friend's near point is 70 cm from her eyes. Is she nearsighted or farsighted? Does she need a converging or a diverging lens to correct this defect?

13. Why do some people wear *bifocals*?

14. Eyeglasses of power +1.25 D are prescribed for a certain person. Is he nearsighted or farsighted?

15. As we have described the astronomical telescope, the final image it forms is virtual. How then is it possible to take telescopic (astronomical) photographs by allowing the image to fall on photographic film?

16. The objectives of the largest astronomical telescopes are always mirrors, not lenses. Why?

17. The simple magnifier is illustrated in Figure 32-13, which is a reproduction of Figure 32-51 on page 1067 of the text. Where is the virtual image that is seen by the eye?

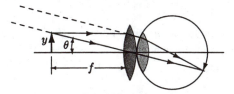

Figure 32-13

**Answers**

1. The index of refraction of the water is greater than that of air, so light leaving the water is refracted away from the normal as shown in Figure 32-14. As a result, the rays appear to emanate from a shallower depth.

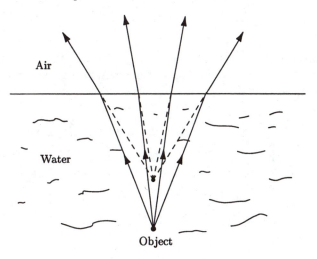

Figure 32-14

2. Actually, a plane mirror does neither, but inverts objects *back* to *front* as illustrated in Figure 32-15. If the mirror inverted right and left, then the object's right hand that points East would appear on the image as a right hand pointing toward the West. Because the image is inverted back to front, the object who faces North is transformed into an Image that faces South. Also, the object's right hand is transformed into a left hand in the Image.

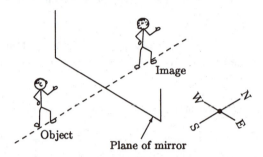

Figure 32-15

3. Chromatic aberration is due to the dispersion of the light by the lens material. The speed of light in any transparent material is a function of the wavelength (color). Thus, for a given lens, the focal length for red light is slightly different than it is for blue light.

4. A converging lens is thicker in the middle than it is at the edges. The converging lenses here are $(a)$, $(b)$, and $(e)$.

5. A plane surface has an infinite radius of curvature, so the focal length ($f = r/2$) is also infinite.

6. Yes. Using the mirror equation to relate the object distance $s$ to the image distance $s'$ and the focal length $f$, we have

$$\frac{1}{s} + \frac{1}{s'} = \frac{1}{f} \quad \text{or} \quad \frac{1}{s} = \frac{1}{f} - \frac{1}{s'}$$

Because a convex mirror has a negative focal length and because a real image has a positive image distance $s'$, the mirror equation implies that $s$ must be negative. That is, a convex mirror can produce a real image if the *object* is virtual. This case is illustrated in Figure 32-16.

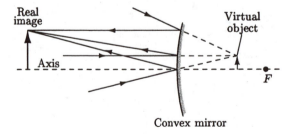

Figure 32-16

7. The focal length of a lens is proportional to the *difference* in index of refraction between the lens material and its surroundings. Thus, if the lens material has a refractive index greater than that of air but less than that of the fluid, its focal length would change sign when you immerse it.

8. Additional information is needed. In accordance with the lens-maker's equation, if the radius of curvature of the front surface is the larger, it is a diverging lens; if the radius of curvature of the front surface is the smaller, it is a converging lens.

9. The mirror has to be half your height as illustrated in Figure 32-17. Perhaps surprisingly, the answer does not depend on how far from the mirror you stand.

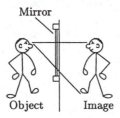

Figure 32-17

10. For a spherical mirror, object and image coincide if $s' = s$. In accordance with the mirror equation,

$$\frac{1}{s} + \frac{1}{s'} = \frac{1}{f} \quad \text{where} \quad f = \frac{r}{2}$$

if $s = s'$, then $s = s' = r$. The image is real because $s'$ is positive and is inverted because $m = -s'/s = -1$ is negative. This image cannot be seen because the reflected light is blocked by the object. If the object is moved off-axis, keeping $s = r$, then the image, also a distance $r$ from the mirror, would be on the opposite side of the axis from the object. Such an image could be viewed either by placing a screen (such as a piece of white cardboard) at the location of the image, or by looking into the mirror from a distance.

11. For a thin lens, the object and image coincide if $s' = -s$. In accordance with thin lens equation

$$\frac{1}{s} + \frac{1}{s'} = \frac{1}{f}$$

if $s' = -s$, then $1/f = 0$. In accordance with the lens-maker's equation

$$\frac{1}{f} = (n-1)\left(\frac{1}{r_1} - \frac{1}{r_2}\right)$$

if $1/f = 0$, then $r_1 = r_2$. The image is virtual because $s'$ is negative and is erect because $m = -s'/s = +1$ is positive. To view this image, you must look back through the lens toward the object. A lens of this type is usually called a window. Several applications have been developed for the device.

12. She can't see objects that are very near; she has to hold a book at arm's length to read it; she is *farsighted.* Her (uncorrected) eye brings light from nearby objects to a focus behind her retina. To correct this, she uses a converging lens in front of her eyes to compensate for the insufficient focusing power of her cornea and lens.

13. The aging eye loses its ability to *accommodate*—that is, the eye's lens is not flexible enough to change the focal length of the cornea–lens system by very much. Bifocals compensate for the eye's poor accommodation by giving it the choice of two corrective lenses of slightly different focal length. (Some of us decrepit types need trifocals!)

14. He is farsighted; his eyes need the help of a converging lens to bring nearby objects to a focus on the retina. Light from nearby objects diverges more strongly at the eye than does light from distant objects.

15. The film is placed at the position of the real image formed by the objective lens (or mirror), and an eyepiece is not used.

16. A mirror does not exhibit chromatic aberration, and it can be ground to achieve a paraboloidal surface, which eliminates spherical aberration. However, the most important advantage is mechanical rather than optical. Because only one surface of a mirror matters, it can be supported from its large back. A lens can be supported only at its edges; so large, heavy lenses sag, losing their shape.

17. The image seen by the eye is a virtual image located an infinite distance to the left of Figure 32-13. The light rays shown are refracted by the magnifier so that they appear to come from the image of the tip of the arrow. The images seen by the eyes in Figure 32-52 and 32-53 of the textbook are also at an infinite distance to the left of the figures.

## VI. Problems

**Example #1.** An object is 25.0 cm in front of a convex spherical mirror. The image of the object in the mirror is exactly half the size of the object. What is the radius of curvature of the mirror?

**Picture the Problem.** The lateral magnification of the object can be used to find the image distance. Then the mirror equation can be used for the radius of curvature.

| 1. Use the expression for the lateral magnification to find the magnitude of the image distance. We are only given the magnitude of the magnification, so we need to remember or value could be positive or negative. | $$m = \frac{-s'}{s}$$ $$\pm 0.5 = \frac{s'}{s}$$ $$s' = \pm 0.5s$$ |
|---|---|

| 2. Use the mirror equation to find the radius of curvature of the mirror. First try $s' = 0.5s$. Here $r$ ends up being positive, but the radius of curvature of a convex mirror should be negative. | $\dfrac{1}{s} + \dfrac{1}{s'} = \dfrac{1}{f} = \dfrac{2}{r}$ <br><br> $\dfrac{1}{s} + \dfrac{2}{s} = \dfrac{2}{r}$ <br><br> $r = 2s/3 = 2(25\,\text{cm})/3 = 16.7\,\text{cm}$ |
|---|---|
| 3. Try $s' = -0.5s$ and see if we get a negative radius of curvature. | $\dfrac{1}{s} + \dfrac{1}{s'} = \dfrac{1}{f} = \dfrac{2}{r}$ <br><br> $\dfrac{1}{s} - \dfrac{2}{s} = \dfrac{2}{r}$ <br><br> $r = -2s = -2(25\,\text{cm}) = -50\,\text{cm}$ |

**Example #2—Interactive.** The opposite walls of a barber shop are covered by plane mirrors, so that multiple images arise from multiple reflections, and you see many reflected images of yourself, receding to infinity. The width of the shop is 6.50 m, and you are standing 2.00 m from the north wall. How far apart are the first two images of you behind the north wall?

**Picture the Problem.** The closest image will be an image of you, a real object. The second closest image will be the image of the first image formed in the south mirror. Remember plane mirrors have a magnification of exactly one. **Try it yourself.** Work the problem on your own, in the spaces provided, to get the final answer.

| 1. Draw a sketch showing the locations of the images. | |
|---|---|
| 2. Determine the distance behind the north wall the first image of you will appear. | |
| 3. Determine the distance behind the south mirror the first image of you will appear. | |
| 4. Determine the distance behind the north mirror the image of the image found in step 3 will be behind the north mirror. | |
| 5. Determine the difference in the distance between the image of step 1 and the image of step 4. | $d = 9.00\,\text{m}$ |

**Example #3.**  An object is 40.0 cm from a concave spherical mirror whose radius of curvature is 32.0 cm. Locate and describe the image formed by the mirror (*a*) by calculating the image distance and lateral magnification and (*b*) by drawing a ray diagram.  On the ray diagram put an eye in a position where it can view the image.

**Picture the Problem.**  Use the appropriate expressions for the calculations, and use the three principle rays for the ray diagram.  Use a ruler to draw the ray diagram as close to scale as possible.

| | |
|---|---|
| 1.  Do the ray diagram first.  It can be used as a sketch of the situation for the calculations.<br><br>Place an arrow, to scale, 40 cm from the mirror.  Locate the center of curvature and the focal point of the mirror. | 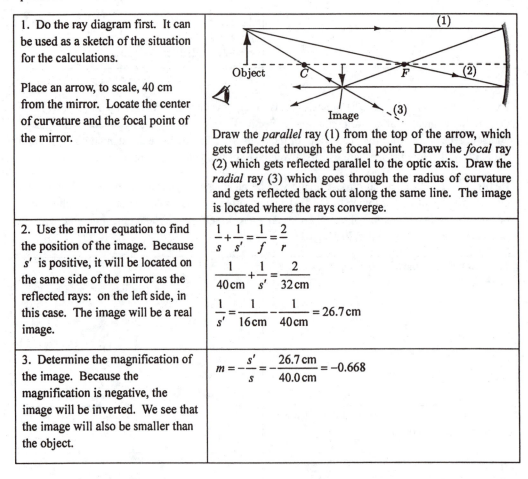<br>Draw the *parallel* ray (1) from the top of the arrow, which gets reflected through the focal point.  Draw the *focal* ray (2) which gets reflected parallel to the optic axis.  Draw the *radial* ray (3) which goes through the radius of curvature and gets reflected back out along the same line.  The image is located where the rays converge. |
| 2.  Use the mirror equation to find the position of the image.  Because $s'$ is positive, it will be located on the same side of the mirror as the reflected rays: on the left side, in this case.  The image will be a real image. | $$\frac{1}{s}+\frac{1}{s'}=\frac{1}{f}=\frac{2}{r}$$ $$\frac{1}{40\,\text{cm}}+\frac{1}{s'}=\frac{2}{32\,\text{cm}}$$ $$\frac{1}{s'}=\frac{1}{16\,\text{cm}}-\frac{1}{40\,\text{cm}}=26.7\,\text{cm}$$ |
| 3.  Determine the magnification of the image.  Because the magnification is negative, the image will be inverted.  We see that the image will also be smaller than the object. | $$m=-\frac{s'}{s}=-\frac{26.7\,\text{cm}}{40.0\,\text{cm}}=-0.668$$ |

**Example #4—Interactive.**  A concave spherical mirror has a radius of curvature of 60.0 cm. Where should an object be placed if its image is to be virtual and three times the size of the object?

**Picture the Problem.**  Because the magnification is three, the image is three times farther from the mirror than the object. If the image is virtual, then $s'$ will be negative. Since the object is real, $s$ will be positive, making $m$ also positive. Use the mirror and magnification expressions to solve for $s$.  **Try it yourself.**  Work the problem on your own, in the spaces provided, to get the final answer.

| 1. Use the expression for the lateral magnification to solve for $s'$ in terms of $s$. | |
| --- | --- |
| 2. Substitute this value for $s'$ into the mirror equation, and solve for the object distance $s$. | |
| | $s = 20.0\,\text{cm}$ (in front of the mirror) |

**Example #5.** A fish is 12.0 cm from the front surface of a spherical fish bowl that has a radius of 20.0 cm. A woman who is 1.00 m outside the surface of the bowl is looking at the fish. (*a*) Where does the woman see the fish? (*b*) Where does the fish see the woman? The index of refraction of water is 1.33.

**Picture the Problem.** Ignore the effects of the very thin glass that makes up the bowl. Thus, the "lens" here is simply the spherical surface of the air-water interface. Use the expression for image and object distances for a single refracting surface to answer both parts of the problem.

| 1. Make a diagram of the situation, drawing the fish, the fishbowl, and the principal rays from the fish. | 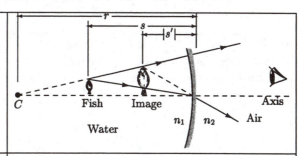 |
| --- | --- |
| 2. Apply the expression for image formation with a single refracting surface. For the case of the woman looking at the fish, the water is medium one, and the air is medium two. The fish is the object. Solve for $s'$. Following the sign convention, the radius of curvature of the bowl is negative. The image is at a negative distance, so the woman sees the fish a distance of 10.6 cm inside the bowl. | $\dfrac{n_1}{s} + \dfrac{n_2}{s'} = \dfrac{n_2 - n_1}{r}$ <br><br> $\dfrac{n_2}{s'} = \dfrac{n_2 - n_1}{r} - \dfrac{n_1}{s} = \dfrac{1 - 1.33}{-0.2\,\text{m}} - \dfrac{1.33}{0.12\,\text{m}} = -9.43\,\text{m}^{-1}$ <br><br> $s' = \dfrac{n_2}{-9.43\,\text{m}} = \dfrac{1}{-9.43\,\text{m}} = -10.6\,\text{cm}$ |

| 3. Apply the expression for image formation with a single refracting surface again, only this time the fish is looking at the woman. The water is medium two, and the air is medium one. In this case the radius of curvature is positive. Solve for $s'$, the distance of the woman's image from the edge of the bowl. | $\dfrac{n_1}{s} + \dfrac{n_2}{s'} = \dfrac{n_2 - n_1}{r}$ <br><br> $\dfrac{n_2}{s'} = \dfrac{n_2 - n_1}{r} - \dfrac{n_1}{s} = \dfrac{1.33 - 1}{0.2\,\text{m}} - \dfrac{1}{1\,\text{m}} = +0.65\,\text{m}^{-1}$ <br><br> $s' = \dfrac{n_2}{+0.65\,\text{m}} = \dfrac{1.33}{+0.65\,\text{m}} = 2.05\,\text{m}$ <br><br> This image would form 2.05 m from the refracting surface, on the side of the fish. However, for most normal fishbowls with a diameter of about 30 cm, this would be outside the bowl, so that the fish would not see a crisp image of the woman, only a blurred image. |

**Example #6—Interactive.** A fish in an aquarium is 12.0 cm below the top surface of the water, and a cat is looking down on the aquarium from a perch located 1.00 m directly above the fish. Where does the cat see the fish? The index of refraction of water is 1.33.

**Picture the Problem.** Follow the solution to part (*a*) of the previous problem. Remember the radius of curvature of a flat surface is infinite. **Try it yourself.** Work the problem on your own, in the spaces provided, to get the final answer.

| 1. Make a diagram of the situation, drawing the fish, the aquarium, the cat, and some rays from the fish directed toward the cat. | |
|---|---|
| 2. Apply the expression for image formation with a single refracting surface. For the case of the cat looking at the fish, the water is medium one, and the air is medium two. The fish is the object. Solve for $s'$. | $s' = 9.02\,\text{cm}$ below the surface |

**Example #7.** The focal length of a glass lens with refractive index 1.55 is +16.0 cm. If the radius of curvature of its front surface is +12.0 cm, what is the radius of curvature of its rear surface? Are the surfaces of this lens convex, concave, or what?

**Picture the Problem.** Use the lens-maker's formula to find the second radius of curvature.

| 1. Determine the curvature of the first surface of the lens. | A positive radius of curvature means the center of curvature of a surface is on the same side of the surface as the refracted light. This first surface must be convex. |
|---|---|
| 2. Use the lens-maker's equation to find the radius of curvature of the second surface. The center of curvature of the second surface is not on the same side as the refracted light, so this surface must bulge outward. This is a biconvex lens, with radii of curvature of 12.0 cm and 33.0 cm. | $\dfrac{1}{f} = (n-1)\left(\dfrac{1}{r_1} - \dfrac{1}{r_2}\right)$ <br> $\dfrac{1}{r_2} = \dfrac{1}{r_1} - \dfrac{1}{f(n-1)} = \dfrac{1}{12\,\text{cm}} - \dfrac{1}{(16\,\text{cm})(1.55-1)} = -.0303\,\text{cm}^{-1}$ <br> $r_2 = -33.0\,\text{cm}$ |

**Example #8—Interactive.** A thin lens made of glass of refractive index 1.60 has surfaces with radii of curvature of magnitude 12.0 and 18.0 cm. What are the possible values for its focal length?

**Picture the Problem.** Use the lens-maker's formula, adopting different sign combinations for each of the surfaces to produce different focal lengths. **Try it yourself.** Work the problem on your own, in the spaces provided, to get the final answer.

| 1. Find the focal length with two positive radii of curvature. | $f = \pm 60.0\,\text{cm}$ |
|---|---|
| 2. Find the focal length with two negative radii of curvature. | $f = \pm 60.0\,\text{cm}$ |
| 3. Find the focal length if one radius of curvature is positive and the other is negative. | $f = \pm 12\,\text{cm}$ |

**Example #9.** A lens of focal length +15.0 cm is 10.0 cm to the left of a second lens of focal length −15.0 cm. Where is the final image of an object that is 30.0 cm to the left of the positive lens?

**Picture the Problem.** The image produced by the first lens will serve as the object for the second lens. Ignoring the second lens, use the thin-lens equation to find the location of the image produced by the first lens. Using this as the object, find the location of the image produced by the second lens.

| | |
|---|---|
| 1. Draw a sketch of the situation to keep track of the location of the lenses and images. The distances shown will be derived as we step through the problem. | 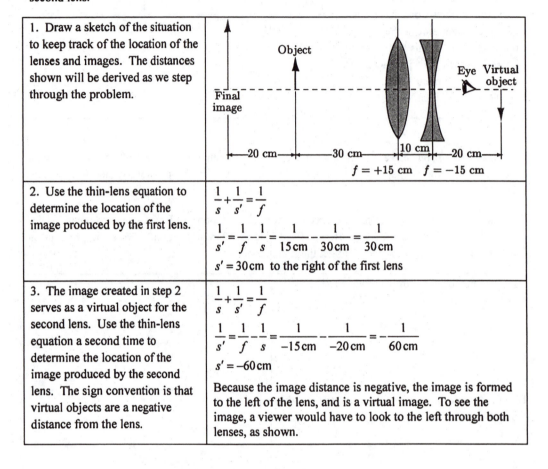 |
| 2. Use the thin-lens equation to determine the location of the image produced by the first lens. | $\dfrac{1}{s}+\dfrac{1}{s'}=\dfrac{1}{f}$ <br><br> $\dfrac{1}{s'}=\dfrac{1}{f}-\dfrac{1}{s}=\dfrac{1}{15\,\text{cm}}-\dfrac{1}{30\,\text{cm}}=\dfrac{1}{30\,\text{cm}}$ <br><br> $s'=30\,\text{cm}$ to the right of the first lens |
| 3. The image created in step 2 serves as a virtual object for the second lens. Use the thin-lens equation a second time to determine the location of the image produced by the second lens. The sign convention is that virtual objects are a negative distance from the lens. | $\dfrac{1}{s}+\dfrac{1}{s'}=\dfrac{1}{f}$ <br><br> $\dfrac{1}{s'}=\dfrac{1}{f}-\dfrac{1}{s}=\dfrac{1}{-15\,\text{cm}}-\dfrac{1}{-20\,\text{cm}}=-\dfrac{1}{60\,\text{cm}}$ <br><br> $s'=-60\,\text{cm}$ <br><br> Because the image distance is negative, the image is formed to the left of the lens, and is a virtual image. To see the image, a viewer would have to look to the left through both lenses, as shown. |

**Example #10—Interactive.** Example #5 involved finding the location of the image of a woman formed by light refracting at the front surface of the bowl. Now find the location of the image of the woman formed by light refracting at the back surface of the bowl.

**Picture the Problem.** From Example #5, we know the location of the real image produced by the front surface of the bowl. This real image serves as a virtual object for the second surface of the bowl. Remember the radius of the spherical fishbowl is 20.0 cm. Think carefully about the signs for $r$ and $s$, as well as what values to use for $n_1$ and $n_2$. **Try it yourself.** Work the problem on your own, in the spaces provided, to get the final answer.

| | |
|---|---|
| 1. Draw a sketch of the situation, showing both surfaces of the bowl, and the location of the virtual object the second surface will image. | |
| 2. Determine the distance of the real image created by the first surface of the bowl from the second surface of the bowl. | |
| 3. Use the spherical boundary formula to obtain the location of the image. | $s = +40.7\,\text{cm}$ beyond the back of the bowl |

**Example #11.** When you place a bright light source 36.0 cm to the left of a lens, you obtain an upright image 14.0 cm from the lens, and also a faint inverted image 13.8 cm to the left of the lens that is due to *reflection* from the front surface of the lens. When the lens is turned around, a faint inverted image is 25.7 cm to the left of the lens. What is the index of refraction of the material?

**Picture the Problem.** Use the thin-lens equation to find the focal length of the lens. Use the reflections to find the radius of curvature of the two surfaces, and the lens-maker's equation to find the index of refraction.

| | |
|---|---|
| 1. We are told the image is upright, so the image distance $s'$ must be negative, which makes it a virtual image. Use the thin-lens equation to find the focal length. | $\dfrac{1}{f} = \dfrac{1}{s} + \dfrac{1}{s'} = \dfrac{1}{36\,\text{cm}} + \dfrac{1}{-14\,\text{cm}} = -0.0437\,\text{cm}^{-1}$ <br> $f = -22.9\,\text{cm}$ |
| 2. Use the mirror equation to determine the radius of curvature of the first surface. | $\dfrac{1}{f} = \dfrac{2}{r_1} = \dfrac{1}{s} + \dfrac{1}{s'} = \dfrac{1}{36\,\text{cm}} + \dfrac{1}{13.8\,\text{cm}} = 0.100\,\text{cm}^{-1}$ <br> $r_1 = 20.0\,\text{cm}$ |
| 3. Repeat for the radius of curvature of the second surface. | $\dfrac{1}{f} = \dfrac{2}{r_2} = \dfrac{1}{s} + \dfrac{1}{s'} = \dfrac{1}{36\,\text{cm}} + \dfrac{1}{25.7\,\text{cm}} = 0.0667\,\text{cm}^{-1}$ <br> $r_2 = 30.0\,\text{cm}$ |

| 4.  Use the lens-maker's equation to find the index of refraction of the lens.  Because both radii of curvature are positive, both surfaces are concave, meaning the lens is thinner in the middle.  The sign conventions for the lens-maker's equations tell us that $r_1$ will be positive, and $r_2$ will be negative. | $\dfrac{1}{f} = (n-1)\left(\dfrac{1}{r_1} - \dfrac{1}{r_2}\right)$ <br><br> $\dfrac{1}{-22.9\,\text{cm}} = (n-1)\left(\dfrac{1}{-20\,\text{cm}} - \dfrac{1}{30\,\text{cm}}\right)$ <br><br> $n = 1.524$ |
| --- | --- |

**Example #12.** The most distant object on which Maria's (uncorrected) eyes focus clearly is about 40.0 cm away, and the nearest is about 14.0 cm away. (*a*) If contact lenses are to correct her vision so that she can focus at infinity when her eye is relaxed, what power is required? (*b*) With corrective lenses of this focal length, what will be her near point?

**Picture the Problem.** The contacts must make images of objects very far away appear to be 40 cm away so the relaxed eye can see them.  For part (*b*), if an image is located 14 cm away, how far away is the object that is imaged by the lens?

| 1.  Use the thin-lens equation to determine the power of the contact lens.  The image distance is negative, because the image must remain upright, and will be on the object side of the lens. | $P = \dfrac{1}{f} = \dfrac{1}{s} + \dfrac{1}{s'} = \dfrac{1}{\infty} + \dfrac{1}{-40\,\text{cm}} = \dfrac{1}{-0.4\,\text{m}} = -2.5\,\text{D}$ |
| --- | --- |
| 2.  Draw a sketch of this situation. | 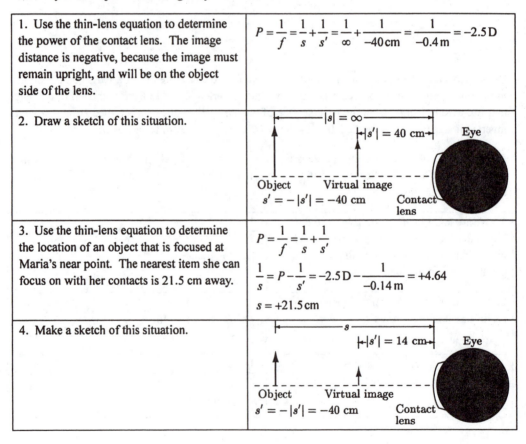 |
| 3.  Use the thin-lens equation to determine the location of an object that is focused at Maria's near point.  The nearest item she can focus on with her contacts is 21.5 cm away. | $P = \dfrac{1}{f} = \dfrac{1}{s} + \dfrac{1}{s'}$ <br><br> $\dfrac{1}{s} = P - \dfrac{1}{s'} = -2.5\,\text{D} - \dfrac{1}{-0.14\,\text{m}} = +4.64$ <br><br> $s = +21.5\,\text{cm}$ |
| 4.  Make a sketch of this situation. | |

**Example #13—Interactive.** A farsighted secretary needs to read from a word-processor screen that is 50.0 cm from his eyes. His uncorrected near point is 110 cm away. What must the power of the lenses in his reading glasses be to form an image of the screen 15.0 cm beyond the near point? Assume that the reading glasses are at a distance of 2.00 cm from the secretary's eye.

**Picture the Problem.** Follow the solution to the previous example. **Try it yourself.** Work the problem on your own, in the spaces provided, to get the final answer.

| | |
|---|---|
| 1. Draw a sketch of the situation, positioning the eye, the glasses, the object, and the image. | |
| 2. Use the thin-lens equation to determine the power of the reading glasses. Remember, $s$ and $s'$ are relative to the lens of the glasses, not the eye. | $P = 1.27\,\mathrm{D}$ |

**Example #14.** A compound microscope has a tube length of 20.0 cm and an objective lens of focal length 8.00 mm. (*a*) If it is to have a magnifying power of 200, what should be the focal length of the eyepiece? (*b*) If the final image is viewed at infinity, how far from the objective should the object be placed?

**Picture the Problem.** The tube length $L$ is the distance between the interior focal points of the two lenses. Use the expressions for a compound microscope. Assume the viewer has the standard near point distance of 25.0 cm.

| | |
|---|---|
| 1. Draw a sketch of the microscope. | <br>Compound microscope |
| 2. Calculate the focal length of the eyepiece from the expression for the magnifying power of a compound microscope. | $M = \left\| -\dfrac{L}{f_o}\dfrac{x_{np}}{f_e} \right\|$<br><br>$f_e = \left\| \dfrac{L}{f_o}\dfrac{x_{np}}{M} \right\| = \dfrac{0.20\,\mathrm{m}}{0.008\,\mathrm{m}}\dfrac{0.25\,\mathrm{m}}{200} = 3.13\,\mathrm{cm}$ |

| 3.  To determine the distance of the object from the objective, realize that the image the objective forms must be positioned at the focal point of the eyepiece if the final image is to be at infinity.  Then use the thin-lens equation to find the actual object location. | $s' = L + f_o$ <br><br> $\dfrac{1}{f_o} = \dfrac{1}{s} + \dfrac{1}{s'}$ <br><br> $\dfrac{1}{s} = \dfrac{1}{f_o} - \dfrac{1}{s'} = \dfrac{1}{f_o} - \dfrac{1}{L + f_o}$ <br><br> $= \dfrac{1}{0.8\,\text{cm}} - \dfrac{1}{(20\,\text{cm}) + (0.8\,\text{cm})} = 1.202\ \text{cm}^{-1}$ <br><br> $s = 0.832\,\text{cm}$ |
| --- | --- |

**Example #15—Interactive.**  A compound microscope has an overall length of 26.0 cm and an overall magnifying power of 40.  If the focal length of the eyepiece is 4.00 cm, what is the focal length of the objective? Assume that the viewer has the standard near point distance of 25.0 cm.

**Picture the Problem.**  The magnifying power depends on the tube length $L$, the near point, and the focal lengths of the lenses.  The tube length is the overall length minus the two focal lengths.  **Try it yourself.**  Work the problem on your own, in the spaces provided, to get the final answer.

| 1.  Determine the tube length in terms of the focal lengths. | |
| --- | --- |
| 2.  Calculate the focal length of the objective from the expression for the magnification of a compound microscope.  Substitute the expression from step one for the tube length. | $f_o = 2.97\,\text{cm}$ |

**Example #16.**  An astronomical telescope with a magnifying power of 20 consists of two converging lenses, the objective and the eyepiece, located 1.00 m apart.  What is the focal length of each lens?

**Picture the Problem.**  In an astronomical telescope the two interior focal points coincide, and the magnification is the ratio of the focal lengths.

| 1. Sketch the telescope. | |
|---|---|
| 2. Relate the separation of the lenses to the focal lengths. | $L = f_o + f_e$ |
| 3. Relate the magnification to the focal lengths. | $M = \dfrac{f_o}{f_e}$ <br><br> $f_o = Mf_e$ |
| 4. Substitute this last expression into that from step 2 to solve for the eyepiece lens focal length. | $L = Mf_e + f_e = f_e(M+1)$ <br><br> $f_e = \dfrac{L}{M+1} = \dfrac{1\,\text{m}}{20+1} = 0.0476\,\text{m}$ |
| 5. Substitute this last result into the magnification expression to find the objective lens focal length. | $f_o = Mf_e = 20(0.0476\,\text{m}) = 0.952\,\text{m}$ |

**Example #17—Interactive.** The objective lens of a large astronomical telescope has a focal length of 14.0 m. It is used to view Jupiter when that planet is at its closest approach to the earth. Under these conditions Jupiter, whose diameter is 143,000 km, is 630,000,000 km away. What is the diameter of the image of Jupiter formed by the objective?

**Picture the Problem.** Use the thin-lens equation to find the image distance. Then compute the magnification. **Try it yourself.** Work the problem on your own, in the spaces provided, to get the final answer.

| 1. Use the thin-lens equation to find the image distance. | |
|---|---|
| 2. Find the magnification of the image. | |
| 3. Use the magnification to determine the diameter of Jupiter in the image. | $d = 3.18\,\text{mm}$ |

# Chapter **33**

# Interference and Diffraction

## I.   Key Ideas

Light is an electromagnetic wave. When two or more light waves superpose (overlap), the electric and the magnetic field vectors add in accordance with the principle of superposition (vector addition). Much of this chapter deals with the distribution of light intensity when the wavefronts are partially blocked and when two or more waves superpose (overlap). The frequency of visible electromagnetic radiation (light) is between $4.3 \times 10^{14}$ Hz and $7.5 \times 10^{14}$ Hz. Neither our eyes nor most other instrument respond rapidly enough to make observations at such frequencies, so we will only discuss observations of light are averaged over many cycles.

*Section 33-1. Phase Difference and Coherence.* When harmonic waves of the same polarization, frequency, and wavelength superpose at a point, the resulting wave, which has the same frequency and wavelength as the original waves, has an amplitude that depends on the phase difference between the component waves. If the component waves are in phase (or equivalently, out of phase by any integer multiple of 360°), they interfere constructively, that is, the amplitude of the resultant wave is the sum of the amplitudes of the component waves. If they are 180° out of phase, they interfere destructively, that is, the amplitude of the resultant wave is the difference of the amplitudes of the component waves.

Suppose light takes times $t_1$ and $t_2$ to travel from a point source to point $P$ along paths 1 and 2, respectively. If the wave source is harmonic with frequency $f$, at $P$ the phase difference $\delta$ between the two wavefronts is $2\pi f (t_2 - t_1)$, and when the wave speed $v$ is the same on both paths, then $t_2 - t_1 = \Delta r / v$, where $\Delta r$ is the difference in the path lengths. This phase difference is

$$\delta = \frac{\Delta r}{\lambda} 2\pi = \frac{\Delta r}{\lambda} 360° \qquad \text{Phase difference due to path difference}$$

where we have eliminated the frequency $f$ using the relation $f\lambda = v$.

Additional phase differences are sometimes produced when waves reflect at a boundary between two transparent media. When light reflects at such a boundary, if the wave is incident from the medium with the higher wave speed (smaller index of refraction), the reflected wave is

180° out of phase with the incident wave.  If the wave is incident from the medium with the lower wave speed (higher index of refraction), there is no phase change on reflection.

If two waves differ in phase by a *constant amount,* they are said to be **coherent.** Coherence in optics is usually achieved by dividing and then recombining light from a single source.

***Section 33-2. Interference in Thin Films.*** We observe interference in the light reflected at the two surfaces of a thin film of transparent material (Figure 33-1). If the phase difference between the two reflected waves, one reflected off the front surface, the other off the back surface, is an even integral multiple of 180°, the interference is constructive and the intensity of the reflected light is a maximum; and if the phase difference is an odd integral multiple of 180°, the interference is destructive and the reflected intensity is a minimum.   Mathematically, this can be expressed by the following general relationships for the phase difference:

$$\delta = 2\pi m \qquad m = 0,1,2,\ldots \quad \text{constructive}$$
$$\delta = 2\pi\left(m+\tfrac{1}{2}\right) \quad m = 0,1,2,\ldots \quad \text{destructive}$$

General interference conditions

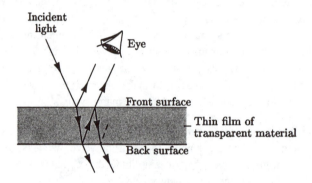

Figure 33-1

One source of phase difference is the different path-lengths two rays of light travel. A path-length difference arises because the light reflected at the back surface travels farther than the light reflected at the front surface. For normal incidence the light reflected at the back surface travels the thickness of the film twice. This additional path length occurs within the film, and the wavelength of the light *in the film* must be used to calculate the associated phase difference.

Additional phase differences can be introduced by phase shifts that occur upon reflection at either surface. If the thickness of the film varies, the phase difference between the two reflected waves varies accordingly. If the thickness varies over several wavelengths, interference fringes (maxima and minima) can be seen in the reflected light. If white light is incident on the film, the interference maxima occur at different thicknesses for different wavelengths, and we see various colors in the reflected light.

For monochromatic light incident nearly normally upon a thin film, we have

*Conditions for thin-film interference with one 180° phase shift due to reflection*

$$\frac{2t}{\lambda'} = \frac{2nt}{\lambda} = m \qquad m = 0, 1, 2, \ldots \quad \text{destructive}$$

$$\frac{2t}{\lambda'} = \frac{2nt}{\lambda} = m + \frac{1}{2} \qquad m = 0, 1, 2, \ldots \quad \text{constructive}$$

*Conditions for thin-film interference with two 180° phase shifts (or no phase shifts at all) due to reflection*

$$\frac{2t}{\lambda'} = \frac{2nt}{\lambda} = m \qquad m = 0, 1, 2, \ldots \quad \text{constructive}$$

$$\frac{2t}{\lambda'} = \frac{2nt}{\lambda} = m + \frac{1}{2} \qquad m = 0, 1, 2, \ldots \quad \text{destructive}$$

where $\lambda'$ is the wavelength in the film, $n$ is the index of refraction of the film, $t$ is the thickness of the film, and $\lambda$ is the vacuum wavelength. The ratio $2t/\lambda'$ is the number of wavelengths in two thicknesses of the film.

***Section 33-3. The Two-Slit Interference Pattern.*** Young's two-slit interference experiment demonstrates conclusively the wave aspect of light. In this experiment two narrow parallel slits $S_1$ and $S_2$ in the opaque screen shown in Figure 33-2 are illuminated by light from a single source. The light passing through each slit spreads out, and the intensity on the screen displays an interference pattern determined by the difference in path lengths from each slit to each point on the screen. The scale of the figure is highly distorted in that both the widths of the slits and the distance $d$ between the slits are actually extremely small compared with the distance $L$ from the slits to the screen. When $\ell_1 = \ell_2$, the slits $S_1$ and $S_2$ can themselves be considered to be coherent sources in phase with one another. The conditions for constructive and destructive interference at points on the screen can be derived as

$$
\begin{array}{lll}
d\sin\theta = m\lambda & m = 0, 1, 2, \ldots & \text{constructive} \\
d\sin\theta = \left(m + \tfrac{1}{2}\right)\lambda & m = 0, 1, 2, \ldots & \text{destructive}
\end{array}
\qquad \text{Two-slit interference}
$$

where $d\sin\theta$ is the path length difference between the waves. Constructive interference occurs at angles where the difference in the path lengths $\Delta r$ from the two slits is an integer number $m$ of wavelengths. The integer $m$ is called the **order of the interference.**

The intensity of any wave is proportional to the square of its amplitude. The intensity at some point in an interference pattern is thus proportional to the square of the net amplitude of the resultant wave produced by the two individual waves. The intensity of any pattern involving the interference of two, in-phase, coherent light sources of equal amplitude can be shown to be

$$I = 4I_0 \cos^2 \tfrac{1}{2}\delta \qquad\qquad \text{Intensity in terms of phase difference}$$

where $\delta$ is the phase difference of the paths of light from the two sources, and $I_0$ is the intensity from either source separately.

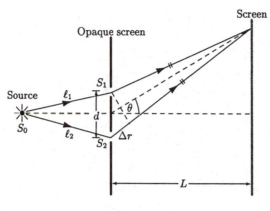

Figure 33-2

If the waves leaving the two sources $S_1$ and $S_2$ differ in phase by 180°, destructive interference occurs at angles where the difference in the path lengths from the two slits is an integral number of wavelengths. That is, if the difference in the phases of the waves as they leave the two slits changes from 0 to 180°, the interference pattern at the screen is shifted by half a fringe compared with the pattern formed when the waves leave $S_1$ and $S_2$ in phase. For a two-slit interference pattern to be observable, the pattern must remain fixed for a time long enough to be observed, which puts a limit on the shortest time for which the difference in the phases of the waves as they leave the two slits must remain constant. This leads to the following definition of coherence: two waves are coherent if at any location the difference in the phases of the two waves remains constant long enough for the interference pattern to be observed. Light waves that are coherent almost always come from a single source as in Figure 33-2. The frequency of visible light waves is $\sim 5 \times 10^{14}$ Hz, and the spread in frequency is $\sim 5 \times 10^5$ Hz, even for light produced by a "monochromatic" laser. Thus, the phase difference between the waves as they leave two independent optical sources rarely remains fixed long enough for an interference pattern to be observed. Radiofrequencies ($\sim 10^8$ Hz) are much lower than optical frequencies so coherent radiofrequency electromagnetic waves can be produced by separate, independent sources.

***Section 33-4.  Diffraction Pattern of a Single Slit.***  According to Huygens' principle, light passing through a single slit can be though of as light emitted from several coherent point sources, rather than light from a single source.  As a result, light traveling through a small slit can actually interfere with itself.  This "self-interference" is called **diffraction.**

Whenever part of a wavefront is limited or blocked by an obstacle, the part or parts of the wavefront that are not blocked spread out in the region beyond the obstacle. Because of this diffraction, light passing through a single slit in an opaque barrier spreads outside the geometrical shadow of the aperture, as shown in Figure 33-3. On the figure, only the rays diffracted at a specified angle $\theta$ are shown. The spreading angle is approximately equal to the ratio of the wavelength and the slit width. The larger this ratio (the smaller the slit), the greater the spreading out (diffraction). The intensity pattern far away from the slit, in the limit of Fraunhofer diffraction, can be derived from Huygens' principle. Diffraction patterns near the aperture, or Fresnel diffraction, which are more difficult to calculate, are not discussed in this study guide. For Fraunhofer diffraction from a slit of width a,

$$a \sin \theta = m\lambda \qquad m = 1, 2, 3, \ldots \qquad \text{(destructive)} \qquad \qquad \text{Single-slit diffraction minima}$$

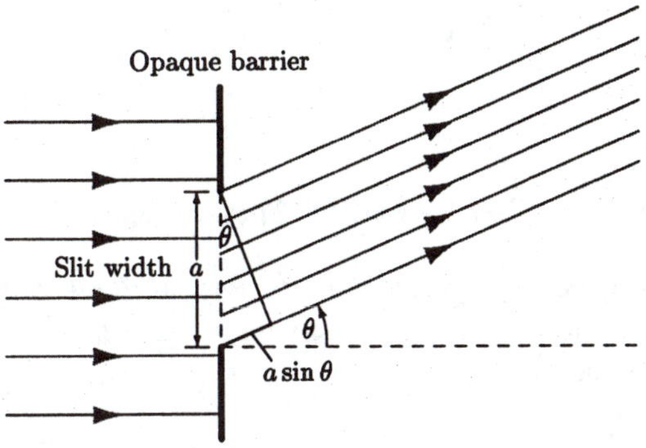

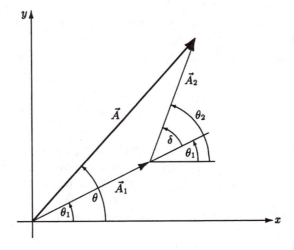

Figure 33-3

*Interference–Diffraction Pattern of Two Slits.* The actual pattern produced in the two-slit experiment is a combination of the two-slit interference pattern and the diffraction pattern of the light passing through each slit.

*Section 33-5. Using Phasors to Add Harmonic Waves.* The superposition of two harmonic waves of the same frequency and wavelength at a point results in a harmonic wave of the same frequency and wavelength. The resultant wave can be found by a geometric construction that is equivalent to ordinary vector addition, as shown in Figure 33-4. Each component wave is represented by a vector (a phasor) whose length is the amplitude of that wave and whose direction is determined by its phase angle. Thus, the phase difference $\delta$ in direction of two phasors $\vec{A}_1$ and $\vec{A}_2$ is the phase difference between the two waves. If the two waves are coherent, that is, if the difference in their phases is constant, the angle between the directions of the two phasors remains constant. Adding two or more phasors vectorally produces a resultant phasor whose length and direction represent the amplitude and phase of the resultant wave.

Figure 33-4

*Calculating the Single-Slit Diffraction Pattern.*  Using phasors, the intensity of the single-slit diffraction pattern can be shown to be

$$I = I_0 \left( \frac{\sin \frac{1}{2}\phi}{\frac{1}{2}\phi} \right)^2$$

Single-slit diffraction intensity pattern

where the phase difference $\phi$ is given by $\phi = (2\pi / \lambda) a \sin \theta$, and $a$ is the width of the slit.

*Calculating the Interference-Diffraction Pattern of Multiple Slits.*  The intensity of the double-slit pattern, which combines both the interference between the slits and the diffraction of the individual slits can be shown to be

$$I = 4I_0 \left( \frac{\sin \frac{1}{2}\phi}{\frac{1}{2}\phi} \right)^2 \cos^2 \frac{1}{2}\delta$$

Interference-Diffraction intensity for two slits

where $\phi$ is the same as for the single-slit pattern, and $\delta = (2\pi / \lambda) d \sin \theta$, where $d$ is the separation of the slits.

*The Interference Pattern of Three or More Equally Spaced Sources.*  For a given spacing of three or more identical coherent sources, the principal interference maxima (there are lesser maxima in between) are at the same directions.  The intensity and sharpness (narrowness) of the principal maxima increase as the number of sources increases.

*Section 33-6.  Fraunhofer and Fresnel Diffraction.*  Diffraction patterns observed at points for which the rays from an aperture or an obstacle are nearly parallel (typically patterns observed at large distances from an aperture) are called Fraunhofer diffraction patterns.  The diffraction pattern observed very close to an aperture or an obstacle, when the rays from that object can no longer be approximated as parallel, are called Fresnel diffraction patterns.  These two patterns, from the same aperture or object, are quite different from each other, as illustrated in the textbook.

*Section 33-7.  Diffraction and Resolution.*  For any optical instrument in which light enters through a finite aperture, diffraction limits the instrument's resolution.  The angle $\theta$ at which the first minimum in the Fraunhofer diffraction pattern of the light from a circular aperture of diameter $D$ occurs is

$$\sin \theta = 1.22 \frac{\lambda}{D}$$

First minimum, circular aperture diffraction

as shown in Figure 33-34 of the textbook.  In the small-angle approximation, $D \gg \lambda$, this becomes $\theta \approx 1.22 \lambda / D$.  This angle is the smallest angle that can be resolved, and is known as the Rayleigh criterion, $\alpha_c$.  The derivation of this result is beyond the scope of both the text and this study guide.  The images of objects separated by angles smaller than this are not distinguishable because they overlap.  An astronomical instrument must have a large aperture for the sake of resolution as well as for light-gathering power.

*Section 33-8.  Diffraction Gratings.*  A diffraction grating consists of a large number of equally spaced lines or slits.  The interference maxima in light that has passed through the grating can be extremely sharp. A diffraction grating is used to analyze the spectrum, the distribution of energy among different wavelengths, of light from various sources. The expression for the resolving power of a grating is

$$R = \frac{\lambda}{|\Delta\lambda|} = mN \qquad\qquad \text{Resolving power of a diffraction grating}$$

where $N$ is the number of slits illuminated, $m$ is the order of the principal interference maxima, and $|\Delta\lambda|$ is the minimum difference in wavelengths that can be resolved.

## II.  Physical Quantities and Key Equations

### Physical Quantities

There are no new physical quantities in this chapter.

### Key Equations

*Phase difference due to path difference*
$$\delta = \frac{\Delta r}{\lambda} 2\pi = \frac{\Delta r}{\lambda} 360°$$

*General interference conditions*
$$\delta = 2\pi m \qquad m = 0,1,2,\dots \quad \text{constructive}$$
$$\delta = 2\pi\left(m + \tfrac{1}{2}\right) \qquad m = 0,1,2,\dots \quad \text{destructive}$$

*Conditions for thin-film interference with one 180° phase shift due to reflection*

$$\frac{2t}{\lambda'} = \frac{2nt}{\lambda} = m \qquad m = 0,1,2,\dots \qquad \text{destructive}$$
$$\frac{2t}{\lambda'} = \frac{2nt}{\lambda} = m + \frac{1}{2} \qquad m = 0,1,2,\dots \qquad \text{constructive}$$

*Conditions for thin-film interference with two 180° phase shifts (or no phase shifts at all) due to reflection*

$$\frac{2t}{\lambda'} = \frac{2nt}{\lambda} = m \qquad m = 0,1,2,\dots \quad \text{constructive}$$
$$\frac{2t}{\lambda'} = \frac{2nt}{\lambda} = m + \frac{1}{2} \qquad m = 0,1,2,\dots \quad \text{destructive}$$

*Two-slit interference*
$$d\sin\theta = m\lambda \qquad m = 0,1,2,\dots \quad \text{constructive}$$
$$d\sin\theta = \left(m + \tfrac{1}{2}\right)\lambda \qquad m = 0,1,2,\dots \quad \text{destructive}$$

*Intensity in terms of phase difference*
$$I = 4I_0 \cos^2 \tfrac{1}{2}\delta$$

*Single-slit diffraction minima*
$$a\sin\theta = m\lambda \qquad m = 1,2,3,\dots \qquad \text{(destructive)}$$

| | | |
|---|---|---|
| *Single-slit diffraction intensity pattern* | $I = I_0 \left( \dfrac{\sin\frac{1}{2}\phi}{\frac{1}{2}\phi} \right)^2$ | $\phi = \dfrac{2\pi}{\lambda} a \sin\theta$ |
| *Interference-Diffraction intensity for two slits* | $I = 4I_0 \left( \dfrac{\sin\frac{1}{2}\phi}{\frac{1}{2}\phi} \right)^2 \cos^2 \frac{1}{2}\delta$ | |
| *First minimum, circular aperture diffraction* | $\sin\theta = 1.22 \dfrac{\lambda}{D}$ | |
| *Resolving power of a diffraction grating* | $R = \dfrac{\lambda}{|\Delta\lambda|} = mN$ | |

## III. Potential Pitfalls

Remember that a difference in path length is only one factor in determining the phase difference between light waves that have traveled from a source by two different paths. One or both of the two waves may have experienced additional phase changes due to reflection.

When computing phase difference, you must divide the path–length difference by the wavelength *in the material where the path difference occurs.*

In thin-film interference, which formula you select to determine whether the reflected light waves interfere constructively or destructively depends not only on the index of refraction of the material, the wavelength of the light, and the thickness of the film, but also on whether the wave undergoes a 180° phase shift on reflection from one or both of the boundaries.

In interference and diffraction calculations, it is easy to confuse phase differences with angles between directions in space. Both quantities are expressed in angle measure, but they are quantities of completely different kinds. Adding to the confusion is that, on phasor diagrams, phase differences are represented as angles between directions.

Do not confuse power with intensity. Power is energy per unit time whereas intensity is power per unit area. Low power can result in high intensity if it is delivered to a small enough area.

When interference is due to three or more equally spaced coherent sources or slits, the *location* on the screen of the principal interference maxima depends only on the spacing of the sources, not on how many of them there are. Both the *sharpness* of the principal maxima and the number of secondary maxima between them increase with the total number of sources or slits.

Do not confuse line sharpness with line spacing. A grating has high *resolution* if you can use it to separate cleanly two light waves of very nearly the same wavelength. That is, after passing through the grating the two waves form two distinct lines. There are two separate issues involved in this—the separation between the centers of the two lines, which is determined by the spacing between slits, and the sharpness (narrowness) of each line, which increases as the number of slits illuminated increases. The resolving power of a grating depends upon both of these properties.

# IV. True or False Questions and Responses

**True or False**

_____ 1.  Only waves from *coherent* sources exhibit observable interference.

_____ 2.  When light traveling in a transparent medium is reflected at a boundary with another medium having a higher index of refraction, the light wave undergoes a 180° phase change on reflection at the boundary.

_____ 3.  Monochromatic light rays reflected from the two surfaces of a film that is exactly one-half wavelength thick always exhibit destructive interference.

_____ 4.  The name *Newton's rings* is given to the diffraction pattern of light passing through a circular aperture.

_____ 5.  When light comes from two coherent point sources that are in phase, constructive interference occurs at locations where the difference in path lengths of the light from each source is an integral number of wavelengths, assuming there are no phase changes due to reflections.

_____ 6.  In the interference pattern of light from two identical slits, the maximum intensity is twice that which one source alone would produce.

_____ 7.  Complete destructive interference of two light waves requires that they be of equal intensity.

_____ 8.  Light from two coherent sources that are not in phase does not produce an interference pattern.

_____ 9.  The principal interference maxima produced by light from four identical slits in an opaque screen a distance $d$ apart occur at angles that are twice those of the interference maxima that would be produced by light from the middle two slits.

_____ 10. If two waves of equal frequency and wavelength are superposed, you can find the amplitude of the resulting wave simply by adding two two-dimensional vectors whose magnitudes and directions are the amplitudes and phases, respectively, of the two waves.

_____ 11. Diffraction of light is important only for apertures and obstructions that are very large compared with the wavelength.

_____ 12. In Huygens' construction, diffraction at a single aperture is equivalent to the interference of an uncountably large number of point sources distributed over the aperture.

_____ 13. In the Fraunhofer diffraction pattern of a single slit, the first *minimum* occurs where the waves from the two edges of the slit are $2\pi$ out of phase.

_____ 14. Fraunhofer diffraction is diffraction that is observed very far from the apertures or obstructions that cause the diffraction.

_____ 15. Light passing through two narrow slits in a barrier exhibits both interference and diffraction.

_____ 16. Diffraction imposes a fundamental limitation on the resolving power of telescopes and microscopes.

_____ 17. The resolving power of a diffraction grating is determined by the spacing between the grating lines.

_____ 18. For a given line *spacing,* a larger diffraction grating is capable of better spectral resolution.

**Responses to True or False**

1. True.

2. True. An example is light that is in air and is reflecting at an air–glass boundary.

3. False. One must consider phase shifts before deciding. It would be destructive if 180° phase shifts occur either at the first or the second surface but not at both. Otherwise, it would be constructive.

4. False. *Newton's rings* refers to the fringe pattern caused by interference in the thin film of air between a plane glass surface and a spherical glass surface. Newton's rings are shown in Figure 33-3 on page 1087 of the text.

5. True.

6. False. It is four times as great. In constructive interference, the resultant amplitude is the sum of the amplitudes, so the amplitude is twice the amplitude due to a single slit. The intensity is proportional to the square of the amplitude, so it is larger by a factor of four.

7. True. To cancel completely, the waves must have the same amplitude. Because, for a given frequency, the intensity of a wave is proportional to the square of its amplitude, coherent waves having the same amplitude also have the same intensity.

8. False. The phase differences affect the locations of the maxima and minima, but interference occurs whenever the two sources are coherent and light waves from the two sources superpose.

9. False. The principal maxima occur at the same angles, but are sharper (narrower and more intense).

10. True. These vectors are called *phasors.*

11. False. Diffraction is important only for apertures or obstructions that are not large compared with the wavelength.

12. True.

13. True. Under this condition, the wave from each point on one half of the aperture is 180° out of phase with, and thus cancels, the wave from the corresponding point on the other half.

14. True.

15. True. Light is diffracted at each slit, and light from one slit interferes with the light from the other slit.

16. True. The usual expression of this is the Rayleigh criterion $\alpha_c \approx 1.22\lambda/D$.

17. False. Resolving power is determined by the total number of slits illuminated.

18. True. The larger the grating, the greater the number of lines that can be illuminated.

## V.  Questions and Answers

**Questions**

1.  What is the origin of the colors we see in the sunlight reflected from thin films?

2.  The maximum intensity of light in the interference pattern produced by two narrow slits is four times, not twice, the intensity that we would see from one narrow slit alone. Where does the extra energy come from?

3.  Antireflecting (AR) coatings for lenses are made with a thickness equal to one-fourth of the wavelength of the light. Is this one-fourth of the wavelength in air, in the coating, or in the glass lens? Is the index of refraction of the coating greater or less than that of glass? Why?

4.  The main reason that some fairly smart guys (like Newton) thought that light is a propagation of particles was that diffraction of light had not been observed. Why not?

5.  What is the advantage of the electron microscope over a light microscope?

6.  It often seems to me (TGR) that a *diffraction grating* should more properly be called an *interference grating*. What do you think?

7.  Is there a maximum wavelength for which both the maxima and minima of the diffraction pattern from a single slit can be observed? Is there a minimum wavelength?

8.  The apparatus sketched in Figure 33-5 is called Fresnel's double mirror. Like Lloyd's mirror, it is a means of realizing two coherent light sources without using slits. How does it work?

9.  Under the most optimistic atmospheric conditions at the premium site for land-based observing (Mauna Kea, Hawaii, elevation $\approx 4.27\,\text{km}$), a telescope can resolve celestial objects that are separated by one-fourth of a second of arc (arc-sec). The viewing never gets any better than this because of atmospheric turbulence, which makes the images jitter. What minimum diameter aperture is necessary to provide $\frac{1}{4}$ arc-sec resolution due to diffraction? Is there ever any point in building a telescope much bigger than this?

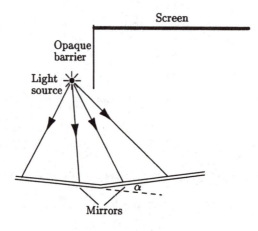

Figure 33-5

10. Describe the two-slit interference pattern you'd get using white light.


**Answers**

1. The colored bands are due to the interference of light reflected at the front and back surfaces of the thin film. The different colors are due to the varying thickness of the film. At one point on the film, for instance, there is an interference maximum in the blue reflected light and a minimum in the red light, so the film looks blue. At another point where the film is slightly thicker, it may reflect red light preferentially. Thus, when we look at sunlight reflecting from an oil slick, we see a shifting rainbow of colors in the reflected light.

2. The power passing through two narrow slits is twice that passing through one narrow slit alone, though the distribution of the power over the observation screen is very different. With one slit the entire screen is approximately uniformly illuminated. This is not so in the two-slit pattern. In the two-slit pattern, the area of the screen that is brightly illuminated is approximately equal to the area that is dimly illuminated. A large intensity over a small area can deliver the same power as a low intensity over a large area. To produce four times the intensity means that twice the power is delivered to an area one-half as large.

3. AR coatings are placed on the front surface of the glass. The index of refraction of the coating is greater than the index of refraction of air and less than the index of refraction of glass. This means that 180° phase shifts are associated with the reflections at both surfaces. Also, the thickness of the coating is one-fourth the wavelength *in the coating*. Thus, the light reflecting at the back surface has a path length that is one-half a wavelength longer than the light reflected at the front surface. Consequently, the reflected light waves are 180° out of phase and interfere destructively.

4. Because the wavelength of visible light is so small, the diffraction of light is usually too small to see.

5. The wavelength of an electron is inversely proportional to its momentum. Electron microscopes use electrons with wavelengths that are much smaller than those of light, so, in

accordance with the Rayleigh criteria, much smaller objects can be resolved distinctly by an electron microscope.

6. Both interference and diffraction are basic to the operation of a grating. The pattern produced by the grating is just a very-many-slit *interference* pattern; on the other hand, the wide-angle pattern wouldn't exist if it weren't for the *diffraction* of the light at each slit.

7. The first minimum in the single-slit diffraction pattern occurs at $\sin\theta = \lambda / a$, where $a$ is the width of the slit. Because $\sin\theta$ cannot be greater than 1, this minimum can be observed only if $\lambda < a$. Thus, the maximum wavelength for which the first minimum in the single-slit diffraction can be observed is $\lambda_{max} = a$. For wavelengths very much smaller than $a$, diffraction effects become unobservably small, and all you see is the geometric shadow of the slit. To say that diffraction effects become unobservably small is to say that the diffraction maxima and minima are so closely spaced that the eye cannot resolve them. The wavelength for which this spacing is just resolvable is the minimum wavelength for which the diffraction pattern can be observed.

8. The two images of the actual light source in the two plane mirrors act as two point sources as seen in Figure 33-6. The two image "sources" are not in phase, because they are at different distances from the source; but they are coherent because all the light actually comes from a single source.

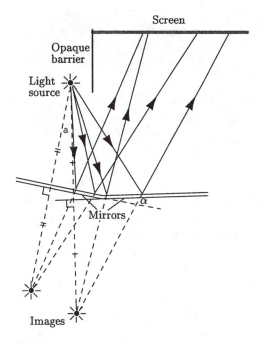

Figure 33-6

9. Since $\frac{1}{4}$ arc-sec is $1.21 \times 10^{-6}$ radian, the Rayleigh criterion is

$$1.22\frac{\lambda}{D} = 1.21 \times 10^{-6}$$

For visible light, $D$ is 55 or 60 cm (about 2 ft). Building a telescope with a diameter much larger than this won't improve the resolution significantly as long as you have to look through the earth's atmosphere. The main reason optical telescopes bigger than this are built is for greater light-collecting power, not resolution. Also, if the telescope is to be placed on a platform above the atmosphere, then telescopes of larger diameters can produce greater resolution.

10. White light is a combination of all the colors of the rainbow. The different colors correspond to different wavelengths, which interfere constructively at different angles $\theta$. Straight ahead the central ($m = 0$) interference maximum is white, but to the sides the pattern dissolves into a few colored bands, after which all evidence of interference disappears. White light can never be very coherent. (Only perfectly monochromatic sources can be completely coherent.)

# VI. Problems, Solutions, and Answers

**Example #1.** Light of wavelength 500 nm falls at near-normal incidence on two flat glass plates 8.00 cm long that are separated at one end by a wire with a 5.00 $\mu$m diameter as in Figure 33-7. Find the spacing of the dark interference fringes along the length of the plates.

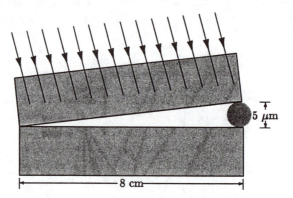

Figure 33-7

**Picture the Problem.** The angle in the figure is greatly exaggerated, so you can treat this problem as if the light is at 100% normal incidence. Interference fringes occur because the thickness of the air gap between the plates varies along the plates.

| 1. Make a sketch of the wedge, showing two paths of light: one in which the light is reflected off the back surface of the top piece of glass, and the other in which light is reflected off the top surface of the lower piece of glass. Because the thickness changes with $x$, interference fringes will occur. The light paths through all the rest of the glass remains constant for all values of $x$. | 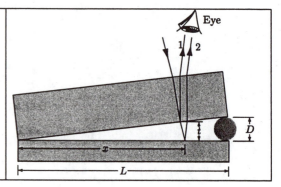 |

| 2. Determine the path length difference required for a dark fringe. | There is no phase shift of the light reflected off the upper air-glass boundary, but there is a 180° phase shift for the light reflected off the lower boundary. As a result, for destructive interference, we have $$2t = m\lambda_{air} \qquad m = 0,1,2,\ldots$$ |
|---|---|
| 3. Relate the thickness of the glass to the distance $x$. Because the distances $t$ and $D$ subtend the same angle, we can relate the tangents. | $$\frac{t}{x} = \frac{D}{L}$$ $$t = \frac{Dx}{L}$$ |
| 4. Substitute the expression for $t$ into the expression obtained in step 2, and solve for $x$, the location of dark fringes. | $$\frac{2Dx}{L} = m\lambda_{air}$$ $$x = \frac{m\lambda_{air}L}{2D} \doteq m\frac{\left(500\times10^{-9}\text{ m}\right)\left(8\times10^{-2}\text{ m}\right)}{2\left(5\times10^{-6}\ \mu\text{m}\right)}$$ $$= m\left(4\times10^{-3}\text{ m}\right)$$ $$= 4\,\text{mm}, 8\,\text{mm}, 12\,\text{mm}, \ldots 80\,\text{mm}$$ |
| 5. Determine the dark fringe spacing. The fringes will appear as shown. | $\Delta x = 4\,\text{mm}$ <br><br> $\vdash\!\!-\!\!-\!\!-\!\!-\!\!-\!\!-8\text{ cm}\!-\!\!-\!\!-\!\!-\!\!-\!\!-\dashv$ |

**Example #2—Interactive.** A wedge-shaped air film is made by placing a small slip of paper between the edges of two glass flats 12.5 cm long. Light of wavelength 600 nm is incident normally on the glass plates. If interference fringes with a spacing of 0.200 mm are observed along the plate, how thick is the paper? This form of "interferometry" is a very practical way of measuring small thicknesses.

**Picture the Problem.** This is a reversal of Example #1. **Try it yourself.** Work the problem on your own, in the spaces provided, to get the final answer.

| 1. Sketch the physical situation, showing the two paths of light that result in the interference pattern. | |
|---|---|
| | |

| | |
|---|---|
| 2. Determine the path length difference required for a bright interference fringe. | |
| 3. Relate the position $t$ to the distance $x$ traveled from the point where the two glass plates touch. | |
| 4. Substitute the expression for $t$ into the expression obtained in step 2. Knowing the separation of the fringes, solve for the thickness of the paper. Use two separate expressions for $D$: one for the $m$th fringe at a position $x$, and one for the $(m+1)$st fringe at a position $x + \Delta x$. This will enable you to solve for the thickness of the paper. | $t = 188\,\mu\text{m}$ |

**Example #3.**  A glass lens with index of refraction $n = 1.57$ is coated with a thin layer of transparent material of index $n = 2.10$. If white light strikes the lens at near-normal incidence, light of wavelengths 495 nm and 660 nm is absent from the reflected light. If the coating is the thinnest possible that meets the conditions given, what is the thickness of the layer?

**Picture the Problem.**  If both wavelengths are absent, then both wavelengths experience destructive interference.

| | |
|---|---|
| 1. Draw a sketch of the situation, including the two paths of light that result in the interference pattern. | 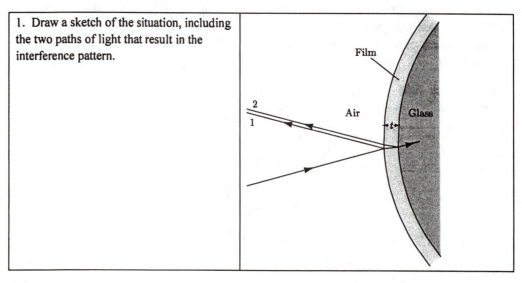 |

| | |
|---|---|
| 2. Determine the path length difference required for destructive interference. | There is a 180° phase change in the ray reflected from the air-film interface, but no phase change in the ray reflected from the film-lens interface. So the destructive interference condition is<br><br>$2t = m\lambda_{film}$    $m = 0, 1, 2, \ldots$ |
| 3. Relate the wavelength in the film to the wavelength in air. | $\lambda_{film} = \dfrac{\lambda}{n_{film}}$ |
| 4. Substitute this expression into that obtained in step 2 to get the condition for destructive interference. | $2t = m\lambda / n_{film}$<br>$t = m\lambda / (2n_{film})$ |
| 5. This expression must be true for both wavelengths given in the problem. Let $\lambda_a = 660\,\text{nm}$ and $\lambda_b = 495\,\text{nm}$. | $t = m_a \lambda_a / (2n_{film})$<br>$t = m_b \lambda_b / (2n_{film})$ |
| 6. Taking the ratio of the two expressions, we can find the smallest possible values for $m_a$ and $m_b$, which will also correspond to the smallest possible thickness. | $\dfrac{m_a \lambda_a}{m_b \lambda_b} = 1$<br><br>$\dfrac{m_a}{m_b} = \dfrac{\lambda_b}{\lambda_a} = \dfrac{495\,\text{nm}}{660\,\text{nm}} = \dfrac{3}{4}$<br><br>$t = m_a \lambda_a / (2n_{film}) = 3(660\,\text{nm}) / (2 \times 2.10) = 471\,\text{nm}$ |

**Example #4—Interactive.** A thin film of soap solution with $n = 1.33$ has air on either side and is illuminated normally with white light. Interference minima are visible in the reflected light only at wavelengths of 400, 480, and 600 nm. What is the minimum thickness of the film?

**Picture the Problem.** Follow the solution to Example #3, above. **Try it yourself.** Work the problem on your own, in the spaces provided, to get the final answer.

| | |
|---|---|
| 1. Draw a sketch of the situation, including the two paths of light that result in the interference pattern. | |
| 2. Determine the path length difference required for destructive interference. | |
| 3. Relate the wavelength in the film to the wavelength in air. | |

| | |
|---|---|
| 4. Substitute this expression into that obtained in step 2 to get the condition for destructive interference. | |
| 5. This expression must be true for all three wavelengths given in the problem. | |
| 6. Taking the ratios of the expressions, we can find the smallest possible values for the integer $m$ for each wavelength, which will also correspond to the smallest possible thickness. | $t = 902\,\text{nm}$ |

**Example #5.**  In a classroom demonstration of Young's experiment, helium-neon laser light of wavelength 632.8 nm passes through two parallel slits and falls on a screen 10.5 m away. Interference maxima on the screen are spaced 4.00 cm apart.  (a) What is the separation of the slits? (b) The instructor slips a piece of clear plastic film 0.1 mil ( 2.54 $\mu$m ) thick, with $n = 1.561$, over one of the slits.  How does the interference pattern change?

**Picture the Problem.**  For part (a), determine the condition for constructive interference at the first maximum beyond the central peak.  In part (b), the plastic film will introduce an additional phase shift in the light that travels through it, which will shift the interference pattern.

| | |
|---|---|
| 1. Sketch the situation. Include the two different paths of light and the angle the paths make with the centerline between the slits. Identify the path difference, $d\sin\theta$. | 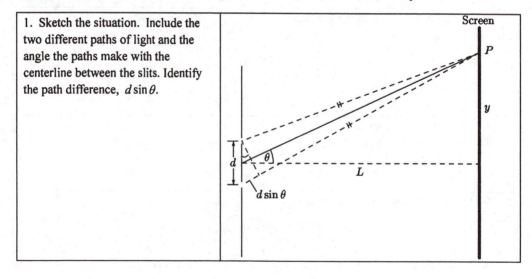 |

| | |
|---|---|
| 2. Determine the tangent of the angle. | $\tan\theta = \dfrac{y}{L}$ |
| 3. Determine the condition for constructive interference at some point. | $d\sin\theta = m\lambda \qquad m = 0,1,2,\ldots$ |
| 4. Because the height $y$ at which the first maximum is observed is much smaller than the distance $L$, the small angle approximation can be used. | $\tan\theta \approx \sin\theta \approx \theta \approx \dfrac{y}{L}$ |
| 5. Substitute this expression into the condition for constructive interference. | $d\dfrac{y_m}{L} = m\lambda$ <br><br> $y_m = Lm\lambda / d$ |
| 6. We know the spacing of subsequent maxima. Use this to solve for $d$. | $\Delta y = y_{m+1} - y_m = (m+1)\lambda L/d - m\lambda L/d = \lambda L/d = 4\,\text{cm}$ <br><br> $d = \lambda L/(4\,\text{cm}) = (632.8\times10^{-9}\,\text{m})(10.5\,\text{m})/(4\times10^{-2}\,\text{m})$ <br><br> $\quad = 166\,\mu\text{m}$ |
| 7. When the plastic is put in front of one of the slits, there will be a phase difference between the two beams leaving the slits, as shown. The light traveling through air will experience some phase change as it travels the thickness $\Delta r$ through the air, but the light traveling through the plastic will experience a different phase change as it travels the distance $\Delta r$ through the plastic. | |
| 8. Calculate the phase difference between the two beams described above as they reach the slits. Let $\delta'$ be the change in phase of the light as it travels through the plastic, and $\delta$ be the change in phase of the light as it travels through the air. <br><br> This phase difference is equivalent to simply $\pi/2$ rad, or one-fourth of a cycle. | $\delta' - \delta = \dfrac{\Delta r}{\lambda_{\text{plastic}}}2\pi - \dfrac{\Delta r}{\lambda_{\text{air}}}2\pi = \left(\dfrac{1}{\lambda_{\text{plastic}}} - \dfrac{1}{\lambda_{\text{air}}}\right)2\pi\,\Delta r$ <br><br> $= \left(\dfrac{n_{\text{plastic}}}{\lambda} - \dfrac{n_{\text{air}}}{\lambda}\right)2\pi\,\Delta r = (n_{\text{plastic}} - n_{\text{air}})\dfrac{2\pi\,\Delta r}{\lambda}$ <br><br> $= (1.561 - 1)\dfrac{2\pi(2.54\times10^{-6}\,\text{m})}{632.8\times10^{-9}\,\text{m}} = 4.50\pi$ <br><br> $= 4\pi + \dfrac{\pi}{2}$ |

| 9.  As a result of the phase difference of the light leaving the slits, the condition for constructive interference is altered by $\frac{1}{4}$ wavelength.<br><br>The result is a shift in the interference pattern.  The spacing is still 4 cm, but the maxima have moved along the $y$ axis. | $d \sin \theta = \left(m + \frac{1}{4}\right)\lambda$<br><br>$y_m = \left(m + \frac{1}{4}\right)\dfrac{\lambda L}{d}$<br><br>$y_0 = \left(0 + \frac{1}{4}\right)\dfrac{\lambda L}{d} = \frac{1}{4}(0.04\,\text{m})$<br><br>$y_1 = \left(1 + \frac{1}{4}\right)\dfrac{\lambda L}{d} = \frac{5}{4}(0.04\,\text{m})$ |
|---|---|

**Example #6—Interactive.**  In a two-slit interference pattern, when light of wavelength 589 nm is used and the screen is 3.00 m from the slits, there are 28 bright fringes per centimeter on the screen.  What is the separation between the slits?

**Picture the Problem.**  The fringe spacing can be calculated from the given information.  This can be used to determine the slit separation as in the previous problem.  **Try it yourself.**  Work the problem on your own, in the spaces provided, to get the final answer.

| 1.  Sketch the situation, including the two different paths of light, and the interference pattern. | |
|---|---|
| 2.  Determine the tangent of the angle the paths of light make with respect to the centerline. | |
| 3.  Determine the condition for constructive interference at some point on the screen. | |
| 4.  Because the height $y$ at which the first maximum is observed is much smaller than the distance $L$, the small angle approximation can be used. | |

| | |
|---|---|
| 5. Substitute this expression into the condition for constructive interference. | |
| 7. Determine the spacing of subsequent interference maxima. | |
| 7. Knowing the spacing of subsequent maxima, solve for $d$. | $d = 4.95\,\text{mm}$ |

**Example #7.** A two-slit interference pattern using 460-nm blue light is thrown on a screen 5.00 m from the slits; bright interference fringes spaced 5.00 cm apart are observed. The fourth maximum in each direction from the central maximum is missing from the pattern. What are the dimensions of the slits?

**Picture the Problem.** The fringe spacing will provide the slit separation as in the previous examples. The fourth fringe is missing because it coincides with the first minimum in the diffraction pattern of a single slit. This can be used to determine the width of the slits.

| | |
|---|---|
| 1. The two-slit configuration is the same as that in Example #5. From the figure in that example, we can find the angle $\theta$. | $\tan\theta = \dfrac{y}{L}$ |
| 2. Determine the condition for constructive interference at the screen. | $d\sin\theta = m\lambda \qquad m = 0,1,2,\dots$ |
| 3. Apply the small angle approximation, and use it to solve for $y_m$, the positions at which interference maxima occur. | $\tan\theta \approx \sin\theta \approx \theta$ <br><br> $d\dfrac{y_m}{L} = m\lambda$ <br><br> $y_m = m\dfrac{\lambda L}{d}$ |
| 4. Use the spacing between subsequent maxima to find the slit separation, $d$. | $\Delta y = y_{m+1} - y_m = (m+1)\dfrac{\lambda L}{d} - m\dfrac{\lambda L}{d} = \dfrac{\lambda L}{d} = 0.05\,\text{m}$ <br><br> $d = \dfrac{\lambda L}{0.05\,\text{m}} = \dfrac{\left(460\times10^{-9}\,\text{m}\right)\left(5\,\text{m}\right)}{0.05\,\text{m}} = 46.0\,\mu\text{m}$ |

| 5. Find an expression for the location of the missing 4th interference maximum. | $y_4 = 4\dfrac{\lambda L}{d}$ |
|---|---|
| 6. Determine the position of the first diffraction maximum from a single slit, and apply the small angle approximation. The slit width is $a$. | $a\sin\theta = \lambda$ <br><br> $a\dfrac{y}{L} = \lambda$ <br><br> $y = \dfrac{\lambda L}{a}$ |
| 7. Because the 4th interference maximum coincides with the first diffraction minimum, we can set the expressions for $y$ in steps 5 and 6 equal to each other to solve for the slit width $a$. | $y_4 = y$ <br><br> $4\dfrac{\lambda L}{d} = \dfrac{\lambda L}{a}$ <br><br> $a = \dfrac{d}{4} = \dfrac{46.0\,\mu m}{4} = 11.5\,\mu m$ |

**Example #8—Interactive.** In a lecture demonstration of diffraction, a laser beam of wavelength 632.8 nm passes through a vertical slit 0.250 mm wide and strikes a screen 12.0 m away. How wide is the central maximum (the distance between the first diffraction minima on each side of the central maximum) of the diffraction pattern on the screen?

**Picture the Problem.** Find the distance of the first diffraction minimum from the central maximum, and multiply by two. **Try it yourself.** Work the problem on your own, in the spaces provided, to get the final answer.

| 1. Determine the position of the first diffraction maximum from a single slit. | |
|---|---|
| 2. Apply the small angle approximation and solve for $y$, the position of the first minimum. | |
| 3. Multiply the position arrived at in step 2 by 2, to find the width of the central maximum. | $y = 6.07\,cm$ |

**Example #9.** Two point sources of light of wavelength 500 nm are photographed from a distance of 100 m using a camera with a 50.0-mm focal length lens. The camera aperture is 1.05 cm in diameter. (*a*) What is the minimum separation of the two sources for them to be resolved in the photograph, assuming the resolution is diffraction-limited? (*b*) What is the separation of the two images when the sources are at this minimum separation?

**Picture the Problem.** The minimum resolution is determined by the Rayleigh criterion. The limiting factor is the size of the aperture. This angular separation corresponds to a linear separation 100 m away. To determine the image separation, assume that the film is at the focal plane of the lens.

| 1. Draw a sketch of the physical situation. | |
|---|---|
| 2. Determine the angular position of the first diffraction minimum for this circular lens, using the small angle approximation. | $\sin\theta = 1.22\dfrac{\lambda}{D}$ <br><br> $\theta \approx 1.22\dfrac{\lambda}{D}$ |
| 3. The critical angular separation requires that the central diffraction maximum of one source is located at the first minimum of the diffraction pattern of the other source. | $\alpha_c = 1.22\dfrac{\lambda}{D}$ |
| 4. Because of the small angle approximation, the critical angle is approximately equal to the tangent of the critical angle. Use this to solve for the source separation distance. | $\dfrac{d}{L} = 1.22\dfrac{\lambda}{D}$ <br><br> $d = 1.22\dfrac{\lambda L}{D} = 1.22\dfrac{\left(500\times10^{-9}\,\text{m}\right)\left(100\,\text{m}\right)}{0.0105\,\text{m}} = 5.81\,\text{mm}$ |
| 5. To find the separation of the images of the two sources, simply use the fact that the object and image must both subtend the same angle $\alpha_c$, so the ratio of object or image separation to the distance from the lens must be equal. | $\dfrac{d'}{f} = \dfrac{d}{L}$ <br><br> $d' = \dfrac{df}{L} = \dfrac{\left(5.81\times10^{-3}\,\text{m}\right)\left(0.05\,\text{m}\right)}{100\,\text{m}} = 2.91\,\mu\text{m}$ |

**Example #10—Interactive.** The world's largest refracting telescope is at the University of Chicago's Yerkes Observatory, located less than 7 miles from the house were I (TGR) grew up. Its objective is 1.02 m in diameter. Suppose you could mount the telescope on a spy satellite 200 km above the ground. (*a*) Assuming that the resolution is diffraction-limited, what minimum separation of two objects on the ground could it resolve? Take 550 nm as a representative wavelength for visible light. (*b*) Due to atmospheric turbulence, objects on the surface of the earth can be distinguished only if their angular separation is at least 1.00 arc-sec. How far apart would two objects on the earth's surface be if they subtend an angle of 1.00 arc-sec? Compare this with your answer to part (*a*).

**Picture the Problem.** Use the Rayleigh criterion to find the minimum angular separation for diffraction-limited resolution. Find the linear separation for this angle and for an angle of 1.00 arc-sec. **Try it yourself.** Work the problem on your own, in the spaces provided, to get the final answer.

| | |
|---|---|
| 1. Draw a sketch of the physical situation. | |
| 2. Determine the angular position of the first diffraction minimum for this circular lens, using the small angle approximation. | |
| 3. The critical angular separation requires that the central diffraction maximum of one source is located at the first minimum of the diffraction pattern of the other source. | |
| 4. Because of the small angle approximation, the critical angle is approximately equal to the tangent of the critical angle. Use this to solve for the source separation distance. | $d = 13.1 \text{ cm}$ |
| 5. Convert 1 arc-sec to radians. | |
| 6. Using the small angle approximation, equate the above angle to the tangent, the separation of two sources divided by the earth-telescope distance, to find the source separation. | $d = 96.9 \text{ cm}$ |

Since the resolution with atmospheric turbulence is larger than the diffraction limited resolution, the atmospheric resolution is the limiting factor, so the telescope can only resolve objects separated by about one meter.

**Example #11.** A diffraction grating with 10,000 lines per centimeter is used to analyze the spectrum of mercury. (*a*) Find the angular separation, in first order, of the two spectral lines of wavelength 577 and 579 nm. (*b*) How wide must the beam of light be on the grating in order to resolve these two spectral lines?

**Picture the Problem.** The angles of the spectrum are the angle for the first-order primary interference maxima from the grating, and depend on the line spacing of the grating. The resolution is determined by the total number of lines involved in producing the interference pattern.

| | |
|---|---|
| 1. Draw a sketch of the physical situation. | 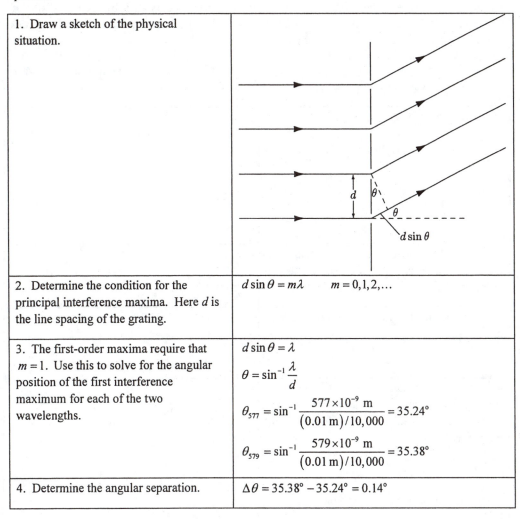 |
| 2. Determine the condition for the principal interference maxima. Here $d$ is the line spacing of the grating. | $d \sin \theta = m\lambda \qquad m = 0, 1, 2, \ldots$ |
| 3. The first-order maxima require that $m = 1$. Use this to solve for the angular position of the first interference maximum for each of the two wavelengths. | $d \sin \theta = \lambda$ <br><br> $\theta = \sin^{-1} \dfrac{\lambda}{d}$ <br><br> $\theta_{577} = \sin^{-1} \dfrac{577 \times 10^{-9}\text{ m}}{(0.01\text{ m})/10,000} = 35.24°$ <br><br> $\theta_{579} = \sin^{-1} \dfrac{579 \times 10^{-9}\text{ m}}{(0.01\text{ m})/10,000} = 35.38°$ |
| 4. Determine the angular separation. | $\Delta\theta = 35.38° - 35.24° = 0.14°$ |

| 5. The resolving power of the grating relates the number of lines illuminated to the wavelength resolution. Use this to determine the number of lines that must be illuminated. We are interested in the first order maxima, where $m = 1$. | $R = \dfrac{\lambda}{\|\Delta\lambda\|} = mN$ <br><br> $N = \dfrac{\lambda}{\|\Delta\lambda\|} = \dfrac{578 \times 10^{-9}\,\text{m}}{2 \times 10^{-9}\,\text{m}} = 289$ |
|---|---|
| 6. Use the line spacing and the number of lines that must be illuminated to find the required beam width. | $\dfrac{0.01\,\text{m}}{10{,}000\,\text{lines}} \times 289\,\text{lines} = 289\,\mu\text{m}$ |

**Example #12—Interactive.**    The spectrum of sodium is dominated by yellow light of wavelengths 589.00 and 589.59 nm. By what angle will these two lines be separated in first order by a diffraction grating of 5000 lines per inch?

**Picture the Problem.**  Follow the solution the part (*a*) of Example #11.  **Try it yourself.**  Work the problem on your own, in the spaces provided, to get the final answer.

| 1. Draw a sketch of the physical situation. | |
|---|---|
| 2. Determine the condition for the principal interference maxima.  Here *d* is the line spacing of the grating. | |
| 3. Calculate the line spacing of the grating in mks units. | |
| 4. The first-order maxima require that $m = 1$. Use this to solve for the angular position of the first interference maximum for each of the two wavelengths. | |
| 5. Determine the angular separation. | $\Delta\theta = 0.0067°$ in first order |

# Chapter 34

# Wave-Particle Duality and Quantum Physics

## I. Key Ideas

The first part of the twentieth century marked one of the most creative times of humankind. Einstein developed the theory of special relativity in 1905 and the theory of general relativity in 1916. The ideas of quantum physics started in 1900 with Planck's explanation of blackbody radiation. Einstein extended Planck's quantum ideas to explain the photoelectric effect in 1905, the same year he put forth the theory of special relativity. Ideas built upon ideas about how nature behaves. The fields of quantum physics, relativity, cosmology, and other areas of modern physics were given birth and nurtured by Einstein, Planck, Bohr, Fermi, Dirac, and many others, most of whom were awarded Nobel prizes for their work. Much of the work in the beginning of modern physics centered around the nature of light.

*Section 34-1. Light.* Light—is it a wave or a particle? The answer depends upon the results of experiments conducted on light. Waves are spread out in space and bend around corners (diffraction) and interfere with each other, producing constructive and destructive interference. The energy of waves is spread out in space. In contrast, particles are localized in space and travel on well-defined lines until they collide with something, after which they again travel on well-defined lines.

Newton (1642–1727) believed that light was composed of particles, a belief that was accepted for over a century. However, many experiments conducted in the 1800s suggested that Newton's particle picture of light was incorrect. Thomas Young in 1801 passed light from a single source through two closely spaced parallel slits and produced an interference pattern on a screen. Augustin Fresnel (1788–1827) developed the mathematical theory of light based on a wave picture, and showed that all experimental results agreed with his wave analysis.

James Clerk Maxwell in 1860 put forth his theory of electromagnetism, which predicted the existence of electromagnetic waves that propagate at the speed of light, $c \approx 3 \times 10^8$ m/s. The implication is that light is an electromagnetic wave, differing from other electromagnetic waves like television and radio only in wavelength. The wavelength of electromagnetic light waves is in the visible region of about 400 nm to about 700 nm ($1 \, \text{nm} = 10^{-9}$ m ). These and many other experiments seemed to show conclusively that light is a wave and not a particle.

***Section 17-2. The Particle Nature of Light: Photons.*** At the beginning of the 1900s, various experiments were performed that could not be explained with a wave picture of light. Instead, the experiments suggested that light energy comes in discrete amounts carried by localized packets called **photons.** Einstein, in explaining the photoelectric effect, first put forth the idea that each photon has an energy $E$ given by

$$E = hf = hc / \lambda \qquad \text{Einstein equation for photon energy}$$

where $f = c / \lambda$ is the frequency of light associated with each photon, and $h$ is **Planck's constant**

$$h = 6.626 \times 10^{-34} \text{ J} \cdot \text{s} = 4.136 \times 10^{-15} \text{ eV} \cdot \text{s} \qquad \textit{Planck's constant}$$

where the standard energy unit Joules (J) is related to the nonstandard energy unit electron volts (eV) by $1 \text{ eV} = 1.602 \times 10^{-19}$ J.

***The Photoelectric Effect.*** The first experiment showing the photon nature of light was the **photoelectric effect.** The apparatus for a photoelectric experiment is shown in Figure 34-1. Light of a variable but known frequency $f$ strikes a metal surface C in an evacuated tube, causing electrons to be emitted from C with various kinetic energies. After traversing a short distance in the tube, the electrons are collected by a second metal plate A. The electrons collected at the plate A constitute a current that is measured by the ammeter.

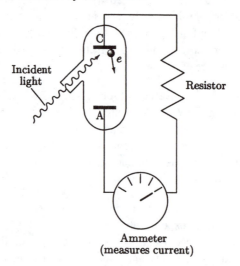

Figure 34-1

In the photoelectric experiment, the current is measured as a function of the frequency and the intensity of the incident light. In addition, the maximum kinetic energy $K_{max}$ of the electron is also measured. The photoelectric experiment has five main results:

**1.** Electrons emitted from the surface C have kinetic energies ranging from zero to a maximum value $K_{max} = \left(\frac{1}{2} mv^2\right)_{max}$.

**2.** $K_{max}$ does not depend on the intensity of the incident light.

**3.** $K_{max}$ does depend linearly on the frequency of the incident light according to a relationship known as **Einstein's photoelectric equation**:

$$K_{max} = \left(\tfrac{1}{2}mv^2\right)_{max} = hf - \phi \qquad \text{Einstein's photoelectric equation}$$

The constant $\phi$, called the **work function,** is equal to the energy needed to remove the least tightly bound electrons from the surface of C and is, therefore, a characteristic of the particular metal composing C.

**4.** Photons with frequencies below a smallest **threshold frequency** $f_t$ (or with wavelengths longer than a longest **threshold wavelength** $\lambda_t$) do not have enough energy to eject electrons from the surface C, no matter how intense the incident light. This means that for $f_t$ and below (and $\lambda_t$ and above), $K_{max} = 0$, which, from Einstein's photoelectric equation, gives

$$\phi = hf_t = \frac{hc}{\lambda_t} \qquad \text{Threshold frequency and wavelength}$$

**5.** The current is observed a very short time, of the order of $10^{-9}$ s, after the incident light strikes C, no matter how low the intensity of the light.

The classical picture of light as being composed of electromagnetic waves does not explain the results of the photoelectric experiment. Classically, the energy in a beam of light depends only on the light's intensity—the frequency of the incident light plays no role at all in energy balance. The electrons in surface C absorb more energy from more intense light and eventually, after a significant time lapse, leave C with kinetic energies ranging from zero to infinity.

Einstein explained *all* the experimental results of the photoelectric experiment by assuming that the light incident on the surface C is composed of photons, all with the same energy $E = hf$, where $h$ is the same constant determined by Planck in his blackbody analysis about five years previously. Each photon interacts with an electron in the emitter C in an all-or-nothing fashion, either giving all its energy to an electron or not interacting at all.

Electrons leave C with varying kinetic energies because some lose energy as they travel through the material of C. An electron at the surface of C requires the smallest amount of energy—equal to the work function $\phi$ of the material—in order to be released. Such an electron is emitted with the maximum kinetic energy $K_{max}$, given in Einstein's photoelectric equation, which is equal to the energy $hf$ of the absorbed photon minus the energy $\phi$ required to release the least tightly bound electron from the surface of C.

Thus, Einstein's quantum picture of light explains how the maximum kinetic energy of an emitted electron depends on the light's frequency, why there is a threshold frequency $f_t$ for which $K_{max} = 0$, and why the experimentally observed threshold frequency depends on the work function of the material through $hf_t = \phi$.

In the picture of light as photons, higher intensity corresponds to more photons in a light beam; and more photons eject more electrons from C. If the frequency of the incident photons stays constant, the maximum kinetic energy of the ejected electrons does not change as the intensity of the photons varies, which agrees with experimental results. The all-or-nothing transfer of energy from a photon to an electron also accounts for the exceedingly short time it takes for a current to be observed after light strikes C.

*Compton Scattering.* In 1923, Compton provided additional evidence that the photon picture of light is correct by his relativistic analysis of a "billiard ball" type of experiment between a photon and an effectively free electron.

In classical theory, the energy and momentum of an electromagnetic wave are related by $E = pc$, or $p = E/c$. Applying the quantum relationship $E = hf$ together with the wave relationship $c = f\lambda$ gives the momentum of a photon as $p = hf/c = h/\lambda$.

$p = h/\lambda$          Momentum of a photon in terms of its associated wavelength

In a Compton experiment, a photon with initial energy $E_1 = hc/\lambda_1$ and momentum $p_1 = h/\lambda_1$ is sent on a collision course with an electron that is at rest. After an elastic collision, the photon moves off at an angle $\theta$ with its original direction with a different energy $E_2 = hc/\lambda_2$ and different momentum $p_2 = h/\lambda_2$, and the electron moves at an angle $\phi$ with the original direction of the photon, as shown in Figure 34-2.

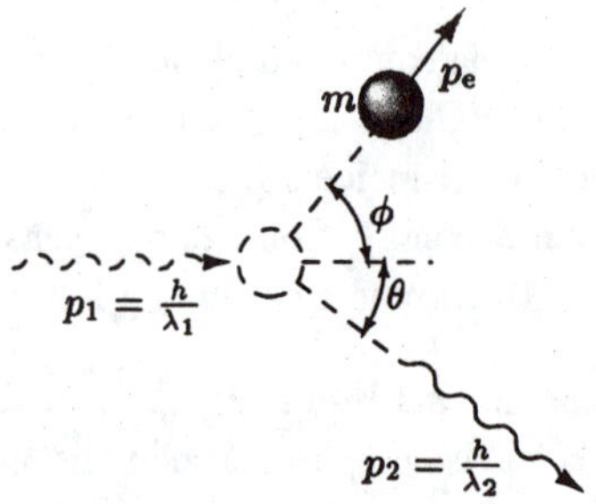

Figure 34-2

Compton applied relativistic conservation of energy and momentum to the collision (see Chapter R). The end result is that the wavelength $\lambda_2$ of the photon scattered at angle $\theta$ after the collision is related to the wavelength $\lambda_1$ of the incident photon before the collision by the **Compton equation**

$$\lambda_2 - \lambda_1 = \frac{h}{m_e c}(1 - \cos\theta)$$          Compton equation

The quantity multiplying the term $(1 - \cos\theta)$ is called the **Compton wavelength**

$$\lambda_C = \frac{h}{m_e c} = \frac{hc}{m_e c^2} = \frac{1240 \text{ eV·nm}}{5.11 \times 10^5 \text{ eV}}$$
$$= 2.43 \times 10^{-3} \text{ nm} = 2.43 \times 10^{-12} \text{ m} = 2.43 \text{ pm}$$          Compton wavelength

This short wavelength means that X-rays or gamma rays are typically needed to observe a $\Delta\lambda$. For Compton scattering from a particle other than an electron, you must use the mass of that particle in the above expressions.

***Section 33-3. Energy Quantization in Atoms.*** In 1913, Niels Bohr postulated that the internal energy of an atom can have only a discrete set of values. That is, the internal energy of an atom is *quantized*. The Bohr model of the atom is still useful today.

***Section 34-4. Electrons and Matter Waves.*** We have seen that a photon has the particle attributes of energy $E = hf$ and momentum $p = h/\lambda$. In addition, light shows wave behavior in its interference and diffraction.

***The de Broglie Hypothesis.*** In 1924, after consulting Einstein, de Broglie boldly suggested in his doctoral dissertation that if photons behave both like waves and particles, then perhaps electrons, which were regarded as "particles," could also have this dual character.

de Broglie used the momentum expression $p = h/\lambda$ for a photon to define the wavelength $\lambda$ of an electron:

$$\lambda = h/p \qquad\qquad \text{de Broglie wavelength of an electron}$$

where $p = mv$ is the (nonrelativistic) momentum of the electron. The expression for the frequency of the wave associated with an electron is the same as that for a photon:

$$f = E/h \qquad\qquad \text{Electron wave frequency}$$

where $E$ is the energy of the electron.

Subsequently, the wave nature of electrons, protons, and neutrons was substantiated experimentally.

***Electron Interference and Diffraction.*** If electrons are waves, it should be possible to perform experiments exhibiting this wave nature. This was first done in 1927 by C. J. Davisson and L. H. Germer at the Bell Telephone Laboratories. Davisson and Germer studied electron scattering from a nickel target. They found that the electron intensity in the scattered beam as a function of scattering angle showed maxima and minima corresponding to the exact de Broglie wavelength associated with the energy of the incident electron beam.

In the same year, G. P. Thomson (son of J. J. Thomson, who discovered the existence of the electron) demonstrated electron diffraction by sending electron beams through thin metal foils. The radii of the resulting diffraction pattern of concentric circles agreed with the de Broglie wavelength associated with the energy of the incident electron beam.

An important application of de Broglie waves associated with electrons is the electron microscope, which employs electrons to "see" objects at scales far smaller than microscopes using visible light. Also, diffraction is now routinely observed with "particles" other than electrons, such as neutrons.

***Standing Waves and Energy Quantization.*** When an object such an electron is confined in a certain spatial region, standing waves occur for only certain wavelengths and frequencies consistent with the boundary conditions. This is similar to the effect of boundary conditions on standing waves in a string. As will be discussed in more detail in Chapter 35, Schrödinger and others showed, around 1928, that the application of boundary conditions to de Broglie waves led to a new fundamental description of nature called **quantum theory, quantum mechanics,** or **wave mechanics,** built around the concept of a wave function that satisfies an equation known as Schrödinger's equation. A consequence of quantum theory is that the energy of a bound system is quantized, that is, the energy of the system cannot be continuous, but can have only certain discrete values.

***Section 34-5. The Interpretation of the Wave Function.*** Classical physics is built on the notion that it is possible to determine the location of a particle such as an electron with unlimited precision; thus, its trajectory—its location as a function of time—can be known exactly.

Quantum mechanics paints an entirely different picture of the entity that was regarded classically as a "particle." According to quantum mechanics, we can determine only the probability of finding a particle, such as an electron, in a small space around a point. In one dimension, quantum mechanics specifies the probability of finding an electron somewhere in a given spatial interval $dx$, rather than exactly at some point $x$.

The probability of finding a particle in an interval is proportional to the size of the interval. If you double the size of $dx$, the probability of finding the particle in the interval doubles. There is never any certainty that you will find an electron in the interval—there is only a probability. The one thing that can be said with certainty is that you will never find an electron in an interval where the probability is zero.

In a probability description, the **probability density** $P(x)$ gives the likelihood of finding a particle in the vicinity of one value of $x$ rather than another value. The probability of finding a particle in a spatial interval from $x$ to $x + dx$ is

$$\text{Probability} = P(x)\,dx \qquad \text{Probability of finding an object in an interval from } x \text{ to } x + dx$$

In the quantum mechanical picture, the quantity that determines the probability density $P(x)$ is called the **wave function** $\psi(x)$. $P(x)$ and $\psi(x)$ are related by

$$P(x) = \psi^2(x) \qquad \text{Relation between the probability density and the wave function}$$

Thus, $\psi^2\,dx$ is the probability of finding a particle in a spatial interval from $x$ to $x + dx$.

$$\text{Probability} = \psi^2(x)\,dx \qquad \text{Probability of finding an object in an interval from } x \text{ to } x + dx$$

The functional form of the wave function $\psi(x)$ is determined from the Schrödinger wave equation, which will be described in Chapter 35. You can think of $\psi(x)$ as the entity that exhibits the wave-like properties of de Broglie waves, which result in the wave properties of objects.

The probability of finding an object in the interval $dx$ is $\psi^2\,dx$, which represents the probability of finding an object in the spatial interval between $x$ and $x + dx$. The object must certainly be somewhere between $x = -\infty$ and $x = +\infty$, so the sum of the probabilities over all intervals must equal 1:

$$\int_{-\infty}^{\infty} \psi^2 \, dx = 1$$    Normalization condition

This defines a **normalization condition** that the wave function $\psi(x)$ must satisfy. In addition, the wave function $\psi(x)$ must approach zero as $x$ approaches plus or minus infinity.

***Section 34-6. Wave–Particle Duality.*** At times objects exhibit wave properties and at other times particle properties. This is known as **wave–particle duality.** Both the wave and particle pictures are necessary for a complete understanding of an object, but you cannot observe both aspects simultaneously in a single experiment, which is the **principle of complementarity,** first put forth by Bohr in 1928.

*The Two-Slit Experiment Revisited.* In Young's two-slit experiment, visible light is incident on two parallel closely-spaced slits, and an interference pattern of alternating bright and dark bands is observed on a screen placed beyond the slits. For a given slit separation, the separation between the bright and dark bands depends upon the wavelength of the light.

If the experiment is repeated with a beam of electrons instead of visible light, a similar interference pattern of bright and dark bands of electrons is observed on a screen or film. The separation between the bright and dark bands depends upon the de Broglie wavelength associated with the energy of the electron beam. One can watch the buildup of the bands as each individual electron strikes the screen. The buildup is determined by the probability of detecting an electron on the screen, which is proportional to $\psi^2(x)$.

*The Uncertainty Principle.* In a thought experiment, when we say we have located a particle at a given position $x$, we really mean we have determined that its position is somewhere in the interval between $x$ and $x + \Delta x$, where $\Delta x$ is the uncertainty in the position $x$. Similarly, if we measure the momentum $p$ of a moving object, we really mean that we have determined that the object's momentum is somewhere in the interval between $p$ and $p + \Delta p$, where $\Delta p$ is the uncertainty in the momentum $p$.

In classical physics, $\Delta x$ and $\Delta p$ can in principle be individually made to be arbitrarily small. In fact, a basic premise in describing the motion of a classical particle is that rigorously $\Delta x = 0$ and $\Delta p = 0$, so that one can talk unambiguously about the *exact* instantaneous position of a particle along with its *exact* instantaneous velocity or momentum.

When uncertainties are analyzed taking quantum considerations into account, one finds that the position and momentum of an object cannot be specified with infinite precision. Suppose, for example, you want to determine the position of an electron by "looking" at it. To do this measurement, you must bounce a photon off the electron. The interaction of the photon with the electron as the bouncing takes place causes the electron's velocity and momentum to change by an unknown amount.

A rigorous treatment of the quantum-mechanical measurement process, first put forth by Werner Heisenberg in 1927, shows that in a simultaneous determination of the position and momentum of a particle such as an electron, the uncertainty in the momentum, $\Delta p$, and position, $\Delta x$, are related by the **uncertainty principle**:

$$\Delta p \, \Delta x \geq \tfrac{1}{2} \hbar \qquad \text{Heisenberg's uncertainty principle for position and momentum}$$

where $\hbar$ (read $h$ bar) $= h / 2\pi$.

The equality gives intrinsic lower limits, and represents the best uncertainty that is possible. If you try to locate the position of a particle more and more precisely by making $\Delta x$ smaller and smaller, the uncertainty $\Delta p$ in the particle's momentum becomes larger and larger, and vice versa. In any actual experiment, additional experimental uncertainties produce values of $\Delta p$ and $\Delta x$ that are larger than the intrinsic lower limits resulting from wave–particle duality, giving rise to the inequality in the uncertainty relationship.

***Section 34-7. A Particle in a Box.*** A "particle in a box" refers to a particle of mass $m$ that is confined to a one-dimensional region of length $L$, and can never be found outside the boundaries of the box. Very importantly, the energies of a particle in a box are quantized—only certain discrete values are possible. Although somewhat artificial, a particle in a box exhibits many properties, such as discrete energies similar to those of an electron bound within an atom, or a proton inside a nucleus.

Let the box be between $x = 0$ and $x = L$. Since the particle is confined to the box, it can never been found outside the boundaries of the box. This means that the wave function $\psi$ is zero outside the boundaries of the box, that is,

$$\psi = 0 \quad \text{for } x \leq 0 \text{ and for } x \geq L \qquad \text{The wave function must be zero outside the box}$$

A consequence of the **boundary conditions** that $\psi = 0$ at $x = 0$ and $x = L$ is that the energy of the particle cannot be continuous, as in classical theory. Rather, the particle's energy can only take on one of the quantized (discrete) values given by

$$E_n = n^2 \frac{h^2}{8mL^2} = n^2 \, E_1 \qquad n = 1, 2, 3, \ldots \qquad \text{Allowed energies for a particle in a box}$$

where the lowest or **ground-state energy** is given by

$$E_1 = \frac{h^2}{8mL^2} \qquad \text{Ground-state energy for a particle in a box}$$

The integers $n = 1, 2, 3, \ldots$ are called **quantum numbers**. The allowed energies $E_n$ are shown in the **energy level diagram** of Figure 34-3.

Thus, the smallest possible value of the energy is not zero, as in classical theory. Rather, the smallest (ground state) energy, called the **zero-point energy**, is given by the expression for $E_1$. This means that a particle in a box can never be at rest, but must always be moving with one of the allowed energies given by $E_n$. Since $E_1$ varies as $1 / L^2$, the smaller the confines of the box, the larger is the zero-point energy.

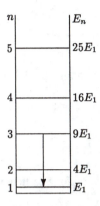

Figure 34-3

The relation of quantized energies to the size of a box is an expression of the uncertainty principle. We know the particle is inside the box, but we do not know exactly where the particle is inside the box, so the uncertainty in the particle's position is $\Delta x = L$. The particle's energy can be any of the quantized values $E_n = n^2 E_1$. A measure of the uncertainty in the particle's energy can be taken as $\Delta E = E_1$. The relationship between a particle's momentum and energy is $p = \sqrt{2mE}$, so the uncertainty $\Delta p$ in momentum corresponding to the uncertainty in energy is $\Delta p = \sqrt{2m\Delta E} = \sqrt{2mE_1} = \sqrt{2m(h^2/8m\,\Delta x^2)} = h/(2\,\Delta x)$, from which $\Delta x\,\Delta p = h/2$, which is a statement of the Heisenberg uncertainty principle for a particle in a box.

It is possible for a system that has quantized energies, such as a particle in a box, to make a transition from an initial energy state $E_i$ to a different final energy state $E_f$. If $E_i$ is higher than $E_f$ ( $E_i > E_f$ ) a photon will be emitted, while if $E_i$ is lower than $E_f$ ( $E_i < E_f$ ) a photon will be absorbed. From conservation of energy, the energy of the emitted or absorbed photon is equal to the energy difference between the initial and final states of the system:

$$hf = hc/\lambda = E_i - E_f \qquad \text{Energy of a photon emitted or absorbed when a system}$$
$$\text{makes a transition between two energy states.}$$

As an example with a particle in a box, a transition from state 3 to the ground state ($n = 1$) is shown by the vertical arrow in Figure 34-3. The frequency of the emitted photon is given by

$$hf = E_i - E_f = E_3 - E_1$$
$$= 3^2 \frac{h^2}{8mL^2} - 1^2 \frac{h^2}{8mL^2} = \frac{h^2}{mL^2}$$

The wavelength of the emitted photon can then be found:

$$\lambda = \frac{c}{f} = \frac{hc}{E_i - E_f} = \frac{hc}{E_3 - E_1} = \frac{mcL^2}{h}$$

**Standing Wave Functions.** The wave functions $\psi_n$ obtained by solving the Schrödinger equation for a particle in a box satisfying the normalization condition, which will be done in Chapter 36, have the same form as for a vibrating string:

$$\psi_n = \sqrt{\frac{2}{L}} \sin\frac{n\pi x}{L} \quad n = 1, 2, 3, \ldots$$

> Wave function for a particle in a box lying between $x = 0$ and $x = L$

The corresponding probability distribution $P_n = \psi_n^2$ is

$$P_n = \psi_n^2 = \frac{2}{L}\sin^2\frac{n\pi x}{L} \quad n = 1, 2, 3, \ldots$$

> Probability distribution for a particle in a box lying between $x = 0$ and $x = L$

Plots of $\psi_n$ and $\psi_n^2$ are shown in Figures 34-4 and 34-5 respectively for the ground state ($n = 1$) and the first two excited states ($n = 2,3$). For each standing wave function, the particle has an energy $E_n = n^2\left(h^2/\left(8mL^2\right)\right) = n^2 E_1$ as given above and shown in Figure 34-3.

For very large values of the quantum number $n$, the maxima of the probability distribution $P_n = \psi_n^2$ are so closely spaced that $\psi^2$ cannot be distinguished from its average value. The fact that the probability distribution is nearly constant across the whole box means that the particle is nearly equally likely to be found anywhere in the box, which is the same as the classical result. An example of this is shown in Figure 34-5 for $n = 10$. This is an example of **Bohr's correspondence principle**:

> In the limit of very large quantum numbers, the classical calculation and the quantum calculation must yield the same results.

*Section 34-8. Expectation Values.* A typical problem in classical mechanics is the specification of the exact position of a particle as a function of time. In contrast, according to quantum mechanics it is intrinsically impossible to specify exactly the position of a particle at any given time. The most we can know about the position of a particle is the relative probability of finding it in one interval or another.

However, if you measure the position $x$ of a particle in a large number of identical experiments, you will obtain some average value of $x$, designated by $\langle x \rangle$. This average value is called the **expectation value** and is related to the wave function $\psi$ by

$$\langle x \rangle = \int_{-\infty}^{+\infty} x\psi^2(x)\,dx$$

> Expectation value of $x$

If, instead of the position $x$, you look at the expectation value of any function $f(x)$, you observe

$$\langle f(x) \rangle = \int_{-\infty}^{+\infty} f(x)\psi^2(x)\,dx$$

> Expectation value of an arbitrary function

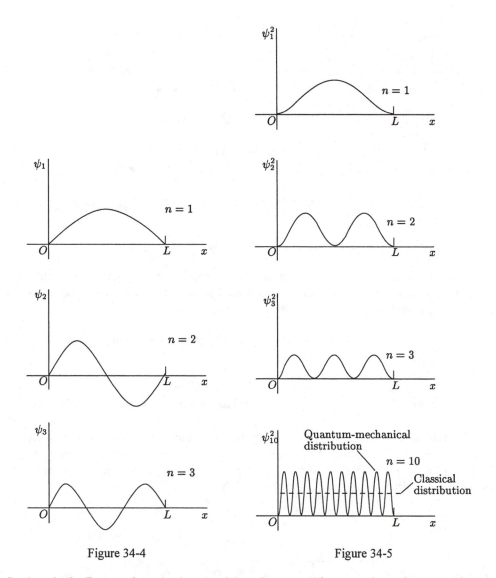

Figure 34-4                    Figure 34-5

***Section 34-9. Energy Quantization in Other Systems.*** When you learn how to solve the Schrödinger equation for quantum mechanical systems in Chapter 35, a key component will be the **potential energy function** $U(x)$ of the system you are looking at. For a particle in a box located between $x = 0$ and $x = L$, which we have looked at above, the potential energy function $U(x)$ is given by

$$U(x) = 0 \qquad 0 < x < L$$
$$U(x) = \infty \qquad x < 0 \text{ or } x > L$$

Potential energy function for a particle in a box

The interpretation of the potential energy function $U(x)$ is that the particle can be found only between the coordinate values $x = 0$ and $x = L$. The particle will never be found outside the box in the region $x < 0$ or $x > L$.

*The Harmonic Oscillator.* The **harmonic oscillator** is a physically more realistic system than a particle in a box. With a harmonic oscillator, a particle of mass $m$ is attached to a spring of force constant $k$ and undergoes small oscillations of amplitude $A$ about a fixed equilibrium point located at $x = 0$. The classical potential energy function $U(x)$ for a harmonic oscillator is

$$U(x) = \tfrac{1}{2}kx^2 = \tfrac{1}{2}m\omega_0^2 x^2 \qquad \text{Potential energy function for a classical harmonic oscillator}$$

where $\omega_0 = \sqrt{k/m}$ is the natural angular frequency in rad/s of the harmonic oscillator. The natural oscillation frequency $f_0$ in vibrations per second is related to $\omega_0$ by $\omega_0 = 2\pi f_0$, so that

$$f_0 = \frac{1}{2\pi}\sqrt{k/m} \qquad \text{Natural oscillation frequency of a classical harmonic oscillator}$$

Classically, the object oscillates between $x = +A$ and $x = -A$.

The allowed energies of a particle in a quantum-mechanical box are quantized. Similarly, a quantum-mechanical analysis of a harmonic oscillator particle (see Chapter 35) shows that the allowed energies are quantized, with the discrete energies $E_n$ given by

$$E_n = (n + \tfrac{1}{2})hf_0 \quad n = 0, 1, 2 \dots \text{ Allowed energies for particles in harmonic oscillator potentials}$$

where

$$E_0 = \tfrac{1}{2}hf_0 \qquad\qquad \text{Ground state energy for a harmonic oscillator particle}$$

is the lowest (ground state) energy of the harmonic oscillator. The allowed energy levels are shown in Figure 34-6, along with the potential energy function $U(x)$. Note that the energy levels for a harmonic oscillator shown in Figure 34-6 are evenly spaced, as compared to the unevenly-spaced energy levels for a particle in a box shown in Figure 34-3.

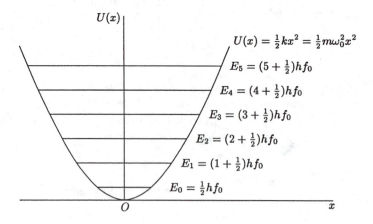

Figure 34-6

The frequency $f$ of a photon emitted when a harmonic oscillator particle undergoes an energy transition from an initial energy state $E_i$ to a final energy state $E_f$ is given by

$$hf = E_i - E_f = (n_i - n_f)hf_0$$

Frequency of an emitted photon when a harmonic oscillator particle undergoes an energy transition

**The Hydrogen Atom.** One main result of a quantum analysis of a particle bound in a box or bound in a harmonic oscillator is that only discrete energies are allowed. Energies are quantized and are described by quantum numbers $n$. A similar situation arises in a hydrogen atom, where an electron is bound to a proton by the electrostatic force of attraction of a nucleus. As will be shown in Chapter 36, the allowed energies of the electron in a hydrogen atom are given by the quantized values

$$E_n = -\frac{13.6\,\text{eV}}{n^2} \quad n = 1, 2, 3, \ldots$$

Allowed energies of an electron in a hydrogen atom

The lowest energy, corresponding to $n = 1$, is the ground-state energy $E_1 = -13.6\,\text{eV}$. Figure 34-7 shows the energy-level diagram for a hydrogen atom.

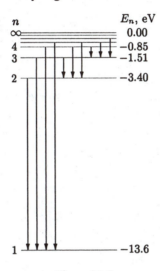

Figure 34-7

As with a particle in a box and a harmonic oscillator, the frequency $f$ of a photon emitted when an electron in a hydrogen atom undergoes an energy transition from an initial energy state $E_i$ to a final energy state $E_f$ is quantized. For the hydrogen atom, it is given by

$$hf = E_i - E_f = \left[-\frac{13.6\,\text{eV}}{n_i^2}\right] - \left[-\frac{13.6\,\text{eV}}{n_f^2}\right]$$

$$= (13.6\,\text{eV})\left[\frac{1}{n_f^2} - \frac{1}{n_i^2}\right]$$

Frequency of an emitted photon when an electron in a hydrogen atom undergoes an energy transition

Transitions from higher to lower energy states are indicated by the vertical arrows on Figure 34-7.

Other atoms are more complicated than hydrogen, but they exhibit discrete energy levels similar to those of hydrogen, with ground-state energies of the order of $-1$ to $-10\,\text{eV}$.

## II.   Physical Quantities and Key Equations

### Physical Quantities

Planck's constant

$$h = 6.626 \times 10^{-34} \text{ J} \cdot \text{s} = 4.136 \times 10^{-15} \text{ eV} \cdot \text{s}$$

$$\hbar = h/2\pi = 1.05 \times 10^{-34} \text{ J} \cdot \text{s}$$
$$= 0.658 \times 10^{-15} \text{ eV} \cdot \text{s}$$

$$hc = 19.865 \times 10^{-26} \text{ J} \cdot \text{m} = 1240 \text{ eV} \cdot \text{nm}$$

Compton wavelength

$$\lambda_C = 2.43 \times 10^{-12} \text{ m} = 2.43 \text{ pm}$$

Ground-state energy for a hydrogen atom

$$E_1 = -13.6\,\text{eV}$$

### Key Equations

Photon energy

$$E = hf = hc/\lambda$$

Einstein's photoelectric equation

$$K_{max} = \left(\tfrac{1}{2}mv^2\right)_{max} = hf - \phi$$

Threshold frequency and wavelength

$$\phi = hf_t = \frac{hc}{\lambda_t}$$

Momentum of a photon in terms of its wavelength

$$p = h/\lambda$$

Energy of a photon

$$E = hc/\lambda$$

Compton equation

$$\lambda_2 - \lambda_1 = \frac{h}{m_e c}\left(1 - \cos\theta\right)$$

de Broglie wavelength of an electron

$$\lambda = h/p$$

Relation between the probability density and the wave function

$$P(x) = \psi^2(x)$$

Probability of finding an object in a spatial interval from $x$ to $x + dx$

$$\text{Probability} = P(x)\,dx = \psi^2(x)\,dx$$

Normalization condition

$$\int_{-\infty}^{\infty} \psi^2\,dx = 1$$

Heisenberg's uncertainty principle for position and momentum

$$\Delta p\,\Delta x \geq \tfrac{1}{2}\hbar$$

*Energy of a photon emitted or absorbed when a system makes a transition from an initial to a final energy state*

$$hf = hc / \lambda = E_i - E_f$$

*Allowed energies for a particle in a box*

$$E_n = n^2 \frac{h^2}{8mL^2} = n^2 E_1 \qquad n = 1, 2, 3, \ldots$$

*Ground-state energy for a particle in a box*

$$E_1 = \frac{h^2}{8mL^2}$$

*Wave function for a particle in a box lying between $x = 0$ and $x = L$*

$$\psi_n = \sqrt{\frac{2}{L}} \sin \frac{n\pi x}{L} \quad n = 1, 2, 3, \ldots$$

*Probability distribution for a particle in a box lying between $x = 0$ and $x = L$*

$$P_n = \psi_n^2 = \frac{2}{L} \sin^2 \frac{n\pi x}{L} \quad n = 1, 2, 3, \ldots$$

*Expectation value of x*

$$\langle x \rangle = \int_{-\infty}^{+\infty} x \psi^2 (x) \, dx$$

*Expectation value of an arbitrary function*

$$\langle f(x) \rangle = \int_{-\infty}^{+\infty} f(x) \psi^2 (x) \, dx$$

*Natural oscillation frequency of a classical harmonic oscillator*

$$f_0 = \frac{1}{2\pi} \sqrt{k/m}$$

*Allowed energies for a quantum-mechanical harmonic oscillator*

$$E_n = (n + \tfrac{1}{2})hf_0 \quad n = 0, 1, 2, \ldots$$

*Ground state energy for a harmonic oscillator particle*

$$E_0 = \tfrac{1}{2} hf_0$$

*Allowed energies of an electron in a hydrogen atom*

$$E_n = -\frac{13.6 \, \text{eV}}{n^2} \quad n = 1, 2, 3, \ldots$$

## III. Potential Pitfalls

Do not simply plug numbers into the photoelectric equation in working photoelectric effect problems. Understand the energy transfers that are going on.

Do not confuse the probability density with probability. The probability density $P(x)$ is a function defined at each point $x$. Probability refers to the probability of finding an object in a small spatial interval $dx$ that straddles a point $x$; for example, the probability of finding an object in an interval between $x$ and $x + dx$. The probability density is related to the probability by $P(x)\,dx$ = probability.

Understand that the probability density $P(x)$ is related to the square of the wave function $\psi(x)$ by $P(x) = \psi^2(x)$.

You might think that there is something unreal about a wave function $\psi$. However, according to quantum mechanics, a wave function is as real a description of an object (such as an electron) as we can get. After all, how real is a point particle in classical physics?

A free electron by itself is not quantized—it can have any arbitrary kinetic energy. The quantization of the energy of an electron occurs only when the electron is confined in some way, say as a particle in a box. When confined, quantization of energy arises from boundary conditions, which can occur only if there are boundaries. The three types of confinement discussed in this chapter are a particle in a box, a harmonic oscillator, and a hydrogen atom.

Be sure you understand the differences in the energy level structures of a particle in a box, a harmonic oscillator, and a hydrogen atom. In particular, make sure you understand how the energy levels depend on the quantum number $n$ for each system.

One main result of quantized systems such as a particle in a box, a harmonic oscillator, and the hydrogen atom is that the energies of the particle in the system is quantized. Make sure you understand that the energy $E = hf$ of an emitted or absorbed photon equals the difference of the energies of the energy levels between which the particle makes a transition.

## IV. True or False Questions and Responses

**True or False**

_____ 1. In a photoelectric effect experiment, as you shine light of a fixed wavelength and increasing intensity on the emitting surface C, the kinetic energies of the electrons emitted from C also increase.

_____ 2. In a photoelectric effect experiment, as you shine light of a fixed wavelength less than the threshold wavelength with larger and larger intensity on the emitting surface C, the current measured at the collecting surface A increases.

_____ 3. For a photoelectric effect experiment with light of a given wavelength less than the threshold wavelength, the larger the work function of the emitting surface C, the smaller the maximum kinetic energy of the electrons ejected from C.

_____ 4. The larger the wavelength of a photon, the more energy it has.

_____ 5. The Compton wavelength of a proton is shorter than the Compton wavelength of an electron.

_____ 6. The larger the scattering angle in a Compton scattering experiment, the larger the change in the wavelength of the scattered photon.

____ 7. The de Broglie wavelength of a neutron is smaller than the de Broglie wavelength of an electron that has the same momentum.

____ 8. As you increase the kinetic energy of an electron, its de Broglie wavelength increases.

____ 9. The probability of finding an object in an interval between $x$ and $x + dx$ is directly proportional to the size $dx$ of the interval.

____ 10. The larger the probability density, the more likely you will find an object in a specified interval $dx$.

____ 11. The probability of finding an object in a given interval $dx$ is proportional to the value of the wave function at the location of $dx$.

____ 12. For a given system, such as a particle in a box or a hydrogen atom, the larger the amplitude of the wave function, the larger will be the quantum number $n$ for an energy level.

____ 13. You cannot detect both the wave and the particle aspects of an object simultaneously.

____ 14. According to the uncertainty principle, the more uncertain that a particle's momentum is, the more uncertain will be its position.

____ 15. Quantized energy levels arise because of confinement of a particle.

____ 16. In a quantum mechanical harmonic oscillator, the frequency of a photon emitted in a transition from $n = 3$ to $n = 2$ is equal to the frequency of a photon emitted in a transition from $n = 2$ to $n = 1$.

____ 17. In a hydrogen atom, the frequency of a photon emitted in a transition from $n = 3$ to $n = 2$ is equal to the frequency of a photon emitted in a transition from $n = 2$ to $n = 1$.

____ 18. On an energy level diagram for a harmonic oscillator, the larger the value of $n$ the higher the energy of the energy level.

____ 19. On an energy level diagram for a hydrogen atom, the larger the value of $n$ the lower the energy of the energy level.

____ 20. When a particle is confined to a certain region of space in a system, the energy of an emitted photon equals the energy difference between the quantized initial and final energy levels of the system through which the particle makes a transition.

## Responses to True or False

1. False. The maximum kinetic energy of the emitted electrons depends only on the frequency of the incident light, as given by Einstein's photoelectric equation. More intense light releases more electrons but does not change their maximum kinetic energy.

2. True.

3.  True.

4.  False. The energy varies as $E = hf = hc/\lambda$, so the larger the wavelength, the smaller the energy.

5.  True.

6.  True.

7.  False. Because the de Broglie wavelength $\lambda$ equals $h/p$, particles with the same momentum have the same wavelength.

8.  False. The larger its kinetic energy, the larger an electron's velocity $v$. Because $\lambda = h/(mv)$, the de Broglie wavelength decreases as an electron's velocity increases with increasing kinetic energy.

9.  True.

10. True.

11. False. The probability is proportional to the *square* of the wave function.

12. False. For a given system, the quantum numbers of the system's energy levels are related to the size of the system and have nothing to do with the amplitude of the wave function.

13. True.

14. False. The more uncertain that a particle's momentum is, the *more precisely* we can locate its position.

15. True.

16. True.

17. False. In a hydrogen atom, the frequency of an emitted photon varies as $f \propto \left[1/n_f^2 - 1/n_i^2\right]$, so the frequency of a photon emitted in a transition from $n = 3$ to $n = 2$ will be different from the frequency of a photon emitted from a transition from $n = 2$ to $n = 1$. (Which frequency will be larger?)

18. True.

19. False. The energy levels in a hydrogen atom are given by $E_n = -(13.6\,\text{eV})/n^2$ (note the minus sign), so the larger the value of $n$, the higher is the value of the energy level (see Figure 34-7).

20. True.

# V. Questions and Answers

## Questions

1. In a photoelectric experiment, why is there a linear relationship between the frequency of incident light and the maximum kinetic energy of emitted electrons?

2. Explain why there is a threshold frequency in a photoelectric experiment.

3. In a photoelectric experiment, what effect does changing the material of the emitting surface C to a new material with a higher work function have on the emitted electrons?

4. In a Compton scattering experiment, how does the change in the wavelength of the scattered photon vary with the scattering angle?

5. How does the de Broglie wavelength of a particle vary with its velocity (neglecting relativistic effects)?

6. An electron is accelerated from rest, acquiring a kinetic energy $K$. How does the de Broglie wavelength of the electron depend on $K$?

7. What two quantities determine the probability of locating an object?

8. How is the wave function related to probability?

9. Describe how energy levels depend upon the quantum number $n$ for (*a*) a particle in a box, (*b*) a harmonic oscillator, and (*c*) a hydrogen atom.

10. Describe an experiment that measures the wave and particle aspects of an object simultaneously.

## Answers

1. In the photoelectric effect, a photon transfers its energy $E = hf$ to an electron in the emitting material C. The electrons at the surface of the emitting material C are least tightly bound and come off with a maximum kinetic energy equal to the energy $hf$ absorbed from the photon minus the work function $\phi$, which is the energy necessary to release an electron from the emitter's surface. This energy balance is described by Einstein's photoelectric equation, $hf = K_{max} + \phi$, which shows there is a linear relationship between the frequency $f$ of the incident light and the maximum kinetic energy $K_{max}$ of the ejected electrons.

2. A threshold photon energy $E_t = hf_t = hc / \lambda_t$ equal to the work function $\phi$ is required to eject the least tightly bound electrons from the surface of the emitting material C: $hc / \lambda_t = \phi$. A photon with less energy than this does not eject any electrons.

3. The larger the work function $\phi$ of the emitting surface C, the more energy is required to release electrons from the emitter's surface. Hence, for incident photons of a given frequency, and therefore a given energy, the larger the work function $\phi$, the lower the maximum kinetic

energy of the emitted electrons. This is seen from the energy balance in Einstein's photoelectric equation $hf = K_{max} + \phi$.

4.    As the scattering angle $\theta$ increases, the change in the wavelength of the scattered photon increases because $\lambda_2 - \lambda_1$ is proportional to $(1 - \cos\theta)$.

5.    The de Broglie wavelength $\lambda$ equals $h/p$. For nonrelativistic situations, $p = mv$, so $\lambda = h/mv$.

6.    The velocity of the electron is related to its kinetic energy by $\frac{1}{2}mv^2 = K$, or $v = (2K/m)^{1/2}$. The de Broglie wavelength is $\lambda = h/mv = h/[m(2K/m)^{1/2}] = h/(2mK)^{1/2}$.

7.    The size $dx$ of the spatial interval, and the probability density $P(x)$. The probability of finding an object in an interval $dx$ is given by $P(x)\,dx$.

8.    The probability density $P(x)$ is the square of the wave function $\psi(x)$: $P(x) = \psi^2(x)$. The probability of finding a particle in an interval $dx$ is $P(x)\,dx = \psi^2(x)\,dx$.

9.    (a) $E_n \propto n^2$, (b) $E_n \propto n$, (c) $E_n \propto -1/n^2$

10.    Such an experiment does not exist. According to Bohr's principle of complementarity, which follows from the uncertainty principle, you cannot measure both the wave and the particle aspects of an object in any single experiment.

# VI. Problems, Solutions, and Answers

**Example #1.**  In a photoelectric effect experiment, it is found that the maximum kinetic energy of emitted electrons for 400-nm light is 2.02 eV. Find the threshold wavelength for the emitter surface.

**Picture the Problem.**  Use the energy balance of Einstein's photoelectric equation to solve for the work function of the material, which can be used to find the threshold wavelength.

| 1. Write down Einstein's photoelectric equation. | $K_{max} = hc/\lambda - \phi$ |
|---|---|
| 2. Substitute the values given into the expression to solve for the work function using conventional units. | $\phi = hc/\lambda - K_{max}$ $= \dfrac{(6.626 \times 10^{-34}\ \text{J·s})(3 \times 10^8\ \text{m/s})}{400 \times 10^{-9}\ \text{m}} - (2.02\ \text{eV})(1.6 \times 10^{-19}\ \text{J/ev})$ $= 1.73 \times 10^{-19}\ \text{J} = 1.08\ \text{eV}$ |
| 3. The same answer can be obtained more easily by using more convenient units introduced in this chapter. | $\phi = hc/\lambda - \phi$ $= \dfrac{1240\ \text{eV·nm}}{400\ \text{nm}} - 2.02\ \text{eV} = 1.08\,\text{eV}$ |

| 4. The work function is related to the threshold wavelength. | $\phi = hc/\lambda_t$ $\lambda_t = hc/\phi = (1240\,\text{eV}\cdot\text{nm})/(1.08\,\text{ev}) = 1148\,\text{nm}$ |

**Example #2—Interactive.** The work function for potassium is 2.21 eV. In a photoelectric effect experiment using a potassium emitter, the maximum kinetic energy of the emitted electrons is observed to be 2.82 eV. What is the wavelength of the incident light?

**Picture the Problem.** Use Einstein's photoelectric equation. **Try it yourself.** Work the problem on your own, in the spaces provided, to get the final answer.

| 1. Write down Einstein's photoelectric equation. | |
| 2. Rearrange the expression to solve for the wavelength of incident light. | |
| 3. Substitute the given values to find the wavelength of incident light. | $\lambda = 247\,\text{nm}$ |

**Example #3.** Find the wavelength of a 0.600-MeV photon after scattering at an angle of 70° in a Compton scattering experiment.

**Picture the Problem.** Use the energy to find the incident wavelength. Then use the Compton relationship to find the scattered wavelength.

| 1. Determine the incident wavelength from the energy. | $E = hc/\lambda$ $\lambda = hc/E = (1240\,\text{eV}\cdot\text{nm})/(0.6\times10^6\,\text{eV})$ $= 2.07\times10^{-3}\,\text{nm} = 2.07\,\text{pm}$ |
| 2. Use the Compton relationship to find the scattered wavelength. | $\lambda_2 - \lambda_1 = \dfrac{h}{m_e c}(1-\cos\theta) = \lambda_C(1-\cos\theta)$ $\lambda_2 = \lambda_C(1-\cos\theta) + \lambda_1$ $= (2.43\,\text{pm})(1-\cos 70°) + 2.07\,\text{pm} = 3.67\,\text{pm}$ |

**Example #4—Interactive.** What is the energy of the electron in Example #3 after the scattering?

**Picture the Problem.**  The energy of the electron will be the difference in the incident and scattered photon energies.  **Try it yourself.**  Work the problem on your own, in the spaces provided, to get the final answer.

| | |
|---|---|
| 1.  Determine the energy of the scattered photon. | |
| 2.  The energy of the electron is the energy difference of the incident and scattered photons. | $E_e = 0.262\,\text{MeV}$ |

**Example #5.**  A typical energy of neutrons in a nuclear reactor is around 0.0400 eV.  What is the de Broglie wavelength of such a neutron?

**Picture the Problem.**  Determine the neutron's momentum, and use that to calculate its wavelength.

| | |
|---|---|
| 1.  Determine the neutron's momentum. | $K = \dfrac{p^2}{2m}$ <br><br> $p = \sqrt{2mK}$ |
| 2.  Use the momentum to determine the de Broglie wavelength of the neutron. | $\lambda = \dfrac{h}{p} = \dfrac{h}{\sqrt{2mK}} = \dfrac{hc}{\sqrt{2mc^2 K}}$ <br><br> $= \dfrac{1240\,\text{eV}\cdot\text{nm}}{\sqrt{2(940\times10^6\,\text{eV})(0.04\,\text{eV})}} = 0.143\,\text{nm}$ |

**Example #6—Interactive.**  What must be the energy of an electron for it to have the same de Broglie wavelength as the neutron in Example #5?

**Picture the Problem.**  Reverse the steps used in Example #5.  **Try it yourself.**  Work the problem on your own, in the spaces provided, to get the final answer.

| | |
|---|---|
| 1.  Use the de Broglie wavelength of the electron to find its momentum. | |
| 2.  From the momentum of the electron, determine its kinetic energy. | $K = 73.6\,\text{eV}$ |

**Example #7.** Find the probability of finding a particle in the central region $L/4 < x < 3L/4$ of a one-dimensional box of length $L$ when the particle is in the second excited stated.

**Picture the Problem.** Determine the probability distribution from the wave function, and integrate over the region of interest.

| | |
|---|---|
| 1. Write down the general expression for the wave functions of a particle in a box. | $\psi_n = \sqrt{\dfrac{2}{L}} \sin \dfrac{n\pi x}{L} \qquad n = 1, 2, 3, \ldots$ |
| 2. Write the wave function for the second excited state; which is the third state overall. | $\psi_3 = \sqrt{\dfrac{2}{L}} \sin \dfrac{3\pi x}{L}$ |
| 3. Determine the probability from the wave function. We need to integrate over the region of interest. | $\text{probability} = \int \psi^2(x)\,dx = \dfrac{2}{L} \int_{L/4}^{3L/4} \sin^2 \dfrac{3\pi x}{L}\,dx$ |
| 4. The integral can be evaluated using a standard integration table. | $\int \sin^2 ax\,dx = \dfrac{x}{2} - \dfrac{1}{4a} \sin 2ax$ |
| 5. This allows us to solve for the probability. There is about a 39% chance of finding the particle in the central region of the box. It spends more time in the turnaround zones. | $\text{probability} = \dfrac{2}{L}\left[ \dfrac{x}{2} - \dfrac{L}{12\pi} \sin \dfrac{6\pi x}{L} \right]_{L/4}^{3L/4}$ $= 0.394$ |

**Example #8—Interactive.** A particle in a one-dimensional box of length $L$ is in its first excited state. If you start from $x = 0$, at what value of $x$ will there be a 25% probability of finding the particle in this interval?

**Picture the Problem.** Follow the same procedure as in Example #7 using $n = 2$. Keeping in mind that the probability equals 1 for finding the particle in the interval from $x = 0$ to $x = L$, draw a picture of the wave function and the probability to help you guess a simple solution to a complicated transcendental equation. **Try it yourself.** Work the problem on your own, in the spaces provided, to get the final answer.

| | |
|---|---|
| 1. Write down the general expression for the wave functions of a particle in a box. | |
| 2. Write down the expression for the first excited state, which is the second state overall. | |
| 3. Sketch the wave function over the range of $0 \le x \le L$. | |

| 4. The probability is the square of the wave function. Sketch the square of the wave function over the range of $0 \leq x \leq L$. Remember, the total integrated probability is 1. | |
| --- | --- |
| 5. From your sketch of the probability, it should be relatively easy to determine the value of $x$ for which the probability of finding the particle in the range from 0 to $x$ is 25%. | $x = L/4$ |

**Example #9.** The size of a nucleus is about $10^{-14}$ m. Treating the nucleus as a one-dimensional particle in a box, find the ground-state energy of a proton and an electron if each particle were bound in the nucleus.

**Picture the Problem.** Substitute the given values into the ground-state energy expression.

| 1. Write down the expression for the ground state energy of a particle in a box. | $E_1 = \dfrac{h^2}{8mL^2} = \dfrac{h^2 c^2}{8mc^2 L^2}$ |
| --- | --- |
| 2. Substitute the given values for a proton. | $E_{1,\text{proton}} = \dfrac{(1240\,\text{eV·nm})^2}{8(938 \times 10^6\,\text{eV})(10^{-5}\,\text{nm})^2} = 2.05\,\text{MeV}$ |
| 3. Substitute the given values for an electron. The energy of an electron is so tremendously large, it is virtually impossible for an electron to exist inside a nucleus. | $E_{1,\text{electron}} = \dfrac{(1240\,\text{eV·nm})^2}{8(0.511 \times 10^6\,\text{eV})(10^{-5}\,\text{nm})^2} = 3,760\,\text{MeV}$ |

**Example #10—Interactive.** For the proton and electron in Example #9, find the wavelength of the photon emitted in a transition from the first excited state to the ground state.

**Picture the Problem.** You already have the ground state energies. Find the energy of the first excited states. The photon energy is the difference between the two states involved in the transition. The wavelength of the emitted photon is related to its energy. **Try it yourself.** Work the problem on your own, in the spaces provided, to get the final answer.

| 1. Determine the energy of the first excited state of the proton. | |
| --- | --- |
| 2. Determine the energy difference of the first excited state and the ground state of the proton. | |

| 3. This energy difference is the same as the energy of the emitted photon. Use this to calculate the wavelength of the emitted photon. | $\lambda = 2.02 \times 10^{-4}$ nm |
|---|---|
| 4. Determine the energy of the first excited state of the electron. | |
| 5. Determine the energy difference of the first excited state and the ground state of the electron. | |
| 6. This energy difference is the same as the energy of the emitted photon. Use this to calculate the wavelength of the emitted photon. | $\lambda = 1.10 \times 10^{-7}$ nm |

**Example #11.** For a harmonic oscillator, find the ratio of the frequency of a photon emitted from an $n = 3$ to $n = 1$ transition to the frequency of a photon emitted from an $n = 2$ to $n = 1$ transition.

**Picture the Problem.** The frequency can be found from the energy of the photon, which is equal to the energy difference of the states involved in the transition.

| 1. Write the general expression for the energy states of a harmonic oscillator. | $E_n = \left(n + \frac{1}{2}\right)hf_0$     $n = 0,1,2,\ldots$ |
|---|---|
| 2. Find the energy difference between the $n = 3$ and $n = 1$ states. | $E_{31} = E_3 - E_1 = \left(3 + \frac{1}{2}\right)hf_0 - \left(1 + \frac{1}{2}\right)hf_0 = 2hf_0$ |
| 3. Find the energy difference between the $n = 2$ and $n = 1$ states. | $E_{21} = \left(2 + \frac{1}{2}\right)hf_0 - \left(1 + \frac{1}{2}hf_0\right) = hf_0$ |
| 4. The energy of a photon is given by $hf$, so we can easily determine the frequency of the two photons, and their ratio. The "harmony" in a harmonic oscillator arises from the fact that the frequency difference between each state is $f_0$. | $f_{31} = 2f_0$ <br> $f_{21} = f_0$ <br> $f_{31} / f_{21} = 2$ |

**Example #12—Interactive.** A 2.00-kg object is attached to the end of a spring with a force constant of 400 N/m and oscillates without friction with an amplitude of 8.00 cm. If the energy of this classical harmonic oscillator were quantized according to quantum mechanical rules, find the value of the corresponding quantum number.

**Picture the Problem.**  Equate the classical expression for the energy of a harmonic oscillator to the quantum mechanical expression for a harmonic oscillator and solve for $n$.  **Try it yourself.** Work the problem on your own, in the spaces provided, to get the final answer.

| | |
|---|---|
| 1.  Find the energy of this classical simple harmonic oscillator. | |
| 2.  Determine the natural frequency of this oscillator. | |
| 3.  Write an expression for the energy of a quantum mechanical harmonic oscillator. | |
| 4.  Equate the expressions from steps 1 and 2 to solve for $n$.  This enormous number is an illustration of Bohr's correspondence principle, where in the limit of very large quantum numbers, classical and quantum calculations must yield the same result. | $n \approx 9 \times 10^{32}$ |

**Example #13.**  The ground-state wave function for a quantum mechanical harmonic oscillator is $\psi(x) = Ce^{-x^2/2A^2}$, where $C$ and $A$ are constants for a given harmonic oscillator.  Find the expectation value of $x^2$.

**Picture the Problem.**  Substitute the given functions into the expression for the expectation value.

| | |
|---|---|
| 1.  Write the general expression for the expectation value. | $\langle f(x) \rangle = \int_{-\infty}^{+\infty} f(x)\psi^2(x)\,dx$ |
| 2.  Substitute the given functions into the expectation value expression. | $\langle x^2 \rangle = \int_{-\infty}^{+\infty} x^2 C^2 e^{-x^2/A^2}\,dx$ |
| 3.  Determine how to calculate this integral from a standard integration table. | $\int_0^{\infty} x^2 e^{-ax^2}\,dx = \frac{1}{4}\sqrt{\frac{\pi}{a^3}}$ |
| 4.  Because this integral is symmetric, we can now easily find the expectation value of $x^2$. | $\langle x^2 \rangle = 2\frac{1}{4}C^2 \sqrt{\frac{\pi}{\left(1/A^2\right)^3}} = \frac{1}{2}C^2 A^3 \sqrt{\pi}$ |

**Example #14—Interactive.**  Find how $C$ is related to $A$ so that the ground-state quantum mechanical oscillator wave function given in Example #13 is normalized.

**Picture the Problem.**  Use the normalization condition.  You will have to use an integral table to evaluate the integral.  **Try it yourself.**  Work the problem on your own, in the spaces provided, to get the final answer.

| | |
|---|---|
| 1.  Write out the normalization condition integral. | |
| 2.  Use an integral table to evaluate the integral. | |
| 3.  Solve for $C$ in terms of $A$. | $C = A^{-1/2}\pi^{-1/4}$ |

# Chapter 35

# Applications of the Schrödinger Equation

## I.   Key Ideas

In Chapter 34 we introduced the notion of a **wave function** $\psi(x)$ which was related to the probability of finding a particle located in a one-dimensional spatial interval $dx$ by

$$\text{Probability} = \psi^2(x)dx$$

> Probability of finding an object in a spatial interval from $x$ to $x + dx$

In Chapter 34 we noted that the wave function $\psi(x)$ must satisfy an equation called the **Schrödinger wave equation.** In this chapter, we discuss the Schrödinger wave equation and its solutions in more detail.

***Section 35-1. The Schrödinger Equation.*** In general, the one-dimensional wave function, which we write as $\Psi(x,t)$, is a function of both the time coordinate $t$ and the one-dimensional position coordinate $x$. The fundamental equation, the wave equation, that gives $\Psi(x,t)$ as a function of $x$ and $t$ cannot be derived, just as Newton's second law, $\Sigma \vec{F} = m\vec{a}$, relating the acceleration of a mass to the forces applied to the mass, cannot be derived.

We postulate the one-dimensional Schrödinger wave equation that determines the probability function for an object of mass $m$:

$$-\frac{\hbar^2}{2m}\frac{\partial^2 \Psi(x,t)}{\partial x^2} + U(x)\Psi(x,t) = i\hbar\frac{\partial \Psi}{\partial t} \qquad \text{Time-dependent Schrödinger equation}$$

In this expression $U(x)$ is the potential energy function for the object, which could be the Coulomb potential energy, the potential energy of a spring, or (prevalent in the following discussions) a constant potential energy that changes its value in sudden jumps at certain locations. It should be noted that the Schrödinger equation involves the imaginary number $i = \sqrt{-1}$.

Often we are not interested in the time behavior of the wave function, but only in its spatial behavior, which determines the probability of finding an object at various locations at a given time. For example, the primary interest may be the most likely location of an electron in a box or in the Coulomb field of a nucleus.

In such a case, we separate the total wave function $\Psi(x,t)$ into two parts. One part $\psi(x)$ depends only on the spatial coordinate (note the lowercase $\psi$). The second part $e^{-i\omega t}$ depends only on the time coordinate and includes the imaginary number $i$. The total wave function is the product of these two parts:

$$\Psi(x,t) = \psi(x)e^{-i\omega t}$$    Separation of wave function

The spatial part of the wave function satisfies the time-independent Schrödinger equation

$$-\frac{\hbar^2}{2m}\frac{d^2\psi(x)}{dx^2} + U(x)\psi(x) = E\psi(x)$$    Time-independent Schrödinger equation

where $E = K + U$ is the total energy of the object. In this book, we will be concerned only with the time-independent Schrödinger equation.

*A Particle in a Box.* One of the easiest problems to solve with the Schrödinger equation is a situation in which a particle is confined in a one-dimensional box of length $L$, which we discussed in Chapter 34. Because the electron can never possess sufficient energy to escape the box, this corresponds to an infinite square-well potential energy of the mathematical form

$$U(x) = 0 \qquad 0 < x < L$$
$$U(x) = \infty \qquad x < 0 \text{ or } x > L$$    Infinite square-well potential energy

The potential energy is shown in Figure 35-1. The problem is to solve the time-independent Schrödinger equation with the square-well potential energy $U(x)$ to obtain the wave function $\psi(x)$, and for each wave function the corresponding energy $E$.

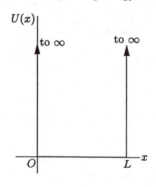

Figure 35-1

The results are as follows: The wave function must satisfy the normalization condition $\int_{-\infty}^{\infty}\psi^2\,dx = 1$, described in Chapter 34, which states that the probability of finding the particle somewhere in the box is 1. This requires the wave function to have the form

$$\psi_n(x) = \sqrt{\frac{2}{L}} \sin\frac{n\pi x}{L} \quad n = 1, 2, 3, \ldots \qquad \text{Wave function for an infinite square-well potential}$$

The integers $n$ are called **quantum numbers.** Plots of $\psi_n$ and the probability distribution function $\psi_n^2$ are shown in Figure 35-2 for the ground state ($n = 1$) and the first two excited states ($n = 2, 3$). Corresponding to each quantum number $n$, there is an energy $E_n$ given by

$$E_n = n^2 E_1 \quad n = 1, 2, 3, \ldots \qquad \text{Allowed energies for an infinite square-well potential}$$

where

$$E_1 = \frac{h^2}{8mL^2} \qquad \text{Ground-state energy for an infinite square-well potential}$$

These energies are shown in the energy-level diagram of Figure 35-3.

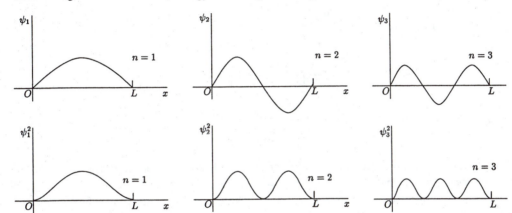

Figure 35-2

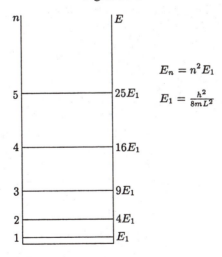

$$E_n = n^2 E_1$$
$$E_1 = \frac{h^2}{8mL^2}$$

Figure 35-3

The preceding results show an important property of the Schrödinger equation, namely, it can generate quantized energy levels that correspond to quantized wave functions. This is in marked contrast with classical ideas which state that energies can have continuous values. Similar quantization appears in the situations that follow.

***Section 35-2. A Particle in a Finite Square Well.*** For a particle in a square well whose walls have a finite (rather than an infinite) height, the potential energy that we put into the Schrödinger equation has the form

$$U(x) = U_0 \quad x < 0$$
$$U(x) = 0 \quad 0 < x < L$$
$$U(x) = U_0 \quad x > L$$

Finite square-well potential energy

The potential energy is shown in Figure 35-4. We shall consider only the case in which $E < U_0$.

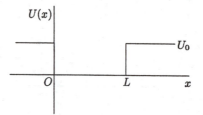

Figure 35-4

Solving the Schrödinger equation with the finite square-well potential energy can be tricky, and matching the boundary conditions at $x = 0$ and $x = L$ is mathematically challenging and somewhat complicated. Consequently we will not give mathematical expressions for $\psi_n$. Instead, we show in Figure 35-5 the end results with graphs of the wave function $\psi_n$ and the corresponding probability distribution $P_n = \psi_n^2$ for the ground state $n = 1$ and for the first two excited states ($n = 2, 3$). These graphs are qualitatively similar to those in Figure 35-2 for an infinite well, except there are exponential "tails" that go into the regions $x < 0$ and $x > L$.

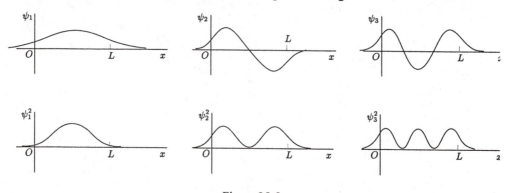

Figure 35-5

For $E < U_0$, the energies corresponding to the allowed wave functions are quantized in a similar way to the energies in an infinite well shown in Figure 35-3, except there are a finite number of quantized energy levels instead of an infinite number. For $E > U_0$ there is a continuum of allowed energies.

Classically, the kinetic energy of a particle is $K = E - U_0$. Because, classically, the kinetic energy must always be positive, a particle can never be found in a region where $E < E_0$. In contrast, the exponential "tail" end of the quantum mechanical results in Figure 35-5 show that there is a finite probability of finding a particle in the classically forbidden regions $x < 0$ and $x > L$.

***Section 35-3. The Harmonic Oscillator.*** An extremely important problem in quantum mechanics is the solution of the Schrödinger equation for a particle whose potential energy is that of a harmonic oscillator. This could, for example, describe small oscillations of atoms in a diatomic molecule oscillating about their equilibrium separation.

Classically, the potential energy of a particle of mass $m$ attached to a spring of force constant $k$ is

$$U(x) = \tfrac{1}{2}kx^2 \qquad \text{Potential energy of a classical object attached to a spring}$$

where $x$ is the amount of stretch or compression of the spring from its normal unstretched position. The angular frequency of an object oscillating at the end of the spring is $\omega_0 = \sqrt{k/m}$, so that the potential energy can also be written as

$$U(x) = \tfrac{1}{2}m\omega_0^2 x^2 \qquad \text{Potential energy of a classical object attached to a spring}$$

The potential energy function is shown in Figure 35-6. Classically, a particle oscillates between $x = \pm A$ with a total energy $E = \tfrac{1}{2}m\omega_0^2 A^2$, and can have any total energy $E$ from zero to any positive value. This is to be contrasted with the quantum mechanical results discussed next, where the total energy of a quantum mechanical harmonic oscillator can only take on certain discrete values.

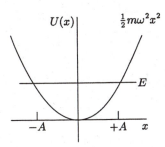

Figure 35-6

When the time-independent Schrödinger equation is solved for a harmonic oscillator potential energy function, the ground-state wave function is

$$\psi_0(x) = A_0 e^{-ax} \qquad a = m\omega_0 /(2\hbar) \quad \text{Ground-state wave function for a harmonic oscillator}$$

where $A_0$ is the normalization constant. This is a Gaussian function whose plot is shown in Figure 35-7.

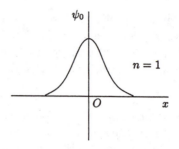

Figure 35-7

The ground-state energy corresponding to the ground state wave function is

$$E_0 = \tfrac{1}{2}\hbar\omega_0 \qquad \text{Ground-state energy for a harmonic oscillator}$$

The energy corresponding to higher-level wave functions is quantized and can take on only the discrete values given by

$$E_n = \left(n + \tfrac{1}{2}\right)\hbar\omega_0, \quad n = 0,1,2,\dots \quad \text{Energy of the nth excited state for a harmonic oscillator}$$

This expression shows that the energy levels of a quantum mechanical harmonic oscillator are evenly spaced by the amount $\hbar\omega_0$, as shown in the energy-level diagram of Figure 35-8. In contrast, the energies of a classical harmonic oscillator are not quantized and can have any values.

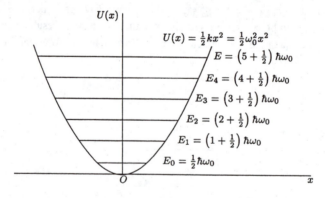

Figure 35-8

***Section 35-4. Reflection and Transmission of Electron Waves: Barrier Penetration.*** In Sections 35-2 and 35-3 we saw that the wave functions for particles with total energies $E$ less than potential energies $U(x)$ could penetrate into classically forbidden regions where $E < U(x)$. Here we look at similar situations where a particle is incident on a step potential energy and a barrier potential energy.

***Step Potential.*** A step potential energy function changes abruptly from zero to a finite value as shown in Figure 35-9 and is represented mathematically by

$$U(x) = 0 \quad x < 0$$

$$U(x) = U_0 \quad x > 0$$

Step potential energy

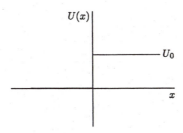

Figure 35-9

If the energy $E$ of the wave function incident from the left is smaller than the step potential energy $U_0$, most of the wave function is reflected at the original wavelength, but a small part is transmitted into the barrier and decays exponentially, as shown in Figure 35-10. If the energy $E$ of the incident wave is larger than the barrier potential energy $U_0$, part of the wave is reflected at the original wavelength and part of the wave is transmitted at a larger wavelength.

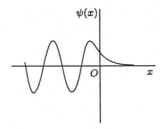

Figure 35-10

The probability of reflection $R$, called the **reflection coefficient,** is given by

$$R = \frac{(k_1 - k_2)^2}{(k_1 + k_2)^2}$$

Reflection coefficient for a wave incident on a step potential energy

Where $k_1$ and $k_2$ are the wave numbers for the incident and transmitted waves, respectively ($k = 2\pi / \lambda$). The probability of transmission $T$ of the transmitted wave, the **transmission coefficient,** is related to the reflection coefficient by

$$T = 1 - R$$

Transmission coefficient for a wave incident on a step potential energy

**Barrier Potential.** A barrier potential energy is like a step potential energy except that after a certain distance $a$ the constant potential energy $U_0$ changes abruptly back to zero, as shown in Figure 35-11. A barrier potential is represented mathematically by

$$U(x) = 0 \quad x < 0$$
$$U(x) = U_0 \quad 0 < x < a$$
$$U(x) = 0 \quad x > a$$

Barrier potential energy function

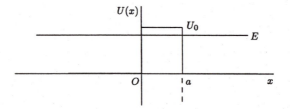

Figure 35-11

Consider a particle, represented by its wave function, incident on a barrier potential at $x = 0$, as shown in Figure 35-12. If the energy of the particle is less than the energy $U_0$ of the barrier energy, the wave function is partially reflected; but there also is exponential transmission of part of the wave function in the region $0 < x < a$, which is forbidden from classical energy considerations. When the exponentially transmitted wave function reaches the other end of the barrier at $x = a$, part of it is reflected again, but another part is transmitted and emerges on the other side with a reduced amplitude but with a wavelength the same as its initial wavelength.

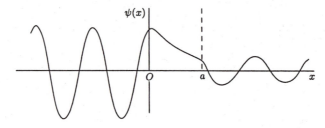

Figure 35-12

This transmission is referred to as **barrier penetration.** Because of the reflections and transmissions at the two sides of the barrier ($x = 0$ and $x = a$) the analysis is somewhat complicated. For the case where the barrier size $a$ and the function $\alpha = \sqrt{2m(U_0 - E)/\hbar^2}$ that depends on the energy difference $(U - E)$ are related such that $\alpha a \gg 1$, the transmission coefficient $T$ follows an exponential expression given by

$$T = e^{-2\alpha a} \qquad \alpha = \sqrt{\frac{2m(U_0 - E)}{\hbar^2}} \quad \text{Transmission coefficient through a barrier where } \alpha a \gg 1$$

The transmission coefficient gives the ratio of the probability distribution function of the transmitted wave function to that of the wave function incident on the front of the barrier: $T = P_{\text{trans}} / P_{\text{incident}}.$

Quantum mechanical "tunneling" through classically forbidden regions explains phenomena such as the emission of $\alpha$ particles from a nucleus and the operation of a **scanning tunneling microscope.**

***Section 35-5. The Schrödinger Equation in Three Dimensions.*** So far, we have looked at a particle that is restricted to motion in one dimension. In three dimensions, the time-independent Schrödinger equation in cartesian coordinates has the form

$$-\frac{\hbar^2}{2m}\left(\frac{\partial^2 \psi}{\partial x^2}+\frac{\partial^2 \psi}{\partial y^2}+\frac{\partial^2 \psi}{\partial z^2}\right)+U\psi = E\psi \qquad \text{Three-dimensional Schrödinger equation}$$

To see the differences between three dimensions and one dimension, imagine a particle confined to a three-dimensional cubical box of side $L$, in which the potential energy is

$$U(x,y,z)=0 \quad 0<x,y,z<L$$
$$U(x,y,z)=\infty \quad x,y,z \text{ outside the box}$$

Three-dimensional infinite square-well potential

For this three-dimensional potential, the solution to the Schrödinger equation has the form

$$\psi(x,y,z)=A\sin\frac{n_1\pi x}{L}\sin\frac{n_2\pi y}{L}\sin\frac{n_3\pi z}{L} \qquad \text{Three-dimensional wave function}$$
$$n_1=1,2,3,\dots \quad n_2=1,2,3,\dots \quad n_3=1,2,3,\dots$$

and the quantized energies are

$$E_{n_1,n_2,n_3}=\frac{\hbar^2\pi^2}{2mL^2}\left(n_1^2+n_2^2+n_3^2\right) \qquad \text{Three-dimensional energy levels}$$

If more than one wave function is associated with the same energy level, that energy level is said to be **degenerate**. For example, the three combinations

$$(n_1,n_2,n_3)=(2,1,1);(1,2,1);(1,1,2)$$

all correspond to the same energy level

$$E_{n_1,n_2,n_3}=6\frac{\hbar^2\pi^2}{2mL^2}=2E_{1,1,1}=6E_1$$

where $E_1$ is the ground-state energy for an infinite one-dimensional square-well potential, as shown in Figure 35-13$a$. However, these three states describe three different wave functions, so there is a threefold degeneracy for this energy level. The degeneracy is removed if the sides of the box are made unequal, as shown in Figure 35-13$b$ for the case $L_1<L_2<L_3$, where $U=0$ for $0<x<L_1$, $0<y<L_2$, and $0<z<L_3$.

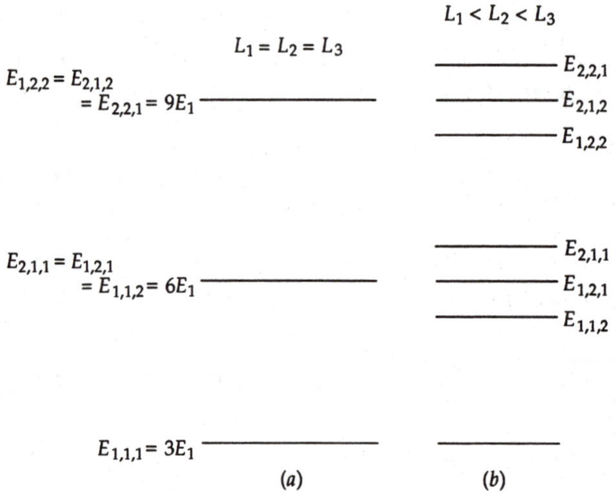

$L_1 < L_2 < L_3$
$L_1 = L_2 = L_3$
$E_{1,2,2} = E_{2,1,2} = E_{2,2,1} = 9E_1$
$E_{2,2,1}$
$E_{2,1,2}$
$E_{1,2,2}$
$E_{2,1,1} = E_{1,2,1} = E_{1,1,2} = 6E_1$
$E_{2,1,1}$
$E_{1,2,1}$
$E_{1,1,2}$
$E_{1,1,1} = 3E_1$
(a)
(b)

Figure 35-13

***Section 35-6. The Schrödinger Equation for Two Identical Particles.*** Classically, the positions of two particles can be followed through any interactions without ambiguity along well-defined distinguishable trajectories, as shown in Figure 35-14*a,b*. In quantum mechanics, when two identical particles come together, interact, and then move apart, it is not possible after the interaction for us to know which particle is which. This is illustrated in Figure 35-14*c*.

①
②
①
②
①
②
②
①
(a)
(b)
①
②
① or ②?
① or ②?
(c)

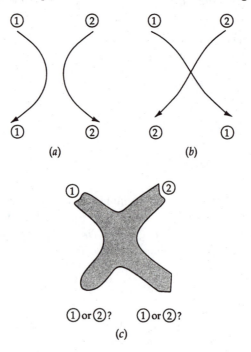

Figure 35-14

Designate by $x_1(t)$ and $x_2(t)$ the coordinates of the trajectories of two identical particles. Let $\psi(x_1, x_2)$ represent the combined wave function of two identical particles that follow one set of trajectories, and let $\psi(x_2, x_1)$ be the wave function for interchanged trajectories. The wave functions are said to be symmetric or antisymmetric upon the exchange of the two identical particles according to

$$\psi(x_2, x_1) = \psi(x_1, x_2) \qquad\qquad \text{Definition of symmetric wave function}$$

$$\psi(x_2, x_1) = -\psi(x_1, x_2) \qquad\qquad \text{Definition of antisymmetric wave function}$$

Both symmetric and antisymmetric wave functions yield the same probability densities because

$$\psi^2(x_2, x_1) = \psi^2(x_1, x_2) \qquad\qquad \text{Probability densities are the same}$$

As an example, consider the wave function describing two identical noninteracting particles in a one-dimensional box. In Section 35-1, we saw that the wave function for one particle in a box is

$$\psi_n(x) = \sqrt{\frac{2}{L}} \sin\frac{n\pi x}{L}$$

For two particles in a box, the symmetric and antisymmetric wave functions are

$$\psi_S(x_1, x_2) = A'\left[\psi_n(x_1)\psi_m(x_2) + \psi_n(x_2)\psi_m(x_1)\right] \qquad\qquad \text{symmetric}$$

$$\psi_A(x_1, x_2) = A'\left[\psi_n(x_1)\psi_m(x_2) - \psi_n(x_2)\psi_m(x_1)\right] \qquad\qquad \text{antisymmetric}$$

Wave functions for two particles in a box

where $A'$ is a constant determined by the normalization condition for the two-particle wave function.

An important difference is seen between symmetric and antisymmetric wave functions when the two quantum numbers are equal. When $n = m$, the antisymmetric wave function is identically zero for all values of $x_1$ and $x_2$, while the symmetric wave function is not zero.

A two-electron system is described by an antisymmetric wave function. If the two wave functions of the individual electrons that make up an overall antisymmetric wave function have the same quantum numbers, the overall wave function is identically zero. This is an example of the **Pauli exclusion principle** for electrons in an atom:

No two electrons in an atom can have the same quantum numbers.

Besides electrons, other particles such as protons and neutrons, called **fermions,** are described by antisymmetric wave functions and obey the Pauli exclusion principle. Another class of particles, called **bosons,** which includes $\alpha$ particles, deuterons, photons, and mesons, are described by symmetric wave functions and do not obey the Pauli exclusion principle.

## II.  Physical Quantities and Key Equations

### Physical Quantities

There are no new physical quantities introduced in this chapter.

### Key Equations

Time-dependent Schrödinger equation
$$-\frac{\hbar^2}{2m}\frac{\partial^2\Psi(x,t)}{\partial x^2}+U(x)\Psi(x,t)=i\hbar\frac{\partial\Psi}{\partial t}$$

Separation of wave function
$$\Psi(x,t)=\psi(x)e^{-i\omega t}$$

Time-independent Schrödinger equation
$$-\frac{\hbar^2}{2m}\frac{d^2\psi(x)}{dx^2}+U(x)\psi(x)=E\psi(x)$$

Infinite square-well potential energy
$$U(x)=0 \quad 0<x<L$$
$$U(x)=\infty \quad x<0 \text{ or } x>L$$

Wave function for a particle in a box
$$\psi_n(x)=\sqrt{\frac{2}{L}}\sin\frac{n\pi x}{L} \quad n=1,2,3,\dots$$

Ground-state energy for a particle in a box     $E_1=h^2/(8mL^2)$

Allowed energies for a particle in a box     $E_n=n^2E_1 \quad n=1,2,3,\dots$

Finite square-well potential energy
$$U(x)=U_0 \quad x<0$$
$$U(x)=0 \quad 0<x<L$$
$$U(x)=U_0 \quad x>L$$

Potential energy of a classical object attached to a spring     $U(x)=\tfrac{1}{2}Kx^2=\tfrac{1}{2}m\omega_0^2x^2$

Ground-state wave function for a harmonic oscillator     $\psi_0(x)=A_0e^{-ax} \quad a=m\omega_0/(2\hbar)$

Ground-state energy for a harmonic oscillator     $E_0=\tfrac{1}{2}\hbar\omega_0$

Energy of the nth excited state for a harmonic oscillator     $E_n=\left(n+\tfrac{1}{2}\right)\hbar\omega_0, \quad n=0,1,2,\dots$

Step potential energy
$$U(x)=0 \quad x<0$$
$$U(x)=U_0 \quad x>0$$

Reflection coefficient for a wave incident on a step potential energy     $R=\dfrac{\left(k_1-k_2\right)^2}{\left(k_1+k_2\right)^2}$

*Transmission coefficient for a wave incident on a step potential energy*     $T = 1 - R$

*Barrier potential energy*

$$U(x) = 0 \quad x < 0$$
$$U(x) = U_0 \quad 0 < x < a$$
$$U(x) = 0 \quad x > a$$

*Transmission coefficient through a barrier*     $T \propto e^{-2\alpha a} \quad \alpha = \sqrt{\dfrac{2m(U_0 - E)}{\hbar^2}}$

*Three-dimensional Schrödinger equation*     $-\dfrac{\hbar^2}{2m}\left(\dfrac{\partial^2 \psi}{\partial x^2} + \dfrac{\partial^2 \psi}{\partial y^2} + \dfrac{\partial^2 \psi}{\partial z^2}\right) + U\psi = E\psi$

*Three-dimensional infinite square-well potential energy (three-dimensional box)*

$$U(x, y, z) = 0 \quad 0 < x < L, 0 < y < L, 0 < z < L$$
$$U(x, y, z) = \infty \quad x, y, z \text{ outside the box}$$

*Wave function for a particle in a three-dimensional box*

$$\psi = A \sin\dfrac{n_1 \pi x}{L} \sin\dfrac{n_2 \pi y}{L} \sin\dfrac{n_3 \pi z}{L}$$
$$n_1 = 1, 2, 3, \ldots \qquad n_2 = 1, 2, 3, \ldots$$
$$n_3 = 1, 2, 3, \ldots$$

*Energy levels for a three-dimensional box*     $E_{n_1, n_2, n_3} = \dfrac{\hbar^2 \pi^2}{2mL^2}\left(n_1^2 + n_2^2 + n_3^2\right)$

*Symmetric wave functions*     $\psi(x_2, x_1) = \psi(x_1, x_2)$

*Antisymmetric wave functions*     $\psi(x_2, x_1) = -\psi(x_1, x_2)$

*Wave functions for two particles in a box*

$$\psi_S(x_1, x_2) = A'\left[\psi_n(x_1)\psi_m(x_2) + \psi_n(x_2)\psi_m(x_1)\right] \qquad \text{symmetric}$$

$$\psi_A(x_1, x_2) = A'\left[\psi_n(x_1)\psi_m(x_2) - \psi_n(x_2)\psi_m(x_1)\right] \qquad \text{antisymmetric}$$

where     $\psi_n(x) \propto \sin\dfrac{n \pi x}{L}$

## III. Potential Pitfalls

Do not confuse the time-dependent wave function $\Psi(x, t)$ with the time-independent wave function $\psi(x)$. They are related by $\Psi(x, t) = \psi(x)e^{-i\omega t}$. You will usually be interested in the time-independent wave function $\psi(x)$.

Do not confuse the wave function $\psi$ with the probability density $P = \psi^2$.

Understand the qualitative differences among the wave functions, energy levels, and tunneling effects of the various types of potential energy functions: (1) particle in a box; (2) particle in a finite square well; (3) harmonic oscillator; (4) step potential; (5) barrier penetration.

Do not confuse the reflection and transmission coefficients for a particle incident on a step potential.

Understand the difference between symmetric and antisymmetric wave functions for two identical particles.

Do not confuse fermions and bosons.

## IV.  True or False Questions and Responses

**True or False**

____ 1.  Quantized energy levels arise because of boundary conditions on the potential energy.

____ 2.  Specification of a potential energy function in the time-independent Schrödinger equation determines a single unique wave function for a particle.

____ 3.  Specification of a potential energy function in the time-independent Schrödinger equation determines a single unique energy for a particle.

____ 4.  A particle has a unique energy for a given one dimensional wave function $\psi_n$ of the particle.

____ 5.  For a given potential energy, the ground-state energy is the lowest energy that a particle can have.

____ 6.  For a particle in a box having infinitely high walls, an object will never be found in the region beyond the extent of the box.

____ 7.  The wave function for a particle in a box varies sinusoidally with position.

____ 8.  An object can never be found in a region where its total energy is less than its potential energy.

____ 9.  The discrete energy levels for a particle in a finite square well range from the ground-state energy to an infinite energy.

____ 10.  The energy levels of a particle in a box are evenly spaced.

____ 11.  The energy levels of a particle in a harmonic energy potential are evenly spaced.

____ 12.  The ground-state wave function for a particle in a harmonic-oscillator potential energy well varies sinusoidally with position.

_____ 13. For a particle incident on a step potential energy function, the sum of the reflection and transmission coefficients equals 1.

_____ 14. When an object penetrates a step potential from the left with $E < U_0$, the wavelength of the wave function on the left side of the step is smaller than the wavelength on the right side.

_____ 15. When an object penetrates a barrier from the left, the wavelength of the wave function on the left side of the barrier is smaller than the wavelength on the right side.

_____ 16. Degeneracy refers to a situation where a wave function becomes infinite.

_____ 17. The probability density for a symmetric wave function is the same as for an antisymmetric wave function (relative to the exchange of two particles).

_____ 18. When two identical particles are described by an antisymmetric wave function, the antisymmetric wave function is identically zero when all of the quantum numbers of each of the particles are the same.

_____ 19. When two identical particles are described by a symmetric wave function, the symmetric wave function is identically zero when all of the quantum numbers of each of the particles are the same.

_____ 20. When a fermion is raised to a higher energy level, it becomes a boson.


**Responses to True or False**

1. True.

2. False. For a given potential energy function, there is a family of wave functions $\psi_n$ that are solutions to the Schrödinger equation.

3. False. For a given potential energy function, there are $n$ quantized energy levels $E_n$ that a particle may have, each one corresponding to the wave functions $\psi_n$ mentioned in 2.

4. True.

5. True.

6. True.

7. True.

8. False. A quantum mechanical analysis predicts tunneling into this classically forbidden region.

9. False. The discrete energy levels of a particle range from the ground-state energy to a maximum energy level that is less than the height $U_0$ of the finite square-well potential. If a particle has an energy above $U_0$, the energy is not quantized and varies continuously.

10. False. The energy levels vary like $E_n = n^2 E_1$.

11. True.

12. False. The ground-state wave function varies like an exponential Gaussian distribution.

13. True.

14. False. On the right side the wave function follows an exponential decay with no associated wavelength.

15. False. Because the kinetic energies on both sides of the barrier are the same, the wavelengths are also the same. The amplitude of the wave function is larger on the left side than on the right side.

16. False. Degeneracy refers to a situation in which a given energy level can be associated with two or more wave functions.

17. True.

18. True. This is the origin of the Pauli exclusion principle.

19. False. A symmetric function is not zero when the quantum numbers of the two particles are the same.

20. False. Fermions and bosons are two different classes of particles. A fermion is described by an antisymmetric wave function, and no two fermions can have the same quantum numbers. Bosons are described by symmetric wave functions, and it is possible for two bosons to have the same quantum numbers.

## V.    Questions and Answers

**Questions**

1.  Classically, why is a region where $E < U_0$ called a "forbidden" region?

2.  What is the main difference between the energy levels of a particle in an infinite square well and in a finite square well?

3.  What is the main difference between the ground-state wave functions for a particle in a box and a harmonic oscillator?

4.  A particle with energy $E < U_0$ is incident from the left on a step potential. On which side of the step potential is the wavelength of the particle longer, and why?

5.  When a particle penetrates a potential barrier, on which side of the barrier is the wavelength of the wave function greater?

6. When a particle penetrates a potential barrier, on which side of the barrier is the amplitude of the wave function greater?

7. What is the main difference between an antisymmetric and a symmetric wave function for two identical particles?

8. What is the main difference between fermions and bosons?

**Answers**

1. In classical physics, the kinetic energy of a particle with total energy $E$ is given by $K = E - U(x)$, where $U(x)$ is the potential energy of the particle. The kinetic energy $K = \frac{1}{2}mv^2$ must always be positive, so it is impossible to have $E < U(x)$. As an example to visualize this, think of a ball having a total energy $E$ rolling up the sloping side of a hill described by a potential energy function $U(x)$. There will be a value $x_0$ where $U(x_0) = U_0 = E$. When the ball reaches the point $E = U_0$ where its total and potential energies are equal, $K = E - U_0$ becomes zero and the ball stops moving upward and rolls back. The ball can never enter the forbidden region where $U(x) > U_0 = E$.

2. There are an infinite number of quantized energy levels in an infinite square well, but there are a finite number of quantized energy levels in a finite square well.

3. The ground-state wave function for a particle in a box varies as a sine function for one-half cycle, while the ground-state wave function for a harmonic oscillator varies like a Gaussian exponential.

4. Once the wave function encounters the step potential, it is no longer characterized by any wavelength. The wave function instead decays exponentially.

5. The wavelength is the same on both sides of the barrier, as long as the barrier is symmetric.

6. The amplitude is greater on the side where the particle is incident. As the particle penetrates the barrier, the amplitude of its wave function decreases exponentially, so it is smaller when it emerges from the barrier.

7. If two particles have the same quantum numbers, the antisymmetric wave function is identically zero. This is not the case for a symmetric wave function.

8. The wave function for two fermions is antisymmetric, while the wave function for bosons is symmetric. This means that fermions satisfy the Pauli exclusion principle, while bosons do not satisfy the Pauli exclusion principle.

## VI. Problems, Solutions, and Answers

**Example #1.** Show that $\psi_n = A_n \sin(n\pi x/L)$ is a solution of the time-independent Schrödinger equation for a particle in a box of length $L$.

**Picture the Problem.**   Calculate the first and second derivatives of $\psi_n$ and substitute into Schrödinger's equation.  If you end up with an energy that satisfies the quantization condition given in the text, then the provided wave function is a solution to Schrödinger's equation.

| | |
|---|---|
| 1.  Write out the time-independent Schrödinger equation as a guide for the problem. | $-\dfrac{\hbar^2}{2m}\dfrac{d^2\psi(x)}{dx^2}+U(x)\psi(x)=E\psi(x)$ |
| 2.  Determine the potential energy function. | Inside the box, we know that $U(x)=0$. |
| 3.  Find the second derivative of $\psi(x)$. | $\dfrac{d\psi(x)}{dx}=A_n\dfrac{n\pi}{L}\cos\dfrac{n\pi x}{L}$ <br><br> $\dfrac{d^2\psi(x)}{dx^2}=-A_n\dfrac{n^2\pi^2}{L^2}\sin\dfrac{n\pi x}{L}$ |
| 4.  Substitute these values into Schrödinger's equation and solve for $E$.  The energy that we arrived at satisfies the quantization condition given in the text, so the given wave function must be a solution to Schrödinger's equation for a particle in a box. | $-\dfrac{\hbar^2}{2m}\left(A_n\dfrac{n^2\pi^2}{L^2}\right)\left(-\sin\dfrac{n\pi x}{L}\right)+0=EA_n\sin\dfrac{n\pi x}{L}$ <br><br> $\dfrac{\hbar^2}{2m}\dfrac{n^2\pi^2}{L^2}=E$ <br><br> $\dfrac{h^2}{4\pi^2}\dfrac{1}{2m}\dfrac{n^2\pi^2}{L^2}=E$ <br><br> $\dfrac{h^2n^2}{8mL^2}=E$ |

**Example #2—Interactive.**   Show that the first excited wave function $\psi_1(x)=A_1xe^{-ax^2}$, shown in Figure 35-15, is a solution of the Schrödinger equation for a harmonic oscillator potential energy function.

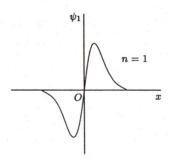

Figure 35-15

**Picture the Problem.**   Follow the solution to Example #1.  **Try it yourself.**   Work the problem on your own, in the spaces provided, to get the final answer.

| | |
|---|---|
| 1.  Write out Schrödinger's equation as a guide for the problem. | |

| | |
|---|---|
| 2. Determine the potential energy function. | |
| 3. Find the second derivative of $\psi_1(x)$. | |
| 4. Substitute these values into Schrödinger's equation and solve for $E$. The energy that we arrived at satisfies the quantization condition given in the text, so the given wave function must be a solution to Schrödinger's equation for a particle in a box. | If $a = m\omega_0/(2\hbar)$ and $E_1 = (3/2)\hbar\omega_0$, then the function is a valid solution. |

**Example #3.** A 5.00-MeV electron strikes a 20.0-MeV potential barrier that is $10^{-13}\,\text{m} = 100\,\text{fm}$ wide. Determine the probability that the electron will tunnel through the barrier.

**Picture the Problem.** Use the transmission coefficient for barrier penetration.

| | |
|---|---|
| 1. Write the expression for the transmission coefficient to guide you through the problem. | $$\frac{P_{\text{trans}}}{P_{\text{incident}}} = T = e^{-2\alpha a} \qquad \alpha = \frac{\sqrt{2m(U_0 - E)}}{\hbar}$$ |
| 2. Determine the value of $\alpha$ by substituting the given values. | $$\alpha = \frac{\sqrt{2m(U_0 - E)}}{\hbar} = \frac{\sqrt{2mc^2(U_0 - E)}}{\hbar c}$$ $$= \frac{\sqrt{2(0.511\,\text{MeV})(20\,\text{MeV} - 5\,\text{MeV})}}{197.4\,\text{MeV}}$$ $$= 1.98 \times 10^{-2}\,\text{fm}^{-1}$$ |
| 3. Substitute into the expression for the transmission coefficient to find the probability the electron will pass through the barrier. | $$T = e^{-2\alpha a} = e^{-2(1.98 \times 10^{-2}\,\text{fm}^{-1})(100\,\text{fm})} = 0.019 \text{ or } 1.9\%$$ |

**Example #4.** In Example #3, what energy of an incident electron will give it a 5% probability of being transmitted through the barrier.

**Picture the Problem.** Follow Example #3, only work it backward to solve for the incident energy.

| | |
|---|---|
| 1. Write the expression for the transmission probability through the barrier. | |
| 2. Rearrange the expression to solve for $\alpha$. | |
| 3. Use the expression for $\alpha$ to solve for the required incident energy. | |
| | $E = 11.5 \, \text{MeV}$ |

# Chapter 36

# Atoms

## I.  Key Ideas

Each element is characterized by a neutral atom that contains a number of protons $Z$, an equal number of electrons, and a number of neutrons $N$. The protons and neutrons are in the nucleus, and the electrons orbit the nucleus. The chemical and physical properties of an element are determined by the number and arrangement of the electrons in the atom.

*Section 36-1. The Nuclear Atom.* Let us first look at the simplest atom—the hydrogen atom, which consists of a central proton (nucleus) around which there is a single orbiting electron.

*Atomic Spectra.* When a hydrogen gas is excited, say by the application of a voltage, the emitted light consists of discrete wavelengths which satisfy the remarkably simple **Rydberg–Ritz formula,**

$$\frac{1}{\lambda} = RZ^2 \left( \frac{1}{n_2^2} - \frac{1}{n_1^2} \right) \qquad n_1 > n_2 \quad n_1, n_2 = 1, 2, 3, \ldots \qquad \textit{Rydberg–Ritz formula}$$

For hydrogen, $Z = 1$, and the **Rydberg constant** is

$$R = 10,977,760 \, \text{m}^{-1} = 10.97776 \, \mu\text{m}^{-1} \qquad \text{Rydberg constant for hydrogen}$$

The Rydberg–Ritz formula also describes the wavelengths in the spectra of **hydrogenlike ions,** which are ions with a single electron and any nuclear charge, such as singly ionized helium $\text{He}^+$ with $Z = 2$ and doubly ionized lithium $\text{Li}^{2+}$ with $Z = 3$. As $Z$ increases, $R$ increases by a small, known amount.

*Section 36-2. The Bohr Model of the Hydrogen Atom.* In 1913 Niels Bohr put forth a model of the hydrogen atom that was in extraordinary agreement with the Rydberg–Ritz formula. Bohr's model is based on a planetary picture of a hydrogenlike atom consisting of a light electron moving in a circular orbit around a heavy nucleus. The attractive Coulomb force $F_c = kZe^2 / r^2$ between the orbiting electron and $Z$ protons in the nucleus produces the necessary centripetal force for the electron to move in the circular orbit.  The total energy of this circular orbit is given by

$$E = -\frac{1}{2}\frac{kZe^2}{r}$$    Orbital energy of an electron circling a nucleus

**Bohr's Postulates.** Bohr's model is built on three postulates.

*Bohr's First Postulate: Nonradiating Orbits (Stationary States).* According to classical electromagnetic theory, an electron that experiences centripetal acceleration in a circular orbit should radiate electromagnetic energy (light). This continuous loss of energy should cause the electron to spiral in to the nucleus. But an electron does not spiral inward; it orbits the nucleus at a fixed radius without radiating. Bohr's "solution" to this dilemma was to postulate that an electron can exist only in discrete stable orbits, called **stationary states,** and in these states there is no radiation.

*Bohr's Second Postulate: Photon Frequency from Energy Conservation (Radiative Transitions).* Bohr built his second postulate on the quantum ideas of Planck and Einstein, postulating that photons are emitted when the electron in a hydrogenlike atom suddenly changes from one stable orbit to another, in what he called a **radiative transition.** The energy of an emitted photon $E = hf$ equals the difference between the energy of an electron in the initial orbit $E_i$ and its energy in the final orbit $E_f$, $hf = E_i - E_f$, giving the frequency of an emitted photon as

$$f = \frac{E_i - E_f}{h}$$    Frequency of a photon emitted in the Bohr model

*Bohr's Third Postulate: Quantized Angular Momentum.* The energy of an electron in one of its stable orbits is quantized in Bohr's third postulate: the angular momentum $L = mvr$ of an electron in an orbit can take on only certain discrete values given by

$$mvr = \frac{nh}{2\pi} = n\hbar \qquad n = 1, 2, 3, \ldots$$    Quantized angular momentum

Application of Bohr's three postulates to a hydrogenlike atom gives the radii of the allowed orbits of the electron as

$$r_n = n^2 \frac{\hbar^2}{mkZe^2} = n^2 \frac{a_0}{Z} \qquad n = 1, 2, 3, \ldots$$    Radii of allowed orbits

where $k$ is the Coulomb constant, and the **first Bohr radius** $a_0$ (when $n = 1$) is

$$a_0 = \frac{\hbar^2}{mke^2} \approx 0.0529 \text{ nm}$$    First Bohr radius

**Energy Levels.** The quantized energy $E_n$ of an electron in one of the quantized orbits is

$$E_n = -\frac{mk^2e^4}{2\hbar^2}\frac{Z^2}{n^2} = -Z^2\frac{E_0}{n^2} \qquad n = 1, 2, 3, \ldots$$    Quantized energy levels

For hydrogen ($Z = 1$) the lowest energy $-E_0$, called the **ground-state energy,** corresponds to $n = 1$, and is given by

$$E_0 = -\frac{mk^2 e^4}{2\hbar^2} = -\frac{ke^2}{2a_0} \approx -13.6 \, \text{eV}$$

Ground-state energy for hydrogen

The magnitude of the ground state energy for hydrogenlike atoms is given by $E_g = Z^2 E_0$.

The Rydberg–Ritz formula can be derived from Bohr's second postulate that the frequency of an emitted photon from a hydrogenlike atom is given by $f = (E_i - E_f)/h = (E_1 - E_2)/h$. Using $1/\lambda = f/c$ and the above expression for $E_n$ gives

$$\frac{1}{\lambda} = Z^2 \frac{mk^2 e^4}{4\pi \hbar^3 c} \left( \frac{1}{n_2^2} - \frac{1}{n_1^2} \right)$$

Bohr expression for $1/\lambda$

Comparison of this expression with the Rydberg–Ritz formula shows that the experimentally determined Rydberg constant can be expressed in terms of fundamental constants of nature.

$$R = \frac{mk^2 e^4}{4\pi \hbar^3 c}$$

Rydberg constant in terms of fundamental constants of nature

When $R$ is evaluated by replacing these constants with their numerical values, very close agreement with the experimental spectroscopic value of $R$ is obtained; this is a major success of the Bohr model.

The specification of $n_2$ in the Rydberg–Ritz formula systematically defines series of spectral lines that lie in the visible and nonvisible regions according to the following scheme (each series is named after the person who first observed it experimentally for hydrogen):

$n_2 = 1, n_1 = 2, 3, 4, \ldots$          Lyman series (ultraviolet region)

$n_2 = 2, n_1 = 3, 4, 5, \ldots$          Balmer series (optical region)

$n_2 = 3, n_1 = 4, 5, 6, \ldots$          Paschen series (infrared region)

$n_2 = 4, n_1 = 5, 6, 7, \ldots$          Brackett series (far infrared region)

A standard way to represent the energies $E_n$ is with an **energy-level diagram,** such as the one shown for hydrogen in Figure 36-1, in which the transitions between energy levels $E_n$ are shown as arrows. When an electron occupies an energy level above its ground state ($n = 1$), it is said to be in an **excited state.**

The **ionization energy** is the energy necessary to remove an electron from an atom. For a hydrogen atom, this is the energy required to remove an electron from the ground state ($n = 1$) to zero energy ($n = \infty$), which is 13.6 eV.

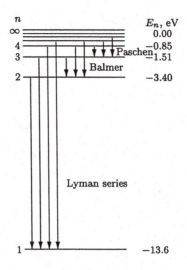

Figure 36-1

## Section 36-3. Quantum Theory of Atoms.

***The Schrödinger Equation in Spherical Coordinates.*** Three-dimensional problems involving electrons orbiting central nuclei are most easily treated in terms of spherical coordinates $(r,\ \theta,\ \phi)$ for which the time-independent Schrödinger equation has the form

$$-\frac{\hbar^2}{2m}\frac{1}{r^2}\frac{\partial}{\partial r}\left(r^2\frac{\partial\psi}{\partial r}\right)-\frac{\hbar^2}{2mr^2}\left[\frac{1}{\sin\theta}\frac{\partial}{\partial\theta}\left(\sin\theta\frac{\partial\psi}{\partial\theta}\right)+\frac{1}{\sin^2\theta}\frac{\partial^2\psi}{\partial\phi^2}\right]+U(r)\psi=E\psi$$

Three-dimensional Schrödinger equation in spherical coordinates

***Quantum Numbers in Spherical Coordinates.*** When the Schrödinger equation in three dimensions is solved using spherical coordinates, and when electron spin is taken into consideration, it is found that there are four quantum numbers that are needed to describe the resultant wave function $\psi$ for electrons in an atom:

| | |
|---|---|
| $n$ | principal quantum number |
| $\ell$ | orbital quantum number |
| $m_\ell$ | magnetic quantum number |
| $m_s$ | intrinsic spin quantum number |

Quantum numbers for spherical coordinates

Specifying these quantum numbers defines an electron's **state**. We now describe these quantum numbers in more detail.

The **principal quantum number $n$,** which takes on the integral values $n = 1, 2, 3, \ldots$, is roughly analogous to the quantum number $n$, discussed in Chapter 35, that determines wave functions and energy levels for a particle in a one-dimensional box. The principal quantum number $n$ specifies the radial part of the wave function, which determines the probability of finding an electron at a radial distance $r$ from the center of an atom. Increasing values of $n$

correspond to increasing values of energy of the quantized energy levels. Electron-shell letters are related to the values of $n$ according to the following scheme.

value of $n$:   1  2  3  4  ...
shell letter:   K  L  M  N  ...

Electron-shell letter notation

The **orbital quantum number** is designated by $\ell$. In a classical picture $\ell$ is related to the angular momentum of an electron as it orbits the nucleus of an atom. It takes on integral values with a range related to the principal quantum number $n$ according to

$$\ell = 0, 1, 2, \ldots, n-1$$ Orbital quantum number range

As an electron orbits a nucleus, its **orbital angular momentum** $L$ is related to its orbital quantum number $\ell$ by

$$L = \sqrt{\ell(\ell+1)}\hbar$$ Orbital angular momentum

For the ground state $n=1$, $\ell=0$, and the orbital angular momentum $L$ is zero, whereas for $n \geq 2$ there are several values of $L$. An orbital letter code is used to indicate the value of $\ell$ for an electron.

value of $\ell$:    0  1  2  3  4  5  ...
letter symbol:   s  p  d  f  g  h  ...

Orbital letter code

*Magnetic Moments.* If an atom is placed in an external magnetic field whose direction is taken as the $z$ direction, we can speak of the $z$ component $L_z$ of an electron's orbital angular momentum $\vec{L}$ along this direction. Quantum-mechanical analysis shows that this component $L_z$ can take on only discrete quantized values given by

$$L_z = m_\ell \hbar$$ Quantized z component of angular momentum

where $m_\ell$ is an integer called the **magnetic quantum** number that can have only the values

$$m_\ell = -\ell, (-\ell+1), (-\ell+2), \ldots, -1, 0, 1, \ldots, (\ell-1), \ell$$ Magnetic quantum number range

For a given value of $\ell$, the magnitude of the orbital angular momentum is $L = \sqrt{\ell(\ell+1)}\hbar$, whereas the maximum value of the $z$ component is $L_z = \ell\hbar$, which is less than $L$. This means that the angular momentum vector $\vec{L}$ never points exactly along the $z$ direction. The angle $\theta$ that $L$ makes with the $z$ direction takes on only discrete values.

The relationships among $\ell$, $L$, $m_\ell$, and $L_z$ apply not only to an individual electron in an atom, but also to the net values for the entire atom as a whole.

*Electron Spin.* In addition to its orbital angular momentum , an electron also possesses an intrinsic angular momentum $\vec{S}$ called its **spin**. The magnitude of the **spin angular momentum** $S$ is quantized according to

$$S = \sqrt{s(s+1)}\hbar \qquad s = \tfrac{1}{2} \qquad\qquad\qquad \text{Quantized spin angular momentum}$$

where $s = \tfrac{1}{2}$ is the electron's **intrinsic spin quantum number.**

Just as $L_z$ is quantized with the values $L_z = m_l \hbar$, the $z$ component of an electron's spin is quantized according to

$$S_z = m_s \hbar \qquad m_s = +\tfrac{1}{2} \quad \text{or} \quad m_s = -\tfrac{1}{2} \qquad\qquad \text{Quantized z component of electron spin}$$

The two values of $S_z$ are usually referred to as "spin up" for the **electron spin quantum number** $m_s = +\tfrac{1}{2}$ and "spin down" for $m_s = -\tfrac{1}{2}$. However, the spin vector $\vec{S}$ never points exactly in the $+z$ or $-z$ direction.

The classical picture of an electron is a ball of charge spinning with an intrinsic angular momentum $\vec{S}$ as it orbits the nucleus with an orbital angular momentum $\vec{L}$, much like a spinning planet orbiting the sun, as shown in Figure 36-2.

Figure 36-2

***Section 36-4. Quantum theory of the Hydrogen Atom.*** In Section 36-2 we discussed Bohr's model of the hydrogen atom, which arose from his three ad-hoc postulates. Here we show how the hydrogen atom appears in Schrödinger's theory.

A hydrogen atom consists of a heavy nucleus composed of a proton around which a light electron orbits. The potential energy that goes into the Schrödinger equation is the Coulomb potential energy of the electron-nucleus system in the hydrogenlike atom:

$$U(r) = -k\frac{Ze^2}{r} \qquad\qquad \text{Coulomb potential energy of an electron in a hydrogenlike atom}$$

The atomic number $Z$, which is 1 for hydrogen, in included in our expressions so that the results are applicable to other hydrogenlike, one-electron, atoms such as singly ionized He⁺, doubly ionized Li²⁺, and so forth.

For the Coulomb potential, the Schrödinger equation can be solved exactly. The total three-dimensional wave function is written as the product of a radial part $R(r)$, a polar part $f(\theta)$, and an azimuthal part $g(\phi)$.

$$\psi(r,\theta,\phi) = R(r)f(\theta)g(\phi) \qquad \text{Three-dimensional wave function}$$

In terms of the quantum numbers $(n, \ell, m_\ell)$, the wave function is usually written as $\psi_{n\ell m}$.

*Energy Levels.* Upon solving the Schrödinger equation for the hydrogen atom Coulomb potential, it is found that the allowed energy levels are quantized according to

$$E_n = -\frac{mk^2e^4}{2\hbar^2}\frac{Z^2}{n^2} = -Z^2\frac{E_0}{n^2} \quad n = 1, 2, \ldots \qquad \text{Allowed energy levels}$$

with

$$E_0 = \frac{mk^2e^4}{2\hbar^2} = \frac{ke^2}{2a_0} \approx 13.6\,\text{eV} \qquad \text{Value of } E_0 \text{ from the Schrödinger equation}$$

Amazingly, these energy levels that arise naturally from solution of the Schrödinger equation are exactly the same quantized energies that result from application of the ad-hoc Bohr theory.

*Wave Functions and Probability Densities.* The **ground state** of a hydrogen atom corresponds to $(n, \ell, m_\ell) = (1, 0, 0)$. The normalized wave function has the form

$$\psi_{100} = \frac{1}{\sqrt{\pi}}\left(\frac{Z}{a_0}\right)^{3/2}e^{-Zr/a_0} \qquad \text{Ground-state hydrogen wave function}$$

where, remarkably, the solution of the Schrödinger equation gives

$$a_0 = \frac{\hbar^2}{mke^2} = 0.0529\,\text{nm} \qquad \text{First Bohr radius}$$

which is equal to the first Bohr radius obtained from Bohr's theory.

Note that $\psi_{100}$ depends only on the radial coordinate $r$, with no dependence on the angles $\theta$ or $\phi$. Also note that since $\ell = 0$, the correct orbital angular momentum is zero, which contrasts with the Bohr model which incorrectly gives the ground-state angular momentum to be $\hbar$.

Because the ground-state wave function has no angular dependence, we do not have to consider angles in determining the probability of finding an electron in the three-dimensional space around the nucleus of the hydrogenlike atom. Thus, we are interested in the radial probability density $P(r)$ for finding an electron between radii $r$ and $r + dr$. To find this, take the volume element of a spherical shell of thickness $dr$, which is $dV = 4\pi r^2\,dr$, and multiply it by the probability density $\psi_{100}^2$ to get the probability of finding an electron in the spherical shell:

$$P(r)\,dr = 4\pi r^2\psi_{100}^2\,dr \qquad \text{Radial probability of finding an electron in a spherical shell}$$

resulting in the probability density of

$$P(r) = 4\pi r^2 \psi_{100}^2 = 4\left(\frac{z}{a_0}\right)^3 r^2\, e^{-2Zr/a_0}$$

Radial probability distribution

$P(r)$ as a function of $r$ is shown in Figure 36-3.

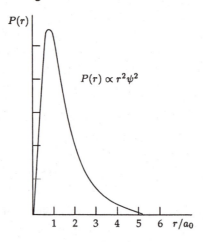

Figure 36-3

For $Z = 1$, the maximum value of $P(r)$ occurs exactly at the first Bohr radius $r = a_0$, showing that the Schrödinger equation predicts that this is the most likely place to find the electron in the ground state of a hydrogen atom. Note, though, the difference between the Bohr model and the Schrödinger theory for the ground state of hydrogen. The Bohr-model electron stays exactly in a well-defined circular orbit of radius $a_0$. In the wave-mechanical picture, the most likely place to find a quantum-theory electron is at the radius $a_0$, but there are finite probabilities for finding it at other radial distances from the nucleus.

The first excited states of hydrogen correspond to $n = 2$, and the wave functions $\psi_{n\ell m}$ have the form

$$\psi_{200} = C_{200}\left(2 - \frac{Zr}{a_0}\right)e^{-Zr/2a_0}$$

$$\psi_{210} = C_{210}\frac{Zr}{a_0}e^{-Zr/2a_0}\cos\theta$$

First excited state hydrogen atom wave functions

$$\psi_{21\pm1} = C_{21\pm1}\frac{Zr}{a_0}e^{-Zr/2a_0}\sin\theta\; e^{\pm i\phi}$$

where $C_{n\ell m}$ is a constant to be determined from the normalization condition.

***Section 36-5.  The Spin–Orbit Effect and Fine Structure.*** The total angular momentum $\vec{J}$ of an atom is equal to the vector sum of the orbital angular momentum $\vec{L}$ and the spin angular momentum $\vec{S}$:

$$\vec{J} = \vec{L} + \vec{S}$$       Total angular momentum

It follows from the vector addition of the orbital and spin angular momenta that the magnitude of $\vec{J}$ is quantized according to the rule

$$\left|\vec{J}\right| = \sqrt{j(j+1)}\hbar$$
$$j = \tfrac{1}{2}, \ell = 0 \quad \text{or} \quad j = \ell \pm \tfrac{1}{2}, \ell > 0$$
Quantized total angular momentum for one electron

The vector addition $\vec{J} = \vec{L} + \vec{S}$ for the case $\ell = 1$ is shown in Figure 36-4. For this case $j$ has the two values $j = \ell + s = \tfrac{3}{2}$ and $j = \ell - s = \tfrac{1}{2}$. The lengths of the various angular momentum vectors are $L = \sqrt{2}\hbar$, $S = \left(\sqrt{3}/2\right)\hbar$, and $J = \left(\sqrt{15}/2\right)\hbar$ or $J = \left(\sqrt{3}/2\right)\hbar$.

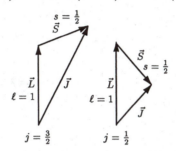

Figure 36-4

***Spectroscopic Notation.*** The notation for specifying the angular momentum quantum numbers, $j$, $\ell$, and $s$ for an atom is to use capital letters S, P, D, and F to designate $\ell$ for an atomic state, and lower-case letters s, p, d, and f to designate $\ell$ for the states of individual electrons in the atom. The letter value for $\ell$ is preceded by $2s + 1$ as a superscript and followed by $j$ as a subscript. Thus, an atom in a state $^2P_{3/2}$ has quantum numbers $\ell = 1$, $s = \tfrac{1}{2}$, and $j = \tfrac{3}{2}$. Sometimes the pre-superscript is omitted, and the letter symbol is multiplied by the value of $n$ of the shell being considered, to get the notation $n\ell_j$. For example, the states $2P_{3/2}$ and $2P_{1/2}$ correspond to a doublet energy level in which $n = 2$, $\ell = 1$, and $j = \tfrac{3}{2}$ and $\tfrac{1}{2}$, with $s = \tfrac{1}{2}$ implied.

***Spin–Orbit Energy Coupling.*** An electron orbiting a nucleus has an orbital angular momentum $\vec{L}$ and a spin angular momentum $\vec{S}$. Associated with the spin angular momentum is a spin magnetic moment $\vec{\mu}_s$.

From the point of view of the electron, the positively charged nucleus orbits around the electron, thereby producing a magnetic field $\vec{B}$ at the location of the electron. The spin magnetic moment $\vec{\mu}_s$ placed in the orbital magnetic field $\vec{B}$ gives rise to a potential energy $U$ given by

$$U = -\vec{\mu}_s \cdot \vec{B}$$       Magnetic potential energy

Thus, the orbital magnetic field interacts with the spin magnetic moment of the electron, giving the electron an additional potential energy that is added or subtracted from its energy without the spin–orbit effect.

The net result of this **spin–orbit effect** is that an energy level for which $\ell \neq 0$ is split into sublevels. For atoms such as hydrogen or sodium, which have one outermost electron, there are two sublevels corresponding to the two values $s = +\frac{1}{2}$ and $s = -\frac{1}{2}$, with the $j = \ell + \frac{1}{2}$ level being slightly higher than the $j = \ell - \frac{1}{2}$ level, as shown in Figure 36-5 for an atom in the states $2P_{3/2}$ and $2P_{1/2}$.

Because spin-orbit coupling produces closely spaced energy levels for each value of $\ell \neq 0$, transitions from these levels to lower levels result in the emission of photons with slightly different wavelengths, as shown in Figure 36-5. This is called **fine structure**, which is easily observed with spectrometers of moderate resolution.

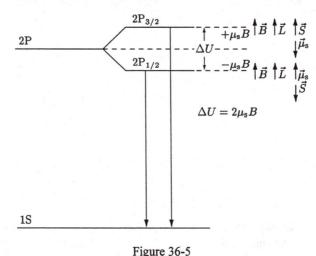

Figure 36-5

***Section 36-6. The Periodic Table.*** The properties and ordering of elements in the periodic table derive largely from application of the Pauli exclusion principle, which states that no two electrons in an atom can have the exact same set of all four quantum numbers ($n$, $\ell$, $m_\ell$, $m_s$). As a consequence, as $Z$ increases from one ground-state atom to the next, the trend is for each added electron to be found in a higher energy level.

As mentioned earlier, the quantum number $n$ designates a shell of electrons in an atom. The pair ($n$, $\ell$) specifies a subshell. the Pauli exclusion principle determines how many electrons can occupy a given subshell. There are $2\ell + 1$ values of $m_\ell$ for each value of $\ell$. Because there are two values of $m_s \left( +\frac{1}{2} \text{ and } -\frac{1}{2} \right)$ for each value of $m_\ell$, the maximum number of electrons that can occupy a subshell is $2(2\ell + 1)$. This gives us a table for specific values of $\ell$:

| $\ell$ value | 0 | 1 | 2 | 3 | |
|---|---|---|---|---|---|
| Subshell letter | | s | p | d | f    Maximum number of electrons in subshells |
| Maximum number of electrons | 2 | 6 | 10 | 14 | |

The subshells of most (but not all) atoms fill from bottom to top in the order shown in Figure 36-6, which also gives a rough indication of the relative values of the energy of each level. When a shell completely fills with electrons, there is a relatively large energy gap to the next subshell. These filled shells correspond to the noble or inert gases at $Z = 2$, 10, 18, 36, 54, and 86, which are difficult to ionize and chemically inactive. Other chemical properties of the elements also depend on the various ways their subshells are filled.

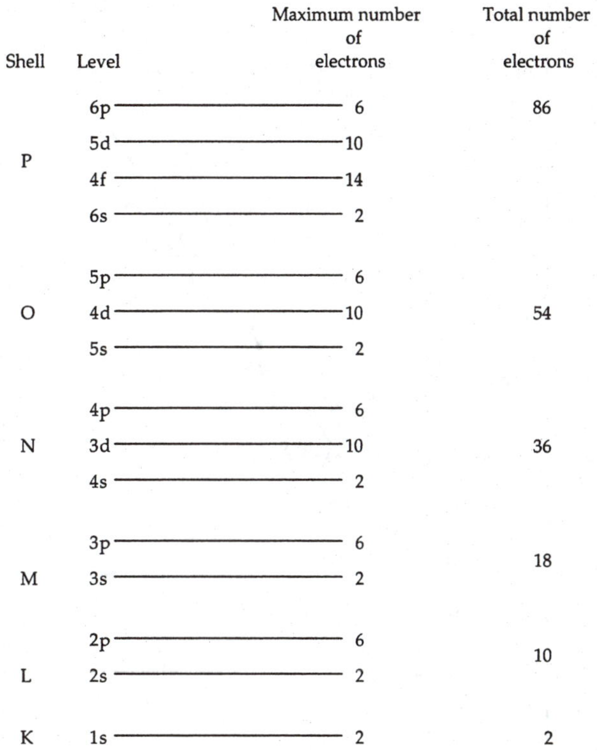

| Shell | Level | Maximum number of electrons | Total number of electrons |
|---|---|---|---|
| P | 6p | 6 | 86 |
| | 5d | 10 | |
| | 4f | 14 | |
| | 6s | 2 | |
| O | 5p | 6 | 54 |
| | 4d | 10 | |
| | 5s | 2 | |
| N | 4p | 6 | 36 |
| | 3d | 10 | |
| | 4s | 2 | |
| M | 3p | 6 | 18 |
| | 3s | 2 | |
| L | 2p | 6 | 10 |
| | 2s | 2 | |
| K | 1s | 2 | 2 |

Figure 36-6

*Section 36-7. Optical and X-Ray Spectra.* One result of Schrödinger's theory is that an atom possesses discrete energy levels into which its electrons can be excited. When an electron in an atom makes a transition to a lower energy level, it emits a photon of frequency $f = (E_i - E_f)/h$ and corresponding wavelength $\lambda = c/f = hc/(E_i - E_f)$. The values of the discrete wavelengths emitted by an atom comprise a characteristic signature of that atom. Except for hydrogenlike atoms, the Schrödinger equation is too complicated to solve analytically, so most energy levels are determined experimentally.

*Optical Spectra.* Transitions involving excitations of outer (valence) electrons result in photons whose wavelengths are in or near the visible or **optical spectrum.** An energy-level diagram for sodium is shown in Figure 36-7. The first column of energy levels corresponds to $^2S_{1/2}$ states, so for these energy levels $\ell = 0$, $s = \frac{1}{2}$, and $j = \frac{1}{2}$.

Except for the S states, where $\ell = 0$, all the states of sodium have two values for $j$, $j = \ell + \frac{1}{2}$ and $j = \ell - \frac{1}{2}$, corresponding to the two values of spin. The spin–orbit effect gives two slightly different values of energy, with $j = \ell + \frac{1}{2}$ energy levels being at a higher energy than $j = \ell - \frac{1}{2}$ levels. This energy difference is on the order of 0.001 eV, which is much too small to be seen on the scale used in Figure 36-7, where they appear as a single line. These states are referred to as **doublets.**

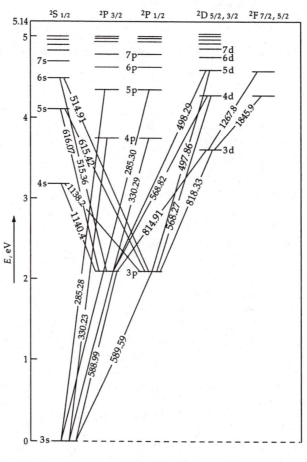

Figure 36-7

Not all transitions between the various energy levels of an atom can actually take place. Only those transitions that obey the following **selection rules** are allowed.

$$\Delta m = \pm 1 \text{ or } 0$$
$$\Delta \ell = \pm 1 \qquad\qquad\qquad \text{Selection rules for transitions}$$
$$\Delta j = \pm 1 \text{ or } 0 \text{ (but } j = 0 \rightarrow j = 0 \text{ is forbidden)}$$

Figure 36-7 shows various wavelengths that result from allowed transitions in sodium.

*X-Ray Spectra.* X rays, first observed by Röntgen in 1895, are high-energy photons that can be produced when a target in a cathode-ray tube is bombarded with electrons that have been accelerated through a voltage $V$, giving them a kinetic energy $K = eV$.

The electrons striking the target lose speed rapidly, emitting photons of continuously varying wavelengths, which give rise to a continuous **bremsstrahlung spectrum** (braking radiation spectrum). This is the curved part of the X-ray spectrum shown in Figure 36-8.

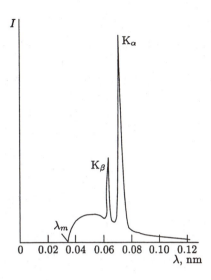

Figure 36-8

The maximum frequency that a photon can have in the bremsstrahlung spectrum is given by $hf_{max} = eV$, where $eV$ is the kinetic energy of the bombarding electrons. This corresponds to a cutoff minimum wavelength, $\lambda_m$, given by

$$\lambda_m = hc / E = hc / eV \qquad \qquad \text{Cutoff X-ray wavelength}$$

as shown in Figure 36-8. The cutoff wavelength $\lambda_m$ depends only on the accelerating voltage $V$ of the X-ray tube, and not on the material of the target.

The sharp $K_\alpha$ and $K_\beta$ spikes in Figure 36-8 show the **characteristic spectrum,** which is superimposed on the continuous bremsstrahlung spectrum. The wavelengths of these peaks do depend on the target material. The characteristic spectrum arises because the bombarding electrons knock out inner-core electrons from atoms. Other electrons in the atom then make transitions from higher energy levels to the vacant energy levels, emitting the characteristic X-ray photons.

The labels of X rays arise from the letter value of the atomic shell from which the inner electron was removed. The $K_\alpha$ and $K_\beta$ lines result from transitions from the $n = 2$ and $n = 3$ levels to the K shell ($n = 1$). Similarly, transitions to the L shell ($n = 2$) are labeled $L_\alpha$ and $L_\beta$. In actuality, these lines show fine structure from the splitting of the energy levels above the K shell.

In 1913 Moseley measured the wavelengths of characteristic X-ray lines for about 40 elements. For K lines, caused by one of the two electrons being knocked out of the $n = 1$ energy level, it was found that his data fitted very well into the theoretical **Moseley relationship,**

$$\lambda = \frac{hc}{E_n - E_1} = \frac{hc}{\left(Z-1\right)^2 \left(13.6\,\text{eV}\right)\left(1 - 1/n^2\right)} \qquad \text{Moseley relationship for X-ray wavelengths}$$

which follows from a Bohr-type model in which electrons in shells outside the K shell feel the charge $+Ze$ of the nucleus and the charge $-e$ of the electron that remains in the K shell, or a net charge of $(Z-1)e$. By measuring wavelengths of characteristic X-ray spectra, Moseley was able to determine atomic numbers $Z$ for various elements. Now, X-ray spectra can be very useful in determining the elements that make up a substance under investigation.

## II.  Physical Quantities and Key Equations

### Physical Quantities

Rydberg constant for hydrogen

$$R = 10,977,760 \text{ m}^{-1} = 10.97776 \ \mu\text{m}^{-1}$$

General Rydberg constant

$$R = \frac{mk^2 e^4}{4\pi \hbar^3 c}$$

First Bohr radius

$$a_0 = \frac{\hbar^2}{mke^2} \approx 0.0529 \text{ nm}$$

Ground-state energy for hydrogen

$$E_0 = -\frac{mk^2 e^4}{2\hbar^2} = -\frac{ke^2}{2a_0} \approx -13.6 \text{ eV}$$

### Key Equations

Rydberg–Ritz formula

$$\frac{1}{\lambda} = RZ^2 \left( \frac{1}{n_2^2} - \frac{1}{n_1^2} \right) \qquad n_1 > n_2$$

Frequency of a photon emitted in the Bohr model

$$f = \frac{E_i - E_f}{h}$$

Radii of allowed orbits

$$r_n = n^2 \frac{\hbar^2}{mkZe^2} = n^2 \frac{a_0}{Z}$$

Quantized energy levels for hydrogenlike atoms

$$E_n = -\frac{mk^2 e^4}{2\hbar^2} \frac{Z^2}{n^2} = -Z^2 \frac{E_0}{n^2}$$

Bohr expression for $1/\lambda$

$$\frac{1}{\lambda} = Z^2 \frac{mk^2 e^4}{4\pi \hbar^3 c} \left( \frac{1}{n_2^2} - \frac{1}{n_1^2} \right)$$

Quantum numbers for spherical coordinates

$n$    principal quantum number
$\ell$    orbital quantum number
$m_\ell$    magnetic quantum number
$m_s$    instrinsic spin quantum number

Orbital quantum number range

$$\ell = 0, 1, 2, \ldots, n-1$$

| | |
|---|---|
| *Orbital angular momentum* | $L = \sqrt{\ell(\ell+1)}\hbar$ |

*Magnetic quantum number range* $\qquad m_\ell = -\ell, (-\ell+1), (-\ell+2), \ldots, -1, 0, 1, \ldots, (\ell-1), \ell$

*Ground-state hydrogen wave function* $\qquad \psi_{100} = \dfrac{1}{\sqrt{\pi}} \left(\dfrac{Z}{a_0}\right)^{3/2} e^{-Zr/a_0}$

*Ground-state radial probability density for hydrogen* $\quad P(r) = 4\pi r^2 \psi_{100}^2 = 4\left(\dfrac{z}{a_0}\right)^3 r^2\, e^{-2Zr/a_0}$

*First excited state hydrogen atom wave functions*

$$\psi_{200} = C_{200}\left(2 - \frac{Zr}{a_0}\right) e^{-Zr/2a_0}$$

$$\psi_{210} = C_{210}\frac{Zr}{a_0} e^{-Zr/2a_0} \cos\theta$$

$$\psi_{21\pm1} = C_{21\pm1}\frac{Zr}{a_0} e^{-Zr/2a_0} \sin\theta\, e^{\pm i\phi}$$

*Magnetic potential energy* $\qquad U = -\vec{\mu}_s \cdot \vec{B}$

*Cutoff X-ray wavelength* $\qquad \lambda_m = hc/E = hc/eV$

*Moseley relationship for X-ray wavelengths* $\qquad \lambda = \dfrac{hc}{E_n - E_1} = \dfrac{hc}{(Z-1)^2 (13.6\ \text{eV})(1 - 1/n^2)}$

## III. Potential Pitfalls

A main result of the Bohr model of hydrogenlike atoms is that the energy of the atom is quantized. Make sure you understand that the energy $E = hf$ of an emitted photon equals the difference of the energies of the initial energy level $E_i$ and final energy level $E_f$ between which an electron makes a transition: $hf = E_i - E_f$.

In considering the probability of locating an electron in a hydrogen atom, do not mistake the probability density $P(r)$, which gives the probability of finding the electron between $r$ and $r + dr$, with $\psi^2$, which gives the probability density *per unit volume*. $P(r) = 4\pi r^2\psi^2$ is the probability density *per unit radius*.

Do not confuse angular momentum, which is a physical quantity, and the quantum number that is associated with that angular momentum, which is an integer or half-integer number. For example, the value of the magnitude of the total angular momentum $\vec{J}$ of an atom is related to the associated quantum number $j$ by $\left|\vec{J}\right| = \sqrt{j(j+1)}\hbar$.

Do not confuse the orbital angular momentum with the orbital magnetic moment. The orbital angular momentum is associated with the mass of an electron orbiting a nucleus, much like the angular momentum of a massive planet as it orbits the sun. The orbital magnetic moment is associated with the magnetic field produced by the charge of an electron that is orbiting the nucleus.

Realize the distinction between the spin angular momentum and the spin magnetic moment. The spin angular momentum is associated with the mass of an electron rotating about an axis, much like the angular momentum of the earth rotating about its axis. The spin magnetic moment is associated with the magnetic field produced by the charge of the electron as it spins.

Understand that $\lambda_m$ of the bremsstrahlung part of an X-ray spectrum depends only on the accelerating voltage of the bombarding electrons and has nothing to do with the material of the target.

Remember also that the wavelengths of the X-ray photons emitted as the target electrons return to lower energy levels depend only on the structure of the target material and have nothing to do with how much energy the bombarding electrons have (as long as this energy is sufficiently large to eject the inner electrons).

## IV. True or False Questions and Responses

**True or False**

_____ 1.  The larger the value of $n$ in the Bohr model, the larger the energy of the energy level.

_____ 2.  In the Bohr model, the wavelengths of a spectral series such as the Lyman series or Balmer series are calculated from the Rydberg–Ritz formula by fixing the value of $n$ of the initial energy level.

_____ 3.  In the Bohr model, the energy of an emitted photon equals the energy difference between the initial and final energy levels of an electron transition in an atom.

_____ 4.  The absolute value of the ionization energy of a hydrogenlike atom is numerically equal to the absolute value of the energy of the $n = 1$ energy level.

_____ 5.  The energy of an energy level in an atom is always related to the principal quantum number $n$ by $E = \text{constant}/n^2$.

_____ 6.  The angular momentum $L$ is related to the orbital quantum number $\ell$ by $L = \ell\hbar$.

_____ 7.  The $z$ component of the orbital angular momentum is related to the magnetic quantum number $m_\ell$ by $L_z = m_\ell\hbar$.

_____ 8.  The spin angular momentum is related to the spin quantum number $s = \frac{1}{2}$ by $S = s\hbar$.

_____ 9.  The $z$ component of the spin angular momentum is related to the quantum number $m_s = \pm\frac{1}{2}$ by $S_z = m_s\hbar$.

_____ 10. Fine structure arises from the interaction of orbital angular momentum and spin angular momentum.

_____ 11. There can be only one electron with a given value of $m_\ell$ for a given value of $n$ and $\ell$.

_____ 12. Elements with filled shells are chemically active.

_____ 13. A transition from one doublet energy level to a lower doublet energy level results in four closely spaced spectral lines.

_____ 14. In a doublet, $j = \ell - \frac{1}{2}$ energy levels are lower than $j = \ell + \frac{1}{2}$ energy levels.

_____ 15. The hydrogen wave function with $n = 2$ is spherically symmetric.

_____ 16. The wavelengths of the K lines in the characteristic X-ray spectrum of an element can be closely predicted from a Bohr model in which outer electrons see an attractive nucleus of charge $(Z-1)e$.

_____ 17. Increasing the accelerating voltage in an X-ray tube increases the frequencies of the characteristic spectral lines.

_____ 18. Photons emitted from an X-ray tube have a maximum energy that depends on the value of the voltage that accelerates the bombarding electrons.

**Responses to True or False**

1. True.

2. False. The wavelengths of spectral series are determined by fixing the value of $n$ of the final energy level.

3. True.

4. True.

5. False. This relationship holds only for hydrogenlike atoms.

6. False. The relationship is $L = \sqrt{\ell(\ell+1)}\hbar$. .

7. True.

8. False. The relationship is $S = \sqrt{s(s+1)}\hbar = \left(\sqrt{3}/2\right)\hbar$.

9. True.

10. True.

11. False. There can be two electrons with the given value of $m_\ell$, one with $m_s = +\frac{1}{2}$ and one with $m_s = -\frac{1}{2}$.

12. False. Elements with filled shells are chemically inactive and difficult to ionize.

13. False. Because of the selection rule on $j$, there are three closely spaced spectral lines.

14. True.

15. False. The hydrogen wave function $\psi_{2\ell m}$ is spherically symmetric only for $\ell = 0$ and $m_{\ell} = 0$.

16. True.

17. False. The frequencies and corresponding wavelengths of lines in the characteristic spectrum depend on the electron energy levels in the atoms of the target material and have nothing to do with the energy of the bombarding electrons.

18. True.

# V.  Questions and Answers

## Questions

1.  How do the energies of an electron in the Bohr model vary with $n$?

2.  According to the Bohr model, how do the energies of the electron in hydrogenlike atoms vary with the number of protons in the nucleus?

3.  Most spectral lines of atoms occur in groups of two or more closely spaced lines. What causes these groupings?

4.  Why are only three closely spaced lines observed in transitions between two doublet energy levels instead of four?

5.  Why doesn't the radial probability distribution function $P(r)$ for the ground state of hydrogen simply equal $\psi^2(r)$?

6.  How does the $(Z-1)e$ term in the Moseley relationship for the wavelengths of K X-ray lines arise?

7.  What effect does increasing the accelerating voltage have on the X-ray spectrum for a given target?

8.  How are the wavelengths of the characteristic peaks in an X-ray spectrum affected by increasing the accelerating voltage?

## Answers

1.  From the expression $E_n = -Z^2 E_0 / n^2$, the energies are seen to vary as $1/n^2$.

2.  From the expression $E_n = -Z^2 E_0 / n^2$, the energies are seen to vary as $Z^2$, where $Z$ is the number of protons in the nucleus.

3.  The interaction between the spin and orbital magnetic moments splits each energy level with $\ell > 0$ into closely spaced energy levels. For a single electron, for example, there is a doublet

splitting corresponding to $j = \ell + \frac{1}{2}$ and $j = \ell - \frac{1}{2}$. Transitions from one pair of closely spaced levels to another pair of closely spaced levels results in the observed fine structure.

4. One of the four possible transitions is ruled out by the selection rule on $j$ that $j = 0 \rightarrow j = 0$ is forbidden.

5. $\psi^2(r)\, dv = \psi^2(r) r^2 \sin\theta\, dr\, d\theta\, d\phi$ is the probability of finding an electron in the volume element $dV = r^2 \sin\theta\, dr\, d\theta\, d\phi$. You get a factor of $4\pi$ when the angular integration is performed, leaving you with $\psi^2(r) r^2 4\pi\, dr = P(r)\, dr$ as the probability for finding an electron between $r$ and $r + dr$. The volume of the corresponding shell is $4\pi r^2\, dr$.

6. When a K-shell ($n = 1$) electron is knocked out of an atom, the outer electrons that can make transitions to the empty K shell see a nucleus with charge $+Ze$,, and a charge $-e$ on the remaining electron in the K shell, or a net charge $(Z-1)e$. The Moseley relation arises when the charge $(Z-1)e$ is used in the Bohr theory.

7. The more energetic the bombarding electrons are, the larger the maximum energy $E_{max} = hf_{max} = hc/\lambda_m$ of the photons emitted in the bremsstrahlung process. This means that the cutoff wavelength $\lambda_m$ decreases as the accelerating voltage of the bombarding electrons is increased, resulting in more and more characteristic wavelengths being observed.

8. The wavelengths of the characteristic peaks are unaffected by increasing the accelerating voltage of the bombarding electrons because the characteristic spectrum depends only on the electron energy levels in the atoms of the target material.

## VI. Problems

**Example #1.** The wavelengths of visible light range from about 400 to 700 nm. Determine the wavelengths in the Balmer series for hydrogen that lie in the visible region.

**Picture the Problem.** Determine the values of $n_2$ and $Z$ that pertain to this problem. Use them in the Rydberg-Ritz formula, varying $n_1$ to determine the wavelengths in the visible.

| | |
|---|---|
| 1. Determine $Z$ and $n_2$ for the Balmer series. | $n_2 = 2$ <br> $Z = 1$ for hydrogen |
| 2. Write out the Rydberg-Ritz formula, and substitute these values in to reduce it to a simple form. | $\dfrac{1}{\lambda} = RZ^2\left(\dfrac{1}{n_2^2} - \dfrac{1}{n_1^2}\right) = (0.010978\,\text{nm}^{-1})(1)\left(\dfrac{1}{2^2} - \dfrac{1}{n_1^2}\right)$ |
| 3. Determine the longest wavelength in the Balmer series by setting $n_1 = 3$. This wavelength is in the visible spectrum. | $\dfrac{1}{\lambda} = (0.010978\,\text{nm}^{-1})\left(\dfrac{1}{4} - \dfrac{1}{9}\right)$ <br> $\lambda = 655.9\,\text{nm}$ |

| | |
|---|---|
| 4. Determine the shortest wavelength in the Balmer series by setting $n_1 = \infty$. This wavelength is outside the visible spectrum, so we must determine the largest value for $n_1$ in a different fashion. | $\frac{1}{\lambda} = \left(0.010978\,\text{nm}^{-1}\right)\left(\frac{1}{4} - \frac{1}{\infty}\right)$ <br><br> $\lambda = 364.4\,\text{nm}$ |
| 5. Determine the largest value of $n_1$ that produces an emission line in the visible by setting $\lambda = 400\,\text{nm}$ and solving for $n_1$. Since $n_1$ must be an integer, $n_1 = 6$ is the largest numbered initial state that results in an emission in the visible. | $\frac{1}{400\,\text{nm}} = \left(0.010978\,\text{nm}^{-1}\right)\left(\frac{1}{4} - \frac{1}{n_1^2}\right)$ <br><br> $n_1 = 6.7$ |
| 6. Determine all the visible emission wavelengths, using $3 \le n_1 \le 6$. The case of $n_1 = 3$ has already been done in step 3. | $\frac{1}{\lambda_4} = \left(0.010978\,\text{nm}^{-1}\right)\left(\frac{1}{4} - \frac{1}{4^2}\right)$ <br><br> $\lambda_4 = 485.8\,\text{nm}$ <br><br> $\frac{1}{\lambda_5} = \left(0.010978\,\text{nm}^{-1}\right)\left(\frac{1}{4} - \frac{1}{5^2}\right)$ <br><br> $\lambda_5 = 433.8\,\text{nm}$ <br><br> $\frac{1}{\lambda_6} = \left(0.010978\,\text{nm}^{-1}\right)\left(\frac{1}{4} - \frac{1}{6^2}\right)$ <br><br> $\lambda_6 = 409.9\,\text{nm}$ |

**Example #2—Interactive.** Electrons in singly ionized helium are excited into the $n = 3$ energy level. Using the Bohr model, determine the wavelengths of photons that are emitted when the electrons return to the ground state.

**Picture the Problem.** These excited electrons can either return directly to the ground state, or first decay to the $n = 2$ state en route to the ground state. **Try it yourself.** Work the problem on your own, in the spaces provided, to get the final answer.

| | |
|---|---|
| 1. Determine Z, $n_1$, and $n_2$ for all possible transitions. | |
| 2. Write out the Rydberg-Ritz formula, and substitute the appropriate values to determine the emission wavelength for a direct transition from the $n = 3$ state to the $n = 1$ state. | $\lambda_{31} = 25.6\,\text{nm}$ |

| 3. Write out the Rydberg-Ritz formula, and substitute the appropriate values to determine the emission wavelength for a transition from the $n = 3$ state to the $n = 2$ state. | |
|---|---|
| | $\lambda_{32} = 164\,\text{nm}$ |
| 4. Write out the Rydberg-Ritz formula, and substitute the appropriate values to determine the emission wavelength for a transition from the $n = 2$ state to the $n = 1$ state. | |
| | $\lambda_{21} = 30.4\,\text{nm}$ |

**Example #3.** How many electrons can fit into the L shell of an atom?

**Picture the Problem.** The value of $n$ can be determined from the letter code. This determines the possible values of $\ell$ which in turn determines the values for $m_\ell$.

| 1. Determine the value of $n$. | $n = 2$ |
|---|---|
| 2. Determine the number of allowed values for the $\ell$ quantum number. There are two allowed values for $\ell$. | $\ell = 0, 1, \ldots, n-1$ <br> $\ell = 0, 1$ |
| 3. Determine the allowed values for $m_\ell$. | $\ell = 0$ <br> $\quad m_\ell = 0$ <br> $\ell = 1$ <br> $\quad m_\ell = -1, 0, 1$ |
| 4. For each value of $m_\ell$ there are two allowed spin states. The list of allowed states is shown at right. There are a total of 8: two s states ($\ell = 0$), and six p states ($\ell = 1$). | $(n, \ell, m_\ell, m_s) = \left(2,0,0,+\tfrac{1}{2}\right), \left(2,0,0,-\tfrac{1}{2}\right),$ <br> $\left(2,1,-1,+\tfrac{1}{2}\right), \left(2,1,-1,-\tfrac{1}{2}\right), \left(2,1,0,+\tfrac{1}{2}\right),$ <br> $\left(2,1,0,-\tfrac{1}{2}\right), \left(2,1,1,+\tfrac{1}{2}\right), \left(2,1,1,-\tfrac{1}{2}\right)$ |

**Example #4—Interactive.** How many electrons can occupy an N shell?

**Picture the Problem.** Follow the steps of Example #3. **Try it yourself.** Work the problem on your own, in the spaces provided, to get the final answer.

| 1. Determine the value of $n$. | |
|---|---|
| 2. Determine the number of allowed values for the $\ell$ quantum number. There are two allowed values for $\ell$. | |

| 3. Determine the allowed values for $m_\ell$. | |
|---|---|
| 4. For each value of $m_\ell$ there are two allowed spin states. List all the allowed states and add them up. | 32 total electrons |

**Example #5.** In a particular atom the two outermost electrons in p states combine to give $\ell = 1$ and $s = 1$. Give the spectroscopic notation of the possible states of the atom.

**Picture the Problem.** Spectroscopic notation is given by $^{2s+1}\ell_j$.

| 1. Determine the letter code for $\ell = 1$. | $\ell = P$ |
|---|---|
| 2. Determine the allowed values of $j$. | $j = \ell + s, \ell + s - 1, \ldots \ell - s$<br>$j = 2, 1, 0$ |
| 3. Convert all allowed states to spectroscopic notation. | $^3P_2, {}^3P_1, {}^3P_0$ |

**Example #6—Interactive.** What are the quantum numbers for an atom in a $^1D_2$ state?

**Picture the Problem.** Spectroscopic notation corresponds to the quantum numbers according to $^{2s+1}\ell_j$. **Try it yourself.** Work the problem on your own, in the spaces provided, to get the final answer.

| 1. Determine the value of $s$ from the preceding superscript value. | $s = 0$ |
|---|---|
| 2. Determine the value of $\ell$ from the letter code. | $\ell = D$ means $\ell = 2$ |
| 3. Determine the value of $j$ from the subscript. | $j = 2$ |

**Example #7.** For $n = 2$, $\ell = 0$, the hydrogen wave function is spherically symmetric and has the form $\psi_{200} = C_{200}(2 - r/a_0)e^{-r/2a_0}$. Find the location(s) of $r/a_0$ where the probability density function is a maximum.

**Picture the Problem.** Find the probability density in terms of the wave function. Maxima and minima are located where the derivative of the probability density function is zero.

| | |
|---|---|
| 1. Determine the probability density function. | $$P(r) = 4\pi r^2 \psi^2 = 4\pi r^2 \left[ C_{200}\left(2 - \frac{r}{a_0}\right)e^{-r/2a_0} \right]^2$$ |
| 2. Find the derivative of the probability density with respect to $r$. | $$\frac{dP(r)}{dr} = 4\pi C_{200}^2 r e^{-r/a_0}\left(2 - \frac{r}{a_0}\right)\left[\left(\frac{r}{a_0}\right)^2 - \frac{6r}{a_0} + 4\right]$$ |
| 3. Set the derivative equal to zero and solve for $r$. | $r/a_0 = 0, r/a_0 = 2,$ <br> $r/a_0 = 0.764, r/a_0 = 5.24$ |
| 4. Using the second derivative test, or plotting out $P(r)$, you discover that the first two values correspond to minima of the function, and the last two values correspond to local maxima, as shown by the solid curve in the figure. The dashed curve shows the probability density for $n = 2$ and $\ell = 1$, which peaks at $r/a_0 = 4$. | |

**Example #8—Interactive.** For a given small interval $\Delta r$, calculate the ratio of the probabilities of finding the electron at the two radii found in Example #7.

**Picture the Problem.** Calculate the probability at the two radii in Example #7 and take their ratio. **Try it yourself.** Work the problem on your own, in the spaces provided, to get the final answer.

| | |
|---|---|
| 1. Calculate $P(r)$ at the first maximum. | |
| 2. Calculate $P(r)$ at the second maximum. | |

| 3. Calculate their ratio. | |
|---|---|
| | 0.272 |

**Example #9.** The potential energy $U$ of an electron with a magnetic moment $\vec{\mu}$ in a magnetic field $\vec{B}$ is $U = -\vec{\mu} \cdot \vec{B}$. For a given magnetic field there are two possible orientations of the magnetic moment of an electron. As a beam of free electrons moves perpendicularly to a 0.200-T magnetic field, determine the energy difference between the electrons that are "aligned" and "anti-aligned" with the field. Use the result that the $z$ component of the magnetic moment of an electron is $\mu_z = -g_s \mu_B m_s$, where $\mu_B = e\hbar/(2m_e) = 5.79 \times 10^{-5}$ eV/T (the Bohr magneton), and $g_s$ (the gyromagnetic ratio for electron spin) has the approximate value $g_s \approx 2$.

**Picture the Problem.** Find the two possible $z$ components of the magnetic moment of an electron. For each component, calculate the potential energy, and then the difference.

| 1. Write out the expression for the potential energy of the electron. | $U = -\vec{\mu} \cdot \vec{B} = -\mu_z B = -g_s \mu_B m_s B$ |
|---|---|
| 2. The two possible values for the spin quantum number are plus and minus one-half. Make this substitution, and solve for the difference in potential energy. | $\Delta U = g_s \mu_B B \left( m_{s,\text{up}} - m_{s,\text{down}} \right) = g_s \mu_B B \left( \tfrac{1}{2} - \left( -\tfrac{1}{2} \right) \right)$ <br> $= 2 \left( 5.79 \times 10^{-5} \text{ eV/T} \right) (0.2 \text{ T}) (1) = 2.32 \times 10^{-5}$ eV |

**Example #10—Interactive.** When an electron in a galactic hydrogen atom "flips" its spin from aligned to anti-aligned with the spin of the proton in an atom, microwave radiation of wavelength 21.0 cm is emitted, which is used in radioastronomy to map the galaxy. Determine the magnetic field felt by the electron in a galactic hydrogen atom.

**Picture the Problem.** The energy of the emitted radiation can be determined from its wavelength. This energy is the same as the energy difference between the aligned and anti-aligned states of the electron. **Try it yourself.** Work the problem on your own, in the spaces provided, to get the final answer.

| 1. Determine the energy of the emitted photon. | |
|---|---|
| 2. Following Example #9, use this energy difference to solve for the magnetic field felt by the electron. | $B = 0.051 \text{T}$ |

**Example #11.** A line in the characteristic X-ray spectrum of a certain element is at 0.0786 nm. If this element is used as the target in an X-ray tube, what is the minimum voltage necessary to observe the line?

**Picture the Problem.** The kinetic energy of the electron must be at least equal to the energy of the line in the characteristic spectrum.

| 1. Determine the energy of the X-ray line. | $E = hc/\lambda = (1240\,\text{eV·nm})/(0.0786\,\text{nm})$ $= 15.78\,\text{keV}$ |
|---|---|
| 2. This energy must equal the kinetic energy of the electron after it has been accelerated through a potential of $V$. | $eV = 15.78\,\text{keV}$ $V = 15.78\,\text{kV} = 15,780\,\text{V}$ |

**Example #12—Interactive.** The minimum wavelength in the continuous X-ray spectrum from a certain target is found to be 0.0322 nm. What is the operating voltage of the X-ray tube?

**Picture the Problem.** The minimum wavelength corresponds to the maximum possible energy of the electron. **Try it yourself.** Work the problem on your own, in the spaces provided, to get the final answer.

| 1. Determine the maximum energy associated with the minimum wavelength. | |
|---|---|
| 2. Equate this energy with the change in the electron's potential energy to find the operating voltage. | $V = 38,500\,\text{V}$ |

**Example #13.** Originally, Ni (atomic weight 58.69) was listed with atomic number $Z = 27$ in the periodic table before Co (atomic weight 58.93), which was assigned $Z = 28$. Moseley measured the $K_\alpha$ line of Co to be 0.179 nm, and the $K_\alpha$ line of Ni to be 0.166 nm. Show from Moseley's data that the ordering of Ni and Co should be reversed from that originally given in the periodic table.

**Picture the Problem.** Substitute the wavelength into Moseley's relationship and solve for $Z$.

| 1. Substitute the wavelength from Ni into Moseley's relationship and solve for Z. For K lines, $n = 2$. | $\lambda = \dfrac{hc}{(Z-1)^2(13.6\,\text{eV})(1-1/n^2)}$<br><br>$0.166\,\text{nm} = \dfrac{1240\,\text{eV} \cdot \text{nm}}{(Z-1)^2(13.6\,\text{eV})(1-0.25)}$<br><br>$Z_{\text{Ni}} = 28.1 \approx 28$ |
|---|---|
| 2. Do the same for Co. | $\lambda = \dfrac{hc}{(Z-1)^2(13.6\,\text{eV})(1-1/n^2)}$<br><br>$0.179\,\text{nm} = \dfrac{1240\,\text{eV} \cdot \text{nm}}{(Z-1)^2(13.6\,\text{eV})(1-0.25)}$<br><br>$Z_{\text{Co}} = 27.1 \approx 27$ |

**Example #14—Interactive.**   For Mo ($Z = 42$) the binding energies of the innermost core electrons are

| Electron | 1s | 2s | 2p | 2p |
|---|---|---|---|---|
| Binding energy (keV) | 20.000 | 2.866 | 2.625 | 2.520 |

Calculate the wavelengths of the fine-structure $K_\alpha$ lines.

**Picture the Problem.** From the energy differences between levels, find the wavelengths of the emitted photons consistent with the selection rule $\Delta\ell = \pm 1$.

| 1. Find the energy difference between the first 2p state and the 1s state. | |
|---|---|
| 2. Find the wavelength associated with this energy difference. | $\lambda = 0.0714\,\text{nm}$ |
| 3. Find the energy difference between the second 2p state and the 1s state. | |
| 4. Calculate the wavelength associate with this energy difference. | $\lambda = 0.0709\,\text{nm}$ |

# Chapter 37

# Molecules

## I.  Key Ideas

*Section 37-1. Molecular Bonding.* The strongest interactive forces among atoms, created by sharing one or more electrons, are attractive forces called bonds, which bind two or more atoms together into a molecule.

*The Ionic Bond.* The ionic bond is the force that holds the atoms of most salt crystals together. In the simplest salts an alkali metal from the first column of the period table (Li, Na, K, . . .) bonds with a halogen from the next-to-last column (Fl, Cl, Br, . . .). Let's look at sodium chloride (NaCl) to illustrate ionic bonding.

Sodium ($Z = 11$) has an atomic configuration $1s^2 2s^2 p^6 3s^1$: a closed inert core plus an outer 3s electron. The last 3s electron is weakly bound to the inner core. A bond is referred to as "weak" or "strong" according to the energy required to break it. The energy needed to remove one electron from a Na atom to form a positive $Na^+$ ion, called the **ionization energy** *I,* is only 5.14 eV. The ionization process is indicated by the reaction

$$Na + I \rightarrow Na^+ + e^- \qquad \qquad \text{Ionization energy reaction}$$

Chlorine ($Z = 17$) has an atomic configuration $1s^2 2s^2 2p^6 3s^2 3p^5$. Adding one more 3p electron would form a closed inert core. It requires only 3.62 eV, called the **electron affinity** *A,* to remove an electron from a $Cl^-$ ion to form a Cl atom, as indicated by the reaction

$$Cl^- + A \rightarrow Cl + e^- \qquad \qquad \text{Electron affinity reaction}$$

The formation of a NaCl molecule from the neutral atoms can be regarded as occurring in two steps. First, an electron is removed from Na and transferred to Cl to form the ions $Na^+$ and $Cl^-$; then the ions are combined to form NaCl.

The energy required for the first step is obtained by adding the ionization energy reaction to the inverse of the electron affinity reaction to obtain

$$Na + Cl + (I - A) \rightarrow Na^+ + Cl^- \qquad \qquad \text{Energetics to form ions from neutral atoms}$$

Thus, the electron transfer from Na to Cl requires energy $I - A = 5.14\text{ eV} - 3.62\text{ eV} = 1.52\text{ eV}$.

The $Na^+$ and $Cl^-$ ions are mutually attracted by a Coulomb force; the corresponding electrostatic potential energy $-ke^2/r$ decreases as the distance $r$ between them diminishes.

When the distance between the two ions becomes very small, the Coulomb attraction between them is overwhelmed by **exclusion-principle repulsion,** which arises because the wave functions of the core electrons of the two ions overlap at small distances. Because, by the exclusion principle, no two electrons in a system can occupy the same state, some of the electrons in the overlap region must go into higher energy states. To accomplish this requires work against the repulsive forces, thereby increasing the potential energy of the ion pair.

The total potential energy of the $Na^+$ and $Cl^-$ ions is the sum of their negative electrostatic potential energy and their positive repulsive potential energy. To express this relationship quantitatively, we first must choose a value for the potential energy when the ions are infinitely separated and there is no overlap region. We let this energy correspond to the energy of neutral Na and Cl atoms; as a result, the energy of a pair of $Na^+$ and $Cl^-$ ions at infinite separation is 1.52 eV, and the total potential energy as a function of the separation $r$ is

$$U(r) = U_{\text{electrostatic}} + U_{\text{repulsive}} + 1.52\text{ eV}$$
$$= -ke^2/r + U_{\text{repulsive}} + 1.52\text{ eV}$$

Total potential energy

The $U(r)$ versus $r$ curve is shown in Figure 37-1.

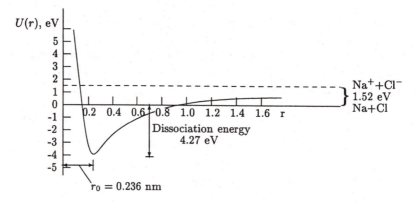

Figure 37-1

The **equilibrium separation** occurs at the minimum of the $U(r)$ curve, which is found experimentally to be at a separation $r_0 = 0.236\text{ nm}$, corresponding to $U(r_0) = -4.27\text{ eV}$. The **dissociation energy** $D$, which is the energy needed to break up the molecule into neutral Na and Cl atoms, is then $D = 4.27\text{ eV}$:

$$NaCl + D \rightarrow Na + Cl$$

Dissociation energy reaction

*The Covalent Bond.* The **covalent bond** arises from quantum-mechanical effects that bind molecules formed from identical or similar atoms, such as hydrogen ($H_2$), nitrogen ($N_2$), carbon monoxide (CO), hydrogen floride (HF), and hydrogen chloride (HCl). To illustrate the quantum-mechanical nature of covalent bonding, let's look at the hydrogen molecule $H_2$, which is formed from two hydrogen atoms whose individual wave functions were described in Chapter 36.

When two hydrogen atoms combine to form $H_2$, the two electrons are either antiparallel or parallel. When the electrons are antiparallel, the total wave function for the $H_2$ molecule, composed of the wave functions of the individual atoms, is symmetrical, and is denoted by $\psi_S$ as shown in Figure 37-2$a$. When the electrons are parallel, the total wave function $\psi_A$ is antisymmetric, as shown in Figure 37-2$b$.

The square of these wave functions, shown in Figures 37-2$c$ and $d$, gives the probability distribution for finding the two electrons in the space around the two protons. Notice in Figure 37-2$c$ that for antiparallel spins the probability is large for finding the electrons between the two protons, whereas in Figure 37-2$d$ the probability is small for finding electrons with parallel spins between the two protons. The "probability cloud" represents the probable locations of the electrons. The negative cloud of electrons with antiparallel spins attracts the protons, producing a bonding force between the protons, whereas electrons with parallel spins create no negative charge between the protons and thus no bonding.

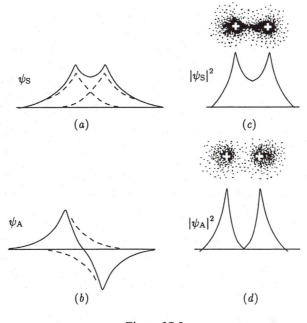

$$\psi_S \qquad |\psi_S|^2$$

$$(a) \qquad\qquad (c)$$

$$\psi_A \qquad |\psi_A|^2$$

$$(b) \qquad\qquad (d)$$

Figure 37-2

Figure 37-3 illustrates these two cases in a different way. The total electrostatic potential energy consists of a positive potential energy that corresponds to repulsive forces between the two electrons and a negative potential energy that corresponds to the attractive force between the electrons and protons. As the distance increases, the antisymmetric $U_A$ potential energy curve, corresponding to parallel spins, decreases steadily but never becomes negative, so there is no equilibrium position—that is, there is no bonding. On the other hand, the symmetric $U_S$ potential energy curve, corresponding to antiparallel spins, has a minimum at $r_0 = 0.074\,\text{nm}$, which is the equilibrium separation of the two protons. Figure 37-3 also shows that the dissociation energy required to separate the hydrogen molecule into two neutral atoms is 4.52 eV.

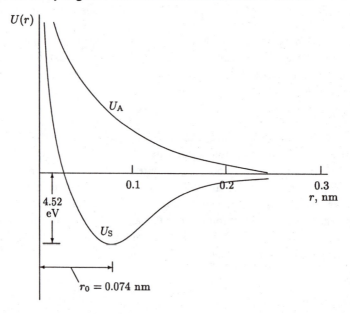

Figure 37-3

The difference in the physical basis for ionic and covalent bonds is that an ionic bond is created when an electron is effectively transferred from one atom to the other, causing the two oppositely charged ions to attract each other. In a covalent bond the protons share the electrons, causing the two positively charged nuclei to be attracted to the cloud of negatively charged electrons between them.

The bond between two identical atoms is totally covalent. In molecules formed from two different atoms, the bonding is usually a mixture of ionic and covalent bonds. In NaCl there is some covalent bonding, and in CO there is some ionic bonding.

The degree (or proportion) of ionic and covalent bonding in a molecule can be determined from the magnitude of the molecule's **electric dipole moment.** The electric dipole moment $\vec{p}$ of two equal and opposite charges $Q$ separated by a distance $d$ has a magnitude $p = Qd$ and a direction that points from the negative to the positive charge.

The distance between two oppositely charged ions in a molecule with ionic bonding is relatively large, resulting in a relatively large dipole moment. The negative charge of the two electrons in a molecule with covalent bonding is located mostly between the two positively charged nuclei, producing two dipole moments that point in opposite directions and therefore tend to cancel, resulting in a relatively small overall dipole moment.

***The van der Waals Bond.*** The **van der Waals bond** arises from weak attractive forces between the dipole moments of molecules. It is the van der Waals force that bonds nearly all substances in the liquid and solid states. Some molecules, such as $H_2O$, have permanent dipole moments that attract each other to produce bonding in the liquid and solid states, as shown in Figure 37-4. Other molecules, which have zero dipole moments, can induce dipole moments in each other, causing a van der Waals force between them.

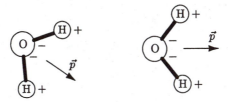

Figure 37-4

*The Hydrogen Bond.* In a **hydrogen bond,** a hydrogen atom is shared between two other atoms. As an example, consider the hydrogen bonding between two water molecules, as shown in Figure 37-5. The electron of a hydrogen atom H in one $H_2O$ molecule is attracted to (spends more time in the vicinity of) an oxygen atom O of an adjacent $H_2O$ molecule. Effectively, the H atom becomes positively charged and the O atom becomes negatively charged. The resulting Coulomb attraction constitutes the hydrogen bond between the two water molecules. Hydrogen bonds often link groups of molecules together and play an important role in enabling giant biological molecules, such as the DNA molecule, to retain their shape.

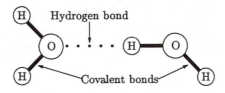

Figure 37-5

*The Metallic Bond.* In a **metallic bond** many electrons—roughly one or two valence electrons per atom—move quite freely as an electron "gas" throughout the metal. The weak attractive force between the positive ions and the electron gas provides the bonding mechanism that holds the ions together.

*Section 37-2. Polyatomic Molecules.* The structure of **polyatomic molecules**—molecules with more than two atoms—follows from the principles of quantum mechanics applied to the bonding between individual atoms. It is primarily the covalent and hydrogen bonds that provide the bonding mechanism of polyatomic molecules.

*Section 37-3. Energy Levels and Spectra of Diatomic Molecules.* The electrons in molecules can be excited into higher energy states, as in atoms. In addition to these electronic excitations, the molecule as a whole rotates and vibrates, and both of these motions are quantized and have energies much lower than the energies of the electronic states. In the following we will restrict ourselves to the simplest case of diatomic (two-atom) molecules.

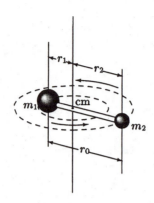

Figure 37-6

*Rotational Energy Levels.* A rotating diatomic molecule can be pictured classically as a dumbbell with unequal masses rotating in a plane about its center of mass, as shown in Figure 37-6. The kinetic energy, with respect to its center of mass, of this molecule, which has moment of inertia $I$ about its center of mass, and rotates at an angular frequency $\omega$, is

$$E = \tfrac{1}{2} I \omega^2 \qquad\qquad \text{Classical kinetic energy of rotation}$$

which can be rewritten in terms of the angular momentum $L = I\omega$:

$$E = \frac{1}{2} I \omega^2 = \frac{1}{2} \frac{(I\omega)^2}{I} = \frac{1}{2} \frac{L^2}{I} \qquad\qquad \text{Classical kinetic energy of rotation}$$

Solution of the Schrödinger equation shows that the rotational angular momentum is quantized according to

$$L^2 = \ell(\ell+1)\hbar^2 \qquad \ell = 0, 1, 2, \ldots \qquad \text{Quantized angular momentum}$$

where $\ell$ is the **rotational quantum number.** We then have quantized rotational energy levels with energies

$$E = \frac{\ell(\ell+1)\hbar^2}{2I} = \ell(\ell+1)E_{0r} \quad \ell = 0, 1, 2, \ldots \qquad \text{Quantized rotational energies}$$

Here

$$E_{0r} = \frac{\hbar^2}{2I} \qquad\qquad \text{Characteristic rotational energy}$$

gives a characteristic rotational energy for a given molecule of moment of inertia $I$. A rotational energy-level diagram is shown in Figure 37-7. Only transitions that satisfy the selection rule $\Delta\ell = \pm 1$ are allowed.

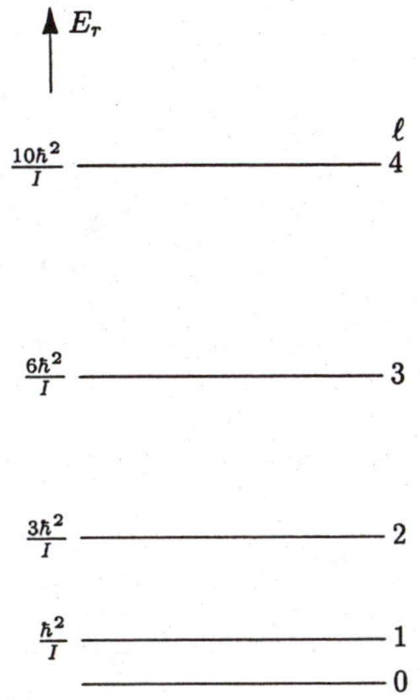

Figure 37-7

If a diatomic molecule is thought of as a dumbbell with masses $m_1$ and $m_2$ located at distances $r_1$ and $r_2$ from the center of mass, as shown in Figure 37-6, the moment of inertia of the dumbbell

$$I = m_1 r_1^2 + m_2 r_2^2$$    Moment of inertia of a dumbbell

can be rewritten in terms of the reduced mass $\mu$

$$\mu = \frac{m_1 m_2}{m_1 + m_2}$$    Reduced mass

as

$$I = \mu r_0^2$$    Moment of inertia of a dumbbell in terms of reduced mass

***Vibrational Energy Levels.*** In addition to rotating, a diatomic molecule can also vibrate in a manner similar to two masses vibrating at the ends of a spring as shown in Figure 37-8. From the Schrödinger equation for a simple harmonic oscillator, it is found that the **vibrational energies** are quantized according to

$$E_v = \left(v + \tfrac{1}{2}\right)hf \quad v = 0, 1, 2, 3, \ldots$$    Quantized vibrational energies

where $v$ is the vibrational quantum number and $f$ is the frequency of vibration.

An energy-level diagram is shown in Figure 37-9. The energy levels are equally spaced by an amount $\Delta E = hf$. Only transitions that satisfy the selection rule $\Delta v = \pm 1$ are allowed, so a photon emitted by a transition between energy levels has the frequency $f$. The lowest energy level is not zero but is $E_0 = \tfrac{1}{2}hf$. Thus, a molecule in its lowest energy state is not at rest but is vibrating with some **zero-point minimum energy** about its equilibrium position.

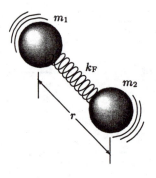

Figure 37-8

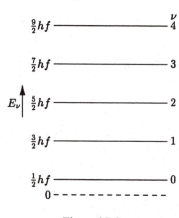

Figure 37-9

The **effective force constant** $k_F$ of a diatomic molecule can be determined by measuring the vibrational frequency and using the relation for two masses vibrating on a spring.

$$f = \frac{1}{2\pi}\sqrt{\frac{k_F}{\mu}}$$

Relationship between frequency and effective force constant

where $\mu$ is the reduced mass described earlier and $k_F$ is the effective force constant.

*Emission Spectra.* The spacing between energy levels of electronic states is much greater than the spacing between vibrational energy levels, which in turn is much greater than the spacing between rotational energy levels. As a consequence we can regard vibrational energy levels as being built up from each electronic energy level, and the smaller rotational energy levels can be treated as being built on each vibrational energy level. The situation is illustrated in Figure 37-10.

A transition from a vibrational energy level $E_U$ of an upper electronic state to a vibrational energy level $E_L$ of a lower electronic state shown in Figure 37-10 results in the emission of a photon whose wavelength is given by $hc/\lambda = E_U - E_L$. The selection rule $\Delta v = \pm 1$ does not hold in this situation.

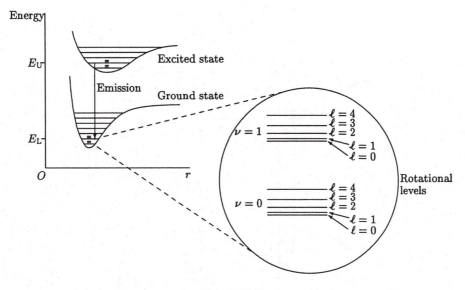

Figure 37-10

The observed **emission spectrum** appears as vibrational bands (one for each transition $\Delta v = v_{\mathrm{U}} - v_{\mathrm{L}}$) rather than isolated lines. With higher resolution, each band shows a **fine structure:** a collection of spectral lines. Each line in the fine structure arises from a transition between the various rotational energy levels that exist at each vibrational energy level.

*Absorption Spectra.* **Absorption spectra** show dark lines where photons have been removed from the incident beam as a result of exciting molecules in the ground electronic state from lower to higher vibrational and rotational energy levels. The predominant transition in the infrared is from the $v = 0$ to the $v = 1$ vibrational energy level. In this transition the atom is excited from various rotational energy levels at the vibrational level $v = 0$ to other rotational levels at the higher vibrational level $v = 1$, subject to the selection rule $\Delta v = \pm 1$, as shown in Figure 37-11.

The frequencies of the **vibrational–rotational absorption lines** fall into two groups,

$$f_{\ell \to \ell+1} = \frac{\Delta E_{\ell \to \ell+1}}{h} = f + \frac{2(\ell+1)E_{0\mathrm{r}}}{h} \quad \ell = 0, 1, 2, \ldots$$

$$f_{\ell \to \ell-1} = \frac{\Delta E_{\ell \to \ell-1}}{h} = f - \frac{2\ell E_{0\mathrm{r}}}{h} \quad \ell = 1, 2, 3, \ldots$$

Frequencies of absorption lines

which result in the lines shown in Figure 37-12. The center of the gap occurs at a photon energy of *hf*, and the separation between the absorption energies of adjacent lines in the two groups on each side of the gap is $2E_{0r}$.

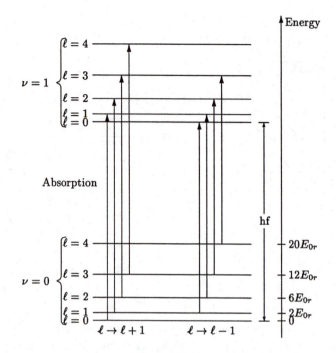

Figure 37-11

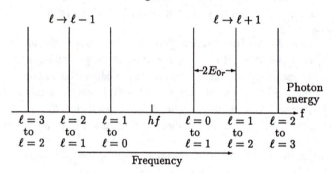

Figure 37-12

# II.  Physical Quantities and Key Equations

### Physical Quantities

There are no new physical quantities introduced in this chapter.

### Key Equations

*Classical kinetic energy of rotation*        $E = \dfrac{1}{2} I \omega^2 = \dfrac{L^2}{2I}$

| | |
|---|---|
| *Quantized angular momentum* | $L^2 = \ell(\ell+1)\hbar^2 \qquad \ell = 0, 1, 2, \ldots$ |
| *Quantized rotational energies* | $E = \ell(\ell+1)\hbar^2/2I = \ell(\ell+1)E_{0r} \qquad \ell = 0, 1, 2, \ldots$ |
| *Characteristic rotational energy* | $E_{0r} = \dfrac{\hbar^2}{2I}$ |
| *Moment of inertia of a dumbbell* | $I = m_1 r_1^2 + m_2 r_2^2$ |
| *Reduced mass* | $\mu = \dfrac{m_1 m_2}{m_1 + m_2}$ |
| *Moment of inertia of a dumbbell in terms of reduced mass* | $I = \mu r_0^2$ |
| *Quantized vibrational energies* | $E = \left(v + \tfrac{1}{2}\right)hf \qquad v = 0, 1, 2, 3, \ldots$ |
| *Selection rules* | $\Delta \ell = \pm 1$ <br> $\Delta v = \pm 1$ |
| *Relationship between frequency and effective force constant* | $f = \dfrac{1}{2\pi}\sqrt{\dfrac{k_F}{\mu}}$ |
| *Frequencies of absorption lines* | $f_{\ell \to \ell+1} = \dfrac{\Delta E_{\ell \to \ell+1}}{h} = f + \dfrac{2(\ell+1)E_{0r}}{h} \qquad \ell = 0, 1, 2, \ldots$ <br> $f_{\ell \to \ell-1} = \dfrac{\Delta E_{\ell \to \ell-1}}{h} = f - \dfrac{2\ell E_{0r}}{h} \qquad \ell = 1, 2, 3, \ldots$ |

## III. Potential Pitfalls

It is easy to confuse ionic and covalent bonds. In an ionic bond, an electron is effectively transferred from one atom to another, producing two oppositely charged ions that attract each other. In a covalent bond, the valence electrons are shared by two nuclei. The two positively charged nuclei are attracted to the negatively charged electron cloud between them.

Do not confuse rotational energies that are specified by the quantum number $\ell$ with the orbital angular momenta of the electrons in an atom of a molecule, which are specified by a quantum number designated by the same letter $\ell$. They are entirely different physical quantities. With rotational energy levels, the quantization condition $L^2 = \ell(\ell+1)\hbar^2$ refers to the angular momentum of the molecule as a whole rotating about its center of mass. This has nothing to do with the orbital angular momenta of the electrons in the molecule.

When dealing with combined rotational and vibrational excitations, do not mix up the different types of energy levels. The smaller-spaced rotational energy levels are built on each larger-spaced vibrational energy level and not the other way around.

## IV. True or False Questions and Responses

**True or False**

_____ 1.  An ionic bond results primarily from the Coulomb attraction between two oppositely charged ions.

_____ 2.  In ionic and covalent bonds there is a repulsive force between the two atoms at short distances because of Coulomb repulsion between the two positively charged nuclei.

_____ 3.  The bonding between two dissimilar atoms is always either totally ionic or totally covalent.

_____ 4.  The degree of ionic or covalent bonding in a molecule can be determined by measuring the electric dipole moment of the molecule.

_____ 5.  The bonding in a metal occurs because of van der Waals forces.

_____ 6.  The van der Waals bond occurs because of dipole attractions.

_____ 7.  The primary bonding mechanism for a polyatomic molecule is the van der Waals bond.

_____ 8.  The reduced mass is larger than the mass of either atom in a diatomic molecule.

_____ 9.  Vibrational energy levels are equally spaced.

_____ 10.  The lowest rotational energy level corresponds to zero energy.

_____ 11.  The lowest vibrational energy level corresponds to zero energy.

_____ 12.  The energy difference between adjacent rotational energy levels is larger than the energy difference between adjacent vibrational energy levels.

_____ 13.  The selection rule $\Delta v = \pm 1$ does not apply when photons are emitted in transitions from vibrational energy levels in one electronic state to vibrational energy levels in another electronic state.

_____ 14.  There are two branches of equally spaced lines in an absorption spectrum of a diatomic molecule.

_____ 15.  Wide bands are seen in the emission spectra of molecules because the vibrational energy levels are not well defined.

**Responses to True or False**

1.  True.

2.  False. The repulsive force occurs because of a quantum-mechanical exclusion-principle repulsion.

3.  False. There is usually a mixture of ionic and covalent bonding.

4.  True.

5.  False. The bonding in a metal arises from the metallic bond between a nearly free gas of electrons that are attracted to the positive ions electrostatically and to each other by quantum-mechanical effects.

6.  True.

7.  False. The covalent bond and the hydrogen bond are the most predominant bonds in a polyatomic molecule.

8.  False. The reduced mass is smaller than the mass of the smaller atom in a diatomic molecule. This can be seen by writing the reduced mass in the form $1/\mu = 1/m_1 + 1/m_2$.

9.  True.

10. True.

11. False. The energy of the lowest vibrational energy level is $E_0 = \frac{1}{2}hf$, called the zero-point energy.

12. False. It's the other way around.

13. True.

14. True.

15. False. Bands are seen in transitions between two vibrational energy levels because the transitions actually occur between closely spaced rotational energy levels that are built on each vibrational energy level. These transitions between the rotational energy levels result in closely spaced spectral lines that can be seen upon higher resolution.

## V.  Questions and Answers

### Questions

1.  In an ionically bound molecule, the two ions are separated by an equilibrium distance $r_0$. Why is the total energy of the molecule not equal to the sum of the coulomb energy $-ke^2/r$ and the difference between the electron affinity of the negative ion and the ionization energy of the positive ion?

2.  Why don't the two oppositely charged ions in an ionically bound molecule continue to attract each other to very small separations?

3.  Why are there no $H_3$ molecules?

4. Which has the larger dipole moment, a molecule with ionic bonding or a molecule with covalent bonding? Why?

5. Briefly explain the mechanism of the van der Waals bond.

6. Briefly explain the mechanism of the hydrogen bond.

7. Briefly explain the mechanism of the metallic bond.

8. What are the primary bonding mechanisms of polyatomic molecules?

9. How can you experimentally measure the moment of inertia of a diatomic molecule?

10. How can you experimentally measure the effective force constant of a diatomic molecule?

11. What is zero-point vibrational energy?

12. What is meant by "bands" in emission spectra of molecules?

13. Explain the origin of the gap in the absorption spectrum shown in Figure 37-12.

14. Explain how you can use an experimental absorption spectrum of a diatomic molecule, such as that in Figure 37-12, to measure the frequency of vibration of the molecule.

**Answers**

1. There is an additional energy due to the exclusion-principle repulsion of the ions at the equilibrium separation.

2. Exclusion-principle repulsion keeps the ions separated.

3. The two electrons in an $H_2$ molecule are antiparallel and form a saturated bond. If a third H atom is brought in, it is not possible for its electron to be also in a 1s state and antiparallel to both of the other two electrons in the $H_2$ molecule, so it is forced into a higher energy state, resulting in repulsion. Essentially there is no room for another electron after an $H_2$ molecule is formed because of the exclusion principle.

4. The ionically bonded molecule has the larger dipole moment. In a molecule that has a covalent bond, the dipole moments, formed from each positively charged nucleus and from the negatively charged electrons between the nuclei, point in opposite directions and tend to cancel, resulting in a small overall dipole moment for the atom.

5. The van der Waals bond arises from the attraction of molecular dipoles for each other. The molecular dipoles can be either permanent dipoles or induced dipoles.

6. The hydrogen bond results from the sharing of a hydrogen atom with two atoms, often two oxygen atoms.

7. An electron gas of nearly free electrons exerts attractive forces on the positive ions of a metal to form a metallic bond.

8. The covalent bond and the hydrogen bond.

9. Measure the energies of the rotational energy levels from the rotational spectrum. Then use

$$E = \frac{\ell(\ell+1)\hbar^2}{2I} \quad \ell = 0, 1, 2, \ldots$$

to find the moment of inertia $I$.

10. Measure the energies of the vibrational energy levels from the vibrational spectrum. From the spacing of the energy levels, find the frequency of oscillation from $\Delta E = hf$. Then find the force constant $k_F$ from

$$f = \frac{1}{2\pi}\sqrt{\frac{k_F}{\mu}}$$

11. From the quantized vibrational energies

$$E_v = \left(v + \tfrac{1}{2}\right)hf \quad v = 0, 1, 2, 3, \ldots$$

the lowest energy corresponding to $v = 0$ has the nonzero value $E_0 = \tfrac{1}{2}hf$. Thus, in its lowest energy state a molecule is not at rest but is vibrating with this zero-point energy about its equilibrium position.

12. In a transition from a vibrational energy level in one electronic state to a vibrational energy level in another electronic state, the resulting spectral line appears as a broad band instead of a sharp line. With higher resolution this band is seen to be composed of many fine-structure lines, which result from transitions between the closely spaced rotational energy levels that exist at each vibrational energy level.

13. The spectral lines in Figure 37-12 result from absorption transitions from a $v = 0$ vibrational state to a $v = 1$ vibrational state, as shown in Figure 37-11. These transitions must satisfy the selection rule $\Delta\ell = \pm 1$. This results in two branches, transitions from $\ell \rightarrow \ell - 1$ and transitions from $\ell \rightarrow \ell + 1$, as shown in Figure 37-12. The gap occurs because there is no $0 \rightarrow 0$ transition.

14. Measuring the energy of the center of the gap, which occurs at $E = hf$, gives the frequency $f$.

## VI.  Problems

**Example #1.**  The following quantities can be measured in an ionically bonded molecule: $r_0$, the equilibrium separation; $I$, the ionization energy of the alkali atom; $A$ the electron affinity of the halogen atom; and $R$ the energy of repulsion at the equilibrium separation.  Obtain an expression for the dissociation energy $D$ of the molecule in terms of these quantities.

**Picture the Problem.** The Coulomb energy can be expressed in terms of $r_0$. The dissociation energy can then be found from the various provided energy terms.

| 1. Find the Coulomb energy. The dissociation energy will be less than this amount because of the energy of repulsion $R$. | $U_C = -ke^2 / r_0$ |
|---|---|
| 2. At infinity, the removal of the electron from the negative halogen ion to neutralize it requires an energy equal to its electron affinity, $A$. Placing this electron on the positive alkali ion to neutralize it gains an energy equal to its ionization energy, $I$. All these combine to give the dissociation energy. | $D = ke^2 / r_0 - R + A - I$ |

**Example #2—Interactive.** The measured dissociation energy of LiF is 5.95 eV, the ionization energy of Li is 5.39 eV, and the electron affinity of F is 3.45 eV. If the repulsive energy at the equilibrium separation of LiF is 1.34 eV, estimate the dipole moment of LiF.

**Picture the Problem.** You can find $r_0$ from the Coulomb energy, which you can find by following Example #1. Use that to find the dipole moment of LiF. **Try it yourself.** Work the problem on your own, in the spaces provided, to get the final answer.

| 1. Determine the Coulomb energy. | |
|---|---|
| 2. From the Coulomb energy, determine $r_0$. | |
| 3. Calculate the ionic dipole moment of LiF. | $p = 2.49 \times 10^{-29}$ C•m |

**Example #3.** The equilibrium separation of KCl is 0.279 nm, and its measured dipole moment is $2.64 \times 10^{-29}$ C•m. What percentage of the bonding of KCl is due to covalent bonding?

**Picture the Problem.** Find the ionic dipole moment corresponding to the equilibrium separation. The percentage of covalent bonding can be found from the ratio of the measured to the calculated ionic dipole moment.

| 1. Assuming the bonding is 100% ionic, find the dipole moment. | $p_{ionic} = er_0 = (1.602 \times 10^{-19}$ C$)(2.79 \times 10^{-10}$ m$)$ <br> $= 4.47 \times 10^{-29}$ C•m |
|---|---|

| 2. Find the percent of ionic bonding. | $\%_{ionic} = p_{measured} / p_{ionic} = 2.64 / 4.47 = 0.59 = 59\%$ |
|---|---|
| 3. The remaining amount is covalent bonding. | $\%_{covalent} = 100\% - \%_{ionic} = 41\%$ |

**Example #4—Interactive.** The equilibrium separation of the two protons in an $H_2$ molecule is 0.074 nm, and the measured binding energy is 4.50 eV. If only Coulomb interactions are taking place, how much negative charge has to be placed at the midpoint between the two nuclei to account for the measured binding energy? Why is this so much smaller than $2e$, the number of electrons in an $H_2$ atom?

**Picture the Problem.** The binding energy is the difference between the positive Coulomb energy between the two protons at equilibrium separation and the negative Coulomb energy between the protons and the unknown negative charge midway between them. **Try it yourself.** Work the problem on your own, in the spaces provided, to get the final answer.

| 1. Determine the positive Coulomb energy between the two protons at their equilibrium separation. | |
|---|---|
| 2. Write an expression for the negative Coulomb energy between the two protons and the unknown negative charge between them. | |
| 3. Solve for the unknown negative charge by equating the difference of the energies in steps 1 and 2 to the binding energy. This charge is much less than $2e$ because the two electrons are not located exactly between the hydrogen atoms. | $q = 0.768e$ |

**Example #5.** Suppose the potential energy curve for an ionically bound molecule, such as that shown in Figure 37-1 for NaCl, is approximated by adding to the Coulomb energy a term to account for exclusion-principle repulsion, giving the total potential energy as a function of the separation of the two ions as $U(r) = -(ke^2/r) + (a/r^9) - b$. Use the values given in Figure 37-1 to find the value of the constants $a$ and $b$ for NaCl.

**Picture the Problem.** The value for $a$ can be determined by finding the minimum of the potential energy function. The value for $b$ can be found from the value of the potential energy function at $r = \infty$.

| 1. Set the derivative of $U(r)$ equal to zero to find the minimum of the curve. From the graph, we know the value of $r$ where the potential energy function is zero. Use this to solve for $a$. | $$\frac{dU(r)}{dr} = \frac{ke^2}{r^2} - \frac{9a}{r^{10}} = 0$$ $$a = \tfrac{1}{9}ke^2 r_0^8 = \tfrac{1}{9}(1.44\,\text{eV·nm})(0.236\,\text{nm})^8$$ $$= 1.54 \times 10^{-6}\ \text{eV·nm}^9$$ |
|---|---|
| 2. Now set $r = \infty$ and use the value for $U(\infty)$ from the graph to find $b$. | $U(\infty) = b = 1.52\,\text{eV}$ |

**Example #6—Interactive.** Use the values of $a$ and $b$ found in Example #5 to find the dissociation energy for NaCl and compare this value with the value given in Figure 37-1.

**Picture the Problem.** The dissociation energy is simply the negative of the potential energy at the equilibrium separation. **Try it yourself.** Work the problem on your own, in the spaces provided, to get the final answer.

| 1. Find the negative of the potential energy function at the equilibrium separation. | $D = 3.90\,\text{eV}$, slightly higher than the value given on the graph |
|---|---|

**Example #7.** The characteristic rotational energy of a CO molecule is $2.39 \times 10^{-4}$ eV. What is the interatomic spacing of the molecule?

**Picture the Problem.** The moment of inertia can be expressed in terms of the reduced mass and the interatomic spacing.

| 1. Find the moment of inertia from the characteristic rotational energy. | $$E_{0r} = \hbar^2 / (2I)$$ $$I = \frac{\hbar^2}{2E_{0r}} = \frac{(1.055 \times 10^{-34}\ \text{J·s})^2}{2(2.39 \times 10^{-4}\ \text{eV})(1.602 \times 10^{-19}\ \text{J/eV})}$$ $$= 1.453 \times 10^{-46}\ \text{kg·m}^2$$ |
|---|---|
| 2. Express the moment of inertia in terms of the reduced mass and interatomic spacing to solve for the spacing. | $$I = \mu r_0^2 = \frac{m_1 m_2}{m_1 + m_2} r_0^2$$ $$r_0^2 = I\frac{m_1 + m_2}{m_1 m_2} = \frac{(1.453 \times 10^{-46}\ \text{kg·m}^2)(12\,\text{u} + 16\,\text{u})}{(1.66 \times 10^{-27}\ \text{kg/u})\ (12\,\text{u})(16\,\text{u})}$$ $$= 1.276 \times 10^{-20}\ \text{m}^2$$ $$r_0 = 0.113\,\text{nm}$$ |

**Example #8—Interactive.** Determine the frequency of radiation corresponding to a transition from the $\ell = 0$ to $\ell = 1$ rotational state of the CO molecule of Example #7.

**Picture the Problem.** The frequency is proportional to the energy difference between the rotational states. **Try it yourself.** Work the problem on your own, in the spaces provided, to get the final answer.

| | |
|---|---|
| 1. Determine the energy of the $\ell = 0$ state. | |
| 2. Determine the energy of the $\ell = 1$ state. | |
| 3. Determine the frequency associated with this energy difference. | $f = 1.16 \times 10^{11}$ Hz |

**Example #9.** The center of the gap of the vibrational-rotational absorption spectrum of HCl gas occurs at 3465 nm. What is the vibrational frequency of a HCl molecule?

**Picture the Problem.** The wavelength is inversely related to the frequency.

| | |
|---|---|
| 1. Relate the wavelength to the frequency and speed of light. | $f = c / \lambda = \left(3 \times 10^8 \text{ m/s}\right) / \left(3.465 \times 10^{-6} \text{ m}\right)$ $= 8.66 \times 10^{13}$ Hz |

**Example #10—Interactive.** What is the effective force constant of a molecule of the HCl gas of Example #9?

**Picture the Problem.** Use the reduced mass and the result of Example #9 in the expression for the effective force constant. **Try it yourself.** Work the problem on your own, in the spaces provided, to get the final answer.

| | |
|---|---|
| 1. Look up the mass of H and Cl. | |
| 2. Determine the reduced mass of the HCl molecule. | |
| 3. Use the reduced mass and the frequency from Example #9 to find the effective force constant. | $k_F = 478$ N/m |

**Example #11.** The vibrational-rotational absorption spectrum for a diatomic molecule is shown in Figure 37-12 with the photon frequencies of the peaks separated by $1.2 \times 10^{12}$ Hz and the center of the gap at $110 \times 10^{12}$ Hz. What is the characteristic rotational energy of the molecule?

**Picture the Problem.** The lines on either side of the gap are separated by $2E_{0r}$.

| 1. Equate the energy spacing to twice the characteristic rotational energy. | $2E_{0r} = hf$ <br><br> $E_{0r} = hf/2 = \left(4.136 \times 10^{-15} \text{ eV} \cdot \text{s}\right)\left(1.2 \times 10^{12} \text{ Hz}\right)/2$ <br><br> $= 2.48 \text{ meV}$ |
|---|---|

**Example #12—Interactive.** Find the lowest vibrational energy of the molecule that produced the absorption spectrum of Example #11.

**Picture the Problem.** Use the relationship between the photon energy of the midpoint of the gap in the absorption spectrum and the lowest molecular vibrational energy. **Try it yourself.** Work the problem on your own, in the spaces provided, to get the final answer.

| 1. The lowest vibrational energy is equal to one-half the energy of the photon in the middle of the gap. | |
|---|---|
| | $E_0 = 0.227 \text{ eV}$ |

# Chapter 38

# Solids

## I.   Key Ideas

*Section 38-1. The Structure of Solids.* By cooling a liquid sufficiently, we can form two kinds of solids: amorphous or crystalline. The molecules of an **amorphous solid,** which is usually formed by rapid cooling, are not arranged in regular arrays but instead are oriented randomly. An amorphous solid does not have a well-defined melting point; it just softens as its temperature increases. Glass is a common amorphous solid.

The molecules of a **crystalline solid,** which has a well-defined melting temperature, are arranged in regular, symmetrical arrays over relatively large regions of the crystal. A single unit structure, called a **unit cell,** repeats regularly throughout the crystal. The makeup of the unit cell depends on the type of bonding between the atoms, ions, or molecules of the crystal, as discussed in Chapter 37.

The structure of a sodium chloride (NaCl) ionic crystal is shown in Figure 38-1. In a NaCl crystal, each $Na^+$ ion has six $Cl^-$ ions as its nearest neighbors, and vice versa; this structure is called a face-centered-cubic crystal. In a body-centered-cubic crystal, each ion has eight nearest neighbor ions of the opposite charge; Figure 38-2 shows the structure of the body-centered-cubic crystal CsCl.

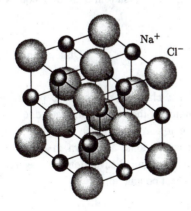

Figure 38-1

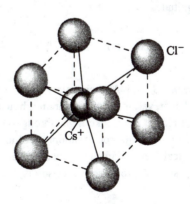

Figure 38-2

The potential energy of a given ion in a crystal due to another ion in the crystal is $U = \pm ke^2 / R$, where $R$ is the distance from the given ion to the other ion, and the sign depends on the charges of the two ions. Adding the potential energies due to the various ions in a crystal results in the following expression for the net attractive potential energy of an ion in a crystal:

$$U_{att} = -\alpha ke^2 / r$$     Net attractive potential energy

where $r$ is the distance between adjacent ions and the **Madelung constant** $\alpha$ depends on the geometry of the structure of a unit cell. For face-centered-cubic structures $\alpha = 1.7476$, and for body-centered-cubic structures $\alpha = 1.7627$.

Besides the Coulomb attractive potential energy between ions in an ionic crystal, there also is an exclusion-principle repulsion potential energy. This is given by the empirical expression

$$U_{rep} = A / r^n$$     Exclusion-principle repulsion potential energy

where $A$ and $n$ are constants. Thus, the total potential energy of an ion in a crystal is

$$U(r) = -\alpha \frac{ke^2}{r} + \frac{A}{r^n}$$     Total potential energy

The equilibrium separation $r_0$ is the value of $r$ at which $U(r)$ is minimum. To find $A$ in terms of $r_0$, set $dU / dr = 0$ and $r = r_0$ to obtain

$$A = \frac{\alpha ke^2 r_0^{n-1}}{n}$$     Constant $A$ in terms of $r_0$

which enables us to write the total potential energy as

$$U(r) = -\alpha \frac{ke^2}{r_0} \left[ \frac{r_0}{r} - \frac{1}{n} \left( \frac{r_0}{r} \right)^n \right]$$     Total potential energy of an ion in a crystal

The dissociation energy (the energy needed to break up the crystal into atoms) equals the magnitude of $U(r)$ at $r = r_0$:

$$\text{Dissociation energy} = |U(r_0)| = \frac{\alpha ke^2}{r_0} \left( 1 - \frac{1}{n} \right)$$     Dissociation energy

***Section 38-2. A Microscopic Picture of Conduction.*** In a classical microscopic picture of a conductor, valence electrons are not bound to particular atoms of the conductor. Instead, electrons move almost freely like a gas throughout the entire conductor. In some given volume $V$, a large number $N$ of electrons randomly collide with the fixed lattice ions, as shown in Figure 38-3. Classical Newtonian theory and the classical Maxwell–Boltzmann distribution can be used to calculate such properties as resistivity, mean free path, and heat conduction.

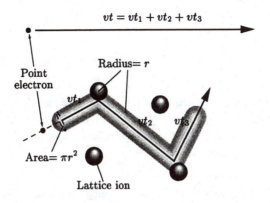

Figure 38-3

Ohm's law for a current-carrying wire segment states that the voltage $V$ across the wire is related to the current $I$ through the wire by

$$V = IR \qquad\qquad \text{Ohm's law}$$

where the resistance $R$ is related to the resistivity $\rho$ by

$$R = \rho L / A \qquad\qquad \text{Relation of resistance to resistivity}$$

with $L$ the length and $A$ the cross-sectional area of the wire segment. When a voltage is applied across a conductor, the resulting electric field inside the conductor causes the electrons to flow, thereby producing a current. As the electrons are accelerated through the wire under the influence of the applied electric field, they collide with the fixed ions in the wire, which slows them down. The **drift velocity** $v_d$ of the electrons, which is superimposed on top of the random velocity of their thermal motion, is related to the electric field $E$ and resistivity by

$$\rho = E / \left( n_e e v_d \right) \qquad\qquad \text{Relation of resistivity to electric field and drift velocity}$$

where $n_e = N / V$ is the number of electrons per unit volume.

It is found that the drift velocity is many orders of magnitude ($\sim 10^{-9}$) smaller than the random thermal velocity. Because of this, as we consider thermal motion of electrons in the following, we are justified in neglecting the extra motion of electrons produced by an applied electric field.

If $\tau$ is the average time between collisions for a given electron in thermal motion, called the **collision time,** the resistivity can be written as

$$\rho = m_e / \left( n_e e^2 \tau \right) \qquad\qquad \text{Relation of resistivity to collision time}$$

If $v_{av}$ is the mean thermal speed of an electron between collisions, the average distance the electron travels between collisions, called the **mean free path** $\lambda$, is

$$\lambda = v_{av} \tau \qquad\qquad \text{Relation of mean free path to collision time}$$

which allows the resistivity to be written as

$$\rho = \frac{m_e v_{av}}{n_e e^2 \lambda}$$     Relation of resistivity to mean free path and mean speed

The classical picture of a point electron colliding with fixed ions does not give a resistivity that agrees with experimental results. Classically, according to the Maxwell–Boltzmann distribution, an electron in an electron gas has a root mean square (rms) speed $v_{rms}$, which is slightly greater than the mean speed $v_{av}$, given by

$$v_{rms} = \sqrt{\frac{3kT}{m_e}}$$     Root mean square speed according to classical Maxwell–Boltzmann theory

where $T$ is the absolute temperature of the electron gas. If this expression for $v_{rms}$ is set equal to $v_{av}$ in the above expression for resistivity, it is seen that classical ideas predict that the resistivity will vary with temperature according to $\rho \propto \sqrt{T}$. Experimentally, though, $\rho$ is found to vary linearly with temperature $T$. Further, at $T = 300\,\mathrm{K}$ the classical calculation for resistivity based on the Maxwell–Boltzmann distribution gives a value for resistivity that is about six times greater than what is measured experimentally. Finally, the classical model does not explain why some materials are conductors, some are insulators, and others semiconductors.

*Section 38-3. The Fermi Electron Gas.* The classical model of electron conduction described above fails because electrons are not Newtonian point particles following Maxwell–Boltzmann statistics having an average kinetic energy equal to $\frac{3}{2}kT$, with zero kinetic energy at $T = 0$. Instead, electrons have wave properties and obey statistics that are described by a quantum-mechanical Fermi–Dirac distribution, obeying the Pauli exclusion principle, which will be described below.

*Energy Quantization in a Box.* To get a feeling for the wave properties of electrons, look at the situation discussed in Chapter 35 where a particle of mass $m$ is confined to a finite region of space such as a one-dimensional box of length $L$. For such a particle, we found that the allowed energies could take on only the quantized values

$$E_n = \frac{n^2 h^2}{8mL^2} = n^2 E_1 \quad n = 1, 2, 3, \ldots$$     Quantized energies of a particle in a 1-D box

where $n$ is called the spatial quantum number, and $E_1$ is the ground state energy. For a given quantum number $n$, the wave function for the particle is given by

$$\psi_n = \sqrt{\frac{2}{L}} \sin \frac{n\pi x}{L}$$     Wave function of a particle in a one-dimensional box

and the wavelength $\lambda_n$ associated with the particle is

$$\lambda_n = 2L / n$$     Wavelength of a particle in a one-dimensional box

***The Pauli Exclusion Principle.*** The electrons in a metal cannot have any arbitrary energy $E_n$ because they must satisfy the **Pauli exclusion principle:**

> No two electrons in a system can be in the same quantum state; that is, they cannot have the same set of values for their quantum numbers.

Electrons are "spin one-half" particles, that is, their *spin* quantum number $m_s$ can have two possible values, $m_s = +\frac{1}{2}$ or $m_s = -\frac{1}{2}$. This means that there can be at most two electrons having a given value of the *spatial* quantum numbers $n, \ell, m_\ell$ :

> There can be at most two electrons with the same set of values for their spatial quantum numbers in any system.

Most of our problems will be one-dimensional, so we will need to consider only one spatial quantum number $n$. Three-dimensional problems involve three spatial quantum numbers, one for each dimension.

Spin one-half particles such as electrons are called **fermions,** and obey the exclusion principle. Other particles with either zero or integral spin quantum numbers, such as $\alpha$ particles, deuterons, and photons, are called **bosons** and do not obey the exclusion principle.

***The Fermi Energy.*** Most systems we will consider involve a large number $N$ of electrons. Let us look at these $N$ electrons in a one-dimensional box at $T = 0$. Only two of the $N$ electrons can be in the lowest energy state corresponding to $n = 1$. There will then be two electrons in the next higher state $n = 2$, two electrons in the state $n = 3$, and so on, until all the electrons are accounted for, and all the available energy states are filled up. For a very large number of electrons, the electrons in the highest state can have an energy much larger than that of the lowest ground state energy $E_1$.

For $N$ electrons, the corresponding value of $n$ of the highest electron state is $n = N/2$. This corresponds to the **Fermi energy** $E_F$ given by setting $n = N/2$ in the above expression for quantized energy levels in a one-dimensional box:

$$E_F = \frac{h^2}{32m_e}\left(\frac{N}{L}\right)^2 \qquad \text{Fermi energy at } T = 0 \text{ in one dimension}$$

For three dimensions, the Fermi energy at $T = 0$ is given by

$$E_F = \frac{h^2}{8m_e}\left(\frac{3N}{\pi V}\right)^{2/3} \qquad \text{Fermi energy at } T = 0 \text{ in three dimensions}$$

where $V$ is the volume of the conductor. When numerical values are substituted, the expression for $E_F$ becomes

$$E_F = \left(0.365 \text{ eV} \cdot \text{nm}^2\right)\left(\frac{N}{V}\right)^{2/3} \qquad \text{Fermi energy at } T = 0 \text{ in three dimensions}$$

These expressions show that the Fermi energy at $T = 0$ depends upon $N/V$, the number of free electrons per unit volume, called the number density of free electrons, which is a characteristic of a given material. In turn, for a given material there is a given Fermi energy $E_F$ at $T = 0$, which can be looked up in standard tables.

The Fermi energy is the highest energy of the $N$ electrons in the three-dimensional conductor under consideration. The $N$ electrons in the conductor of volume $V$ have energies ranging from the ground state energy $E_1$ to the Fermi energy $E_F$. The average energy of this system can be shown to be

$$E_{av} = \tfrac{3}{5} E_F$$
Average energy of electrons in a Fermi gas at $T = 0$

For typical conductors, $E_{av}$, which is of the order of several eV, is much larger than thermal energies of about $kT = 0.026 \, \text{eV}$ at room temperature of $T = 300 \, \text{K}$.

***The Fermi Factor at $T = 0$.***    The **Fermi factor** $f(E)$ gives the probability of finding an electron in a given electron state. At $T = 0$, all the states below $E_F$ are filled, and all the states above $E_F$ are empty. Therefore, the probability is equal to one for finding an electron in a state below $E_F$, and is equal to zero for finding an electron in a state above $E_F$, as shown in Figure 38-4. Thus, at $T = 0$ the Fermi factor is

$$f(E) = 1 \quad E \le E_F$$
$$f(E) = 0 \quad E > E_F$$
Fermi factor at $T = 0$

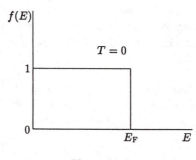

Figure 38-4

***The Fermi Factor for $T > 0$.***    At temperatures greater than $T = 0$, the picture changes, but not by much. Electrons with energies around the Fermi energy $E_F$ gain extra energy of the order of $kT$, which at 300 K is only 0.026 eV. Thus, there will be a certain probability of finding these electrons with an energy slightly higher than the Fermi energy at $T = 0$, and the Fermi factor at $T > 0$ looks as shown in Figure 38-5. Since at $T > 0$ there is no distinct energy separating filled and unfilled states, the Fermi energy at $T > 0$ is defined to be that energy for which the probability of being occupied is one-half. Except for very high temperatures, the Fermi energy at $T > 0$ is very close to the Fermi energy at $T = 0$.

The **Fermi temperature** $T_F$ is defined by

$$kT_F = E_F$$
Fermi temperature

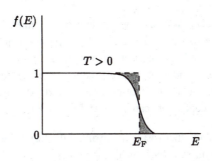

Figure 38-5

Note that even at $T = 0$ there is a finite Fermi temperature. For most metals, the Fermi temperature is so large that the metal would not be a solid at actual temperatures around $T_F$.

Related to the Fermi energy is the **Fermi speed** $u_F$ given by

$$u_F = \sqrt{\frac{2E_F}{m_e}}$$    Relation between Fermi speed and Fermi energy

*Contact Potential.* When two different metals are placed in contact with each other, a potential difference $V_{contact}$ called the **contact potential** develops between them. The contact potential depends upon the work function $\phi$ of each of the metals. Recall from the photoelectric effect of Chapter 34 that $\phi$ is the energy required to remove the least tightly bound electrons from a metal. The contact potential is related to the work functions $\phi_1$ and $\phi_2$ of the two metals in contact by

$$V_{contact} = (\phi_1 - \phi_2)/e$$    Relation of contact potential to work function

*Section 38-4.  Quantum Theory of Electrical Conduction.* If we use the Fermi speed in place of the average speed, the expression for the resistivity of a metal becomes

$$\rho = \frac{m_e u_F}{n_e e^2 \lambda}$$    Resistivity

Unfortunately, the Fermi speed is considerably larger than the average speed, which means the calculated resistivity above is now about 100 times greater than the experimentally determined value.   To solve both this problem, as well as the temperature dependence problem of the resistivity, we need to re-interpret the mean-free-path.

Quantum-mechanically, electron waves do not scatter at all in a perfectly ordered crystal. It is only imperfections in the crystal ordering that cause scattering to occur.  As a result, we can still use

$$\lambda = \frac{1}{n_{ion} A} = \frac{1}{n_{ion} \pi r_0^2}$$    Quantum-mechanical mean free path

for the mean free path, but only if we interpret $r_0$ to be the amplitude of thermal vibrations, rather than the radius of the ions in the lattice.

***Section 38-5. Band Theory of Solids.*** The free-electron picture of a Fermi gas of electrons interacting with atoms in the lattice does not explain why some materials (conductors) are good conductors of electricity, why other materials (insulators) are poor conductors of electricity, and why still other materials (semiconductors) have conductive properties somewhere between those of conductors and insulators. Also, the free-electron picture does not explain why the ratio of resistivities of insulators and conductors can be much larger than $10^{25}$.

An explanation of these and other results is given by the **band theory of solids.** The electrons in a single atom in its ground state occupy the lowest-lying energy levels, with higher energy levels being unoccupied. Figure 38-6*a* shows the first three energy levels of an atom (designated as 1s, 2s, and 2p). If two of these atoms are brought close together, the interaction between them shifts the energy of each level for each atom; the result is that each of the three previously single energy levels is split into two slightly different levels, as shown in Figure 38-6*b*. When $N$ atoms are brought close together, their interactions cause each previously single energy level to split into $N$ separate but closely spaced levels. Figure 38-6*c* shows the splittings when six atoms are brought together.

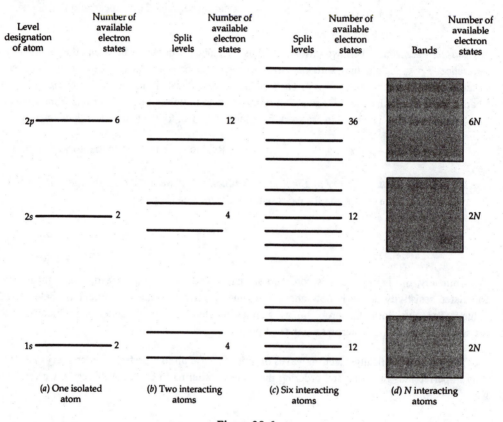

Figure 38-6

Now consider a macroscopic solid, in which a large number of atoms are very close together and $N$ is of the order $10^{23}$. The $N$ energy levels are now so close together that they can be regarded as forming continuous **bands,** one band for each of the previously single energy levels, as shown in Figure 38-6*d.*

The positions of these energy-level bands relative to each other and how they fill with electrons determine whether a material is a conductor, insulator, or semiconductor. A band can be filled with electrons, partially filled with electrons, or empty, as shown in Figure 38-7. The band occupied by the highest-energy valence electrons is called the **valence band.** The lowest band in which there are unoccupied energy states is called the **conduction band.** An energy gap between allowed bands is called a **forbidden energy band.**

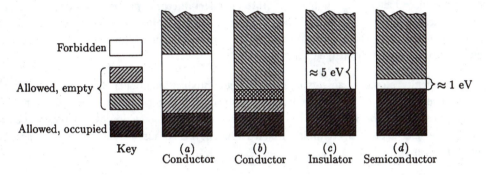

Figure 38-7

A *conductor* can result in two ways. (a) If the valence band is only partially filled (Figure 38-7*a*), it is easy to excite valence electrons into the energy levels in the unfilled part of the band. We simply apply an electric field to give these electrons the kinetic energy they need to participate in electrical conduction. In this case, the valence band is also the conduction band. (b) In another situation (Figure 38-7*b*), the conduction band overlaps the partially or totally filled valence band, resulting in unoccupied energy levels into which valence electrons can move to participate in electrical conduction.

The valence band in an *insulator* is completely filled with electrons, and the conduction band is separated from the valence band by a large energy gap, of the order of 5 eV (Figure 38-7*c*). At ordinary temperatures around 300 K, the thermal energy $kT$ is of the order of 0.026 eV, so only a few electrons have enough energy to be in the conduction band. Even if we apply an electric field to an insulator, we cannot accelerate the electrons to higher kinetic energies because there are no nearby energy levels for them to occupy. Thus, little electrical conduction results.

In a *semiconductor* the energy gap between a filled valence band and the conduction band is small, of the order of 1 eV (Figure 38-7*d*). At ordinary temperatures, thermal excitation has promoted a sizable number of electrons into the conduction band, and each of these thermally excited electrons has left behind it an unfilled **hole** in the valence band. This type of material is called an **intrinsic semiconductor.** When we apply an electric field, the thermally excited electrons acquire kinetic energy and participate in electrical conduction because many energy levels are available to them; electrons in the valence band also can acquire kinetic energy and participate in electrical conduction because there are holes in the energy levels into which they can move. The movement of electrons into holes is equivalent to positive charges moving in a direction opposite to that of the electrons.

For conductors, resistivity increases with increasing temperature and conductivity (the reciprocal of resistivity) decreases. But for semiconductors, resistivity decreases with increasing temperature and conductivity increases. When we raise the temperature of a semiconductor, we excite more electrons into the conduction band, leaving more holes in the valence band. Because both electrons and holes can participate in electrical conduction, conductivity increases and resistivity decreases.

***Section 38-6. Semiconductors.*** The properties of intrinsic semiconductors can be changed in a controllable way by introducing impurities into the semiconductor—that is, by **doping** the intrinsic semiconductor to turn it into an **impurity semiconductor.** By appropriate doping, we can turn an intrinsic semiconductor into either an ***n*-type semiconductor,** in which conduction is due primarily to negatively charged electrons, or into a ***p*-type semiconductor,** in which conduction is due primarily to positively charged holes.

The impurity atoms of an *n*-type semiconductor produce additional electrons that occupy energy levels just below the conduction band, as shown in Figure 38-8*a*. These electrons can easily move into the conduction band to participate in electrical conduction, without leaving extra holes in the valence band. Such levels are called **donor levels** because they donate electrons to the conduction band.

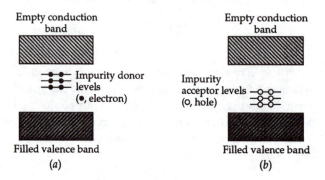

Figure 38-8

The impurity atoms of a *p*-type semiconductor introduce a deficiency of electrons and produce unoccupied energy levels, or holes, just above the nearly filled valence band, as shown in Figure 38-8*b*. Electrons in the valence band can easily move into these holes, thereby producing extra holes in the valence band to participate in electrical conduction, without producing extra electrons in the conduction band. Such levels are called **acceptor levels** because they accept electrons from the valence band.

***Section 38-7. Semiconductor Junctions and Devices.*** A *p*-type semiconductor has a larger concentration of positively charged holes that are available for electrical conduction than of available negatively charged electrons; an *n*-type semiconductor has a larger concentration of available electrons. A ***pn* junction** is formed when a *p*-type and an *n*-type semiconductor are placed in contact. Because of the unequal concentrations of electrons and holes, electrons diffuse from the *n* to the *p* side and holes diffuse from the *p* to the *n* side, until the attractive forces between the holes and electrons result in an equilibrium configuration. (Both the holes and the electrons are called charge carriers.)

At equilibrium there is a double layer of charges at the junction: negative charges on the *p* side and positive charges on the *n* side (Figure 38-9). Between the negative and positive layers is the **depletion region,** which has no charge carriers. As a consequence there is a potential difference *V* across the junction, with the *n* side being at the higher voltage.

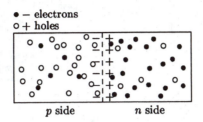

Figure 38-9

***Diodes.*** Now look at what happens when we connect an external voltage source across the junction. If we connect the positive terminal of the source to the *p* side of the junction, we obtain a forward bias as shown in Figure 38-10*a*. That is, the external potential difference reduces the potential difference across the junction, producing an increased diffusion of electrons and holes; this increased diffusion results in a current that increases exponentially with the applied external voltage. If we connect the positive terminal of the source to the *n* side of the junction, we obtain a **reverse bias** (Figure 38-10*b*). In this case the external potential difference adds to the original junction potential difference *V,* producing a very small reverse current that eventually reaches a saturation value as the external voltage is increased. The junction thus acts as a **diode,** through which conduction takes place essentially in only one direction.

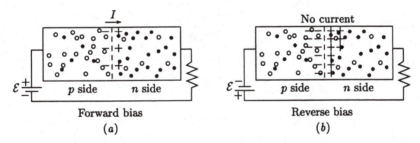

Figure 38-10

Other types of devices built on *pn* junctions are **tunnel diodes, solar cells, light-emitting diodes** (LEDs), and semiconductor lasers.

***Transistors.*** A junction transistor consists of a very thin semiconductor region sandwiched between two semiconductor regions of the opposite type. A ***pnp* transistor** and an ***npn* transistor** are shown with their circuit symbols in Figures 38-11*a* and *b*. The narrow central region of a transistor is the **base,** and the outer regions are the **emitter** and **collector.** Thus, a transistor consists of two *pn* junctions, one between emitter and base, and the other between base and collector. The emitter is much more heavily doped than the base and collector. Here we discuss only the operation of a *pnp* transistor; the operation of an *npn* transistor is essentially the same.

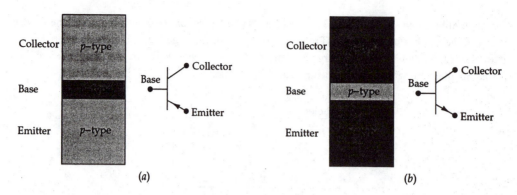

Figure 38-11

The basic operation of a transistor can be seen from an analysis of Figure 38-12. The voltage $V_{ec}$ causes forward biasing of the emitter–base junction and reverse biasing of the base–collector junction. Because of the forward bias, the heavily doped $p$-type emitter emits holes that flow across the emitter–base junction into the base, resulting in the emitter current $I_e$. A small number of the holes that flow across the emitter–base junction will tend to recombine in the base, but this would result in an undesirable accumulation of positive charge in the base that would prevent other holes from crossing the junction. To prevent this accumulation, an alternate path is provided to draw off these holes by connecting a battery $V_{eb}$ between the base and emitter and letting a small base current $I_b$ pass. The collector current is then $I_c = I_e - I_b$.

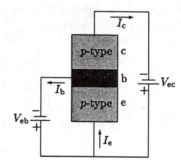

Figure 38-12

In practice $I_b$ is much smaller than either $I_c$ or $I_e$. The current gain $\beta$ of the transistor is defined by

$$I_c = \beta I_b \qquad \qquad \text{Current gain}$$

Typically $\beta$ is in the range from 10 to 100.

A common use for a transistor is the amplification of small time-varying signals. The operation of a typical amplifier circuit can be seen from an analysis of Figure 38-13. The small time-varying input signal $v_s$ produces a time-varying current $i_b$ that is added to the steady-state current $I_b$. This results in a large time-varying output current $i_c = \beta i_b$ that is added to the steady-state current $I_c$ in the collector. The current $i_c$ produces a time-varying output voltage $v_L$ across the load resistance $R_L$.

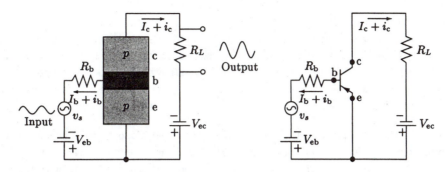

Figure 38-13

The ratio of the output voltage $v_L$ to the input voltage $v_s$ is the **voltage gain** of the amplifier. If $r_b$ is the internal resistance of the transistor between the base and emitter, and $R_b$ is the resistance in series with the bias voltage $V_{eb}$, the voltage gain is

$$\text{Voltage gain} = \frac{v_L}{v_s} = \beta \frac{R_L}{R_b + r_b}$$    Voltage gain

An **integrated circuit** (IC) chip is a collection of resistors, capacitors, diodes, and transistors, interconnected through circuits, that is fabricated on a single semiconductor crystal, usually silicon. It is possible to produce ICs that contain several hundred thousand components in an area smaller than $1\,\text{cm}^2$. The development of ICs since the early 1960s has revolutionized the electronics industry and has led to thousands of applications in computers, cameras, watches, communication networks, and many other areas.

***Section 38-8. Superconductivity.*** The resistivity of a **superconductor** drops suddenly to zero when its temperature is decreased below a **critical temperature** $T_c$, as shown in Figure 38-14, which varies from one superconductor to another. Below $T_c$ the resistivity is truly zero. If a current is established in a superconducting loop, it persists for years with no measurable decay.

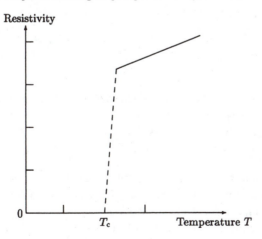

Figure 38-14

When superconductivity was first discovered in the early 1900s, materials became superconducting only at very low temperatures of the order ~ 0.1–10 K. At present, though (April 2003), many **high-temperature superconductors** have been found with critical temperatures as high as 138 K at atmospheric pressure. This has revolutionized the study and applications of superconductors because high-temperature superconductors can be cooled with inexpensive liquid nitrogen, which boils at 77 K.

***The BCS Theory.*** The **BCS theory** is named after Bardeen, Cooper, and Schrieffer, who first put forth an explanation of superconductivity in 1957, and who received the Nobel prize for their work in 1972. In the BCS theory, two electrons in a superconductor at low temperature form a bound state, called a **Cooper pair,** as a result of interactions of the electrons with the crystal lattice of the material. This can be seen in an intuitive manner as follows. Figure 38-15 shows positive ions in the lattice being attracted toward an electron. The effect of this interaction is to produce a net positive charge in the region of these ions, which can now attract another electron toward the region. Thus, the lattice acts as a mediator that produces an attractive force between the two electrons in a Cooper pair that exceeds their Coulomb repulsion.

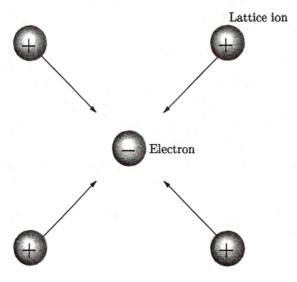

Figure 38-15

Electrons in a Cooper pair have opposite spins and have equal and opposite linear momenta. Thus, a Cooper pair has the properties of a single particle with zero momentum and zero spin. This means that a Cooper pair behaves like a boson, which does not obey the Pauli exclusion principle, so there can be any number of Cooper pairs in the same quantum state at the same energy.

When a material is in a superconducting state, its Cooper pairs can absorb (or emit) energy only by breaking up (or forming). (This is analogous to molecules that must absorb energy before they can break up into their constituent atoms.) The energy required to break up a Cooper pair is called the **superconducting energy gap** $E_g$, which is of the order of $10^{-3}$ eV. In the BCS theory, $E_g$ at $T = 0$ is related to the critical temperature $T_c$ by

$$E_g = 3.5\,kT_c \qquad\qquad \text{Superconducting energy gap at } T = 0$$

As the temperature is increased from $T = 0$, $E_g$ decreases, reaching zero at $T = T_c$ when superconductivity ceases.

When electrical conduction occurs in a superconductor, all the Cooper pairs have the same momentum, which remains constant because individual pairs cannot be scattered by the lattice ions. Because the lattice ions cannot scatter Cooper pairs, there is no resistance.

***The Josephson Effect.*** Consider a junction consisting of two metals separated by a layer of insulating material that is only a few nanometers thick. If a voltage is applied across this junction, charged particles will tunnel through it due to barrier penetration (see Chapter 35). As a result, we obtain a current that can be measured as a function of the voltage applied across the junction.

If the two metals are superconductors, the junction is called a **Josephson junction.** Even when no voltage is applied across this junction, Cooper pairs tunnel through it. The resulting dc current, called the **dc Josephson effect current,** is given by

$$I = I_{max} \sin\left(\phi_2 - \phi_1\right) \qquad \text{DC Josephson effect current}$$

where the maximum current $I_{max}$ depends on the thickness of the junction. The angles $\phi_1$ and $\phi_2$ are the phases of the wave functions of the Cooper pairs in the two superconductors.

If a dc voltage $V$ is applied across the junction, the result is, somewhat surprisingly, an ac current called the **ac Josephson effect current,** with frequency

$$f = \frac{2e}{h} V \qquad \text{AC Josephson effect current frequency}$$

Because we can measure frequencies extremely accurately, this effect provides an experimental method for measuring the ratio $e/h$ very precisely.

***Section 38-9. The Fermi–Dirac Distribution.*** The **Fermi–Dirac distribution function** $n(E)$ gives the number of electrons $dN$ having energies in the interval between $E$ and $E + dE$ according to

$$dN = n(E)\, dE \qquad \text{Number of electrons with energies between } E \text{ and } E + dE$$

The expression for $n(E)$ is composed of two parts, the density of states and the Fermi factor.

The number of states between $E$ and $E + dE$ is $g(E)dE$, where $g(E)$ is the **density of states** given by

$$g(E) = \frac{8\pi\sqrt{2}m_e^{3/2}V}{h^3}E^{1/2} = \frac{3N}{2E_F^{3/2}}E^{1/2} \qquad \text{Density of states}$$

The density of states does not depend on temperature. A plot of $g(E)$ versus $E$ is shown in Figure 38-16.

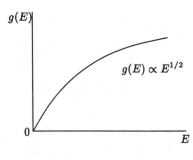

Figure 38-16

The other part of the overall distribution comes from the Fermi factor $f(E)$, which gives the probability of finding an electron in a given state. We have seen in Section 38-3 that at $T = 0$ the Fermi factor is

$$f(E) = 1 \quad E \le E_F$$
$$f(E) = 0 \quad E > E_F$$

Fermi factor at $T = 0$

At $T > 0$, the Fermi factor takes on a more complicated form given by

$$f(E) = \frac{1}{e^{(E - E_F)/(kT)} + 1}$$

Fermi factor at $T > 0$

Figure 38-5 shows the Fermi factor at $T > 0$ compared with the Fermi factor at $T = 0$ shown in Figure 38-4. The shaded region shows there is a finite probability of finding electrons in states with energies larger than $E_F$.

The number of electrons in a given energy interval $dE$ is the number of states in the energy interval, $g(E)dE$, multiplied by the probability of finding an electron in a given state, $f(E)$:

$$n(E)\,dE = g(E)\,dE\,f(E) \qquad \text{Number of electrons with energies between E and } E + dE$$

Using the expressions for $g(E)$ and $f(E)$, we obtain $n(E)$, which is the Fermi–Dirac distribution function:

$$n(E) = \frac{8\pi\sqrt{2}m_e^{3/2}V}{h^3}\frac{E^{1/2}}{e^{(E - E_F)/(kT)} + 1}$$
$$= \frac{3N}{2E_F^{3/2}}\frac{E^{1/2}}{e^{(E - E_F)/(kT)} + 1}$$

Fermi–Dirac distribution function

A plot of $n(E)$ is shown in Figure 38-17, which is the product of Figures 38-5 and 38-16. The dashed curve shows $n(E)$ at $T = 0$, where there are no electrons with energies larger than the Fermi energy. At higher temperatures, some electrons with energies near the Fermi energy are excited above the Fermi energy, as indicated by the shaded region. Because only those electrons within about $kT$ of the Fermi energy can be excited to higher energy states, the difference between $n(E)$ at temperature $T$ and at $T = 0$ is very small except for extremely high temperatures.

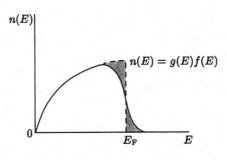

Figure 38-17

# II. Physical Quantities and Key Equations

## Physical Quantities

*Madelung constant for face-centered cubic structures*    $\alpha = 1.7476$

*Madelung constant for body-centered cubic structures*    $\alpha = 1.7627$

## Key Equations

*Total potential energy of an ion in a crystal*

$$U(r) = -\alpha \frac{ke^2}{r_0}\left[\frac{r_0}{r} - \frac{1}{n}\left(\frac{r_0}{r}\right)^n\right]$$

*Dissociation energy*

$$\text{Dissociation energy} = |U(r_0)| = \frac{ake^2}{r_0}\left(1 - \frac{1}{n}\right)$$

*Relation of resistance to resistivity*    $R = \rho L / A$

*Relation of resistivity to electric field and drift velocity*    $\rho = E/(n_e e v_d)$

*Relation of resistivity to collision time*    $\rho = m_e/(n_e e^2 \tau)$

*Relation of mean free path to collision time*    $\lambda = v_{av}\tau$

*Relation of resistivity to mean free path and mean speed*    $\rho = \dfrac{m_e v_{av}}{n_e e^2 \lambda}$

*Root mean square speed according to classical Maxwell–Boltzmann theory*    $v_{rms} = \sqrt{\dfrac{3kT}{m_e}}$

*Quantized energies of a particle in a one-dimensional box*

$$E_n = n^2 \frac{h^2}{8mL^2} = n^2 E_1 \quad n = 1, 2, 3, \ldots$$

Fermi energy at $T = 0$ in one dimension

$$E_F = \frac{h^2}{32m_e}\left(\frac{N}{L}\right)^2$$

Fermi energy at $T = 0$ in three dimensions

$$E_F = \frac{h^2}{8m_e}\left(\frac{3N}{\pi V}\right)^{2/3}$$

Fermi energy at $T = 0$ in three dimensions

$$E_F = \left(0.365\ \text{eV·nm}^2\right)\left(\frac{N}{V}\right)^{2/3}$$

Average energy of electrons in a Fermi gas at $T = 0$

$$E_{av} = \tfrac{3}{5}E_F$$

Fermi factor at $T = 0$

$$f(E) = 1 \quad E \leq E_F$$
$$f(E) = 0 \quad E > E_F$$

Fermi temperature

$$kT_F = E_F$$

Relation between Fermi speed and Fermi energy

$$u_F = \sqrt{\frac{2E_F}{m_e}}$$

Relation of contact potential to work function

$$V_{contact} = \left(\phi_1 - \phi_2\right)/e$$

Quantum-mechanical mean free path

$$\lambda = \frac{1}{n_{ion}A} = \frac{1}{n_{ion}\pi r_0^2}$$

Current gain of a transistor

$$I_c = \beta I_b$$

Voltage gain of a transistor

$$\text{Voltage gain} = \frac{v_L}{v_s} = \beta\frac{R_L}{R_b + r_b}$$

Superconducting energy gap at $T = 0$

$$E_g = 3.5kT_c$$

DC Josephson effect current

$$I = I_{max}\sin\left(\phi_2 - \phi_1\right)l$$

AC Josephson effect current frequency

$$f = 2eV/h$$

Number of electrons with energies between $E$ and $E + dE$

$$dN = n(E)\,dE$$

Density of states

$$g(E) = \frac{8\pi\sqrt{2}m_e^{3/2}V}{h^3}E^{1/2} = \frac{3N}{2E_F^{3/2}}E^{1/2}$$

Fermi factor at $T > 0$

$$f(E) = \frac{1}{e^{(E - E_F)/(kT)} + 1}$$

*Fermi–Dirac distribution function*

$$n(E) = \frac{8\pi\sqrt{2}m_e^{3/2}\,V}{h^3}\frac{E^{1/2}}{e^{(E-E_F)/(kT)}+1}$$

$$= \frac{3N}{2E_F^{3/2}}\frac{E^{1/2}}{e^{(E-E_F)/(kT)}+1}$$

## III. Potential Pitfalls

Do not confuse the drift velocity of electrons with their random thermal mean velocity.

Understand the difference between the general Pauli exclusion principle and the Pauli exclusion principle for spatial states.

Understand the difference between fermions and bosons.

Understand the band structure that leads to the differences between conductors, insulators, and semiconductors.

Do not confuse Cooper pairs in the BCS theory with free electrons in the free-electron theory.

Understand the differences between the dc and ac Josephson effects.

Do not confuse the number of particles that have an energy between $E$ and $E + dE$ with the number of states that have an energy between $E$ and $E + dE$.

Do not confuse the number of particles that have an energy between $E$ and $E + dE$ with the Fermi–Dirac distribution function $n(E)$.

Do not mix up the structures of face-centered-cubic crystals, body-centered-cubic crystals, and hexagonal close-packed crystals.

Understand how the Madelung constant is related to the geometry of a crystal and why it is of the order of 1.8.

Remember the difference between $n$-type and $p$-type semiconductors.

## IV. True or False Questions and Responses

**True or False**

_____ 1.  A good electrical conductor is always a good heat conductor.

_____ 2.  The classical free-electron theory can explain the observed properties of electrical conduction but cannot explain the observed properties of heat conduction.

_____ 3.  In a conductor, the drift speed and mean speed of electrons are comparable in magnitude.

_____ 4.  In a conductor, the mean speed and root mean square speed of electrons are comparable in magnitude.

_____ 5.  In the ground state, $N$ bosons in a one-dimensional box will have the same energy as $N$ fermions in the same box.

_____ 6.  Free-electron theory can explain the observed properties of electrical conduction and heat conduction if the classical Maxwell–Boltzmann distribution function is replaced with the quantum-mechanical Fermi–Dirac distribution function.

_____ 7.  A conductor is a material whose valence band is completely filled.

_____ 8.  The difference between an insulator and a semiconductor is in the size of the energy gap between the valence and conduction bands.

_____ 9.  In a quantum-mechanical picture, the contribution of the electron gas to heat conduction is small because only electrons near the value $kT$ of the Fermi energy level can be excited.

_____ 10. The band theory arises because atoms in metals attract each other as though rubber bands were stretched between them.

_____ 11. A Cooper pair has the properties of a spin-$\frac{1}{2}$ fermion.

_____ 12. The resistance of a superconductor is very small, but it is not exactly zero.

_____ 13. The zero resistance of a superconductor results because electrons in Cooper pairs cannot be scattered by the lattice ions.

_____ 14. Tunneling through a Josephson junction occurs because Cooper pairs punch holes in the material at the junction separating the two superconductors.

_____ 15. An ac current results when a dc voltage is applied across a Josephson junction.

_____ 16. The current in a Josephson junction is zero when the voltage across the junction is zero.

_____ 17. A new class of high-temperature superconductors operates at around room temperature.

_____ 18. At $T = 0$ the probability is 0 that all energy levels above the Fermi energy level are occupied.

_____ 19. At $T > 0$ the probability is less than $\frac{1}{2}$ that all energy levels above the Fermi energy level are occupied.

_____ 20. The number of electrons occupying energy levels between $E$ and $E + dE$ is equal to the number of states in the same energy interval.

_____ 21. The density of states $g(E)$ is the same at a higher temperature as at $T = 0$.

_____ 22. The Fermi factor $f(E)$ is the same at a higher temperature $T$ as at $T = 0$.

_____ 23. The Fermi factor is never greater than 1.

_____ 24. The Fermi energy at a higher temperature $T$ is very close to the Fermi energy at $T = 0$.

_____ 25. The Fermi energy depends primarily on the density of free electrons in the material.

_____ 26. For a particular type of crystal structure (say, a face-centered-cubic structure), the Madelung constant depends on the distance between the atoms of the crystal.

_____ 27. The repulsive force between atoms at small distances in a crystal results from the exclusion principle.

_____ 28. In a crystal that has a face-centered-cubic structure, each positive ion has six nearest negative-ion neighbors, and vice versa.

_____ 29. In a crystal with a body-centered-cubic structure, each ion has six nearest neighbors of the opposite charge.

_____ 30. As a general rule, it is advantageous to get rid of impurities in a semiconductor.

_____ 31. A semiconductor diode rectifier is formed by joining $p$-type and $n$-type semiconductors together to form a $pn$ junction.

_____ 32. The current in a semiconductor diode is very large when the diode is reverse biased.

_____ 33. A transistor is equivalent to a $pn$ junction joined to an $np$ junction.

_____ 34. It is possible to interconnect several hundred thousand electronic devices in an area smaller than $1 \text{cm}^2$ on an integrated circuit chip.


**Responses to True or False**

1. True.

2. False. The classical free-electron theory cannot explain the observed properties of either electrical conduction or heat conduction.

3. False. The mean speed is of the order of $10^9$ times larger than the drift speed.

4. True.

5. False. In the ground state, the $N$ bosons will all occupy the lowest energy level. Because of the exclusion principle, however, only two fermions can occupy the lowest energy level, and the other fermions will occupy higher energy levels. Thus, in the ground state the energy of $N$ fermions will be larger than the energy of $N$ bosons.

6. True.

7. False. A conductor is a material whose valence band is only partially filled.

8. True.

9. True.

10. False. It is called the band theory because the energy levels of many closely packed atoms split into groups of nearly continuous closely spaced energy levels called bands.

11. False. A Cooper pair behaves like a spin-0 boson and does not satisfy the Pauli exclusion principle.

12. False. The resistance of a superconductor is exactly zero.

13. True.

14. False. Tunneling occurs because of wave properties of Cooper pairs.

15. True.

16. False. There is a current in a Josephson junction even when there is no voltage across the junction. The current results from the tunneling of Cooper pairs through the junction.

17. False. At present the highest high-temperature superconductors at atmospheric pressure have critical temperatures of around 138 K, which is above the boiling point of liquid nitrogen (77 K) but is far below room temperature (around 300 K).

18. True.

19. True.

20. False. The number of electrons $n(E)dE$ with energies between $E$ and $E+dE$ equals the number of states $g(E)dE$ between $E$ and $E+dE$ multiplied by the Fermi factor $f(E)$, which gives the probability that a state is occupied: $n(E)dE = g(E)dE\,f(E)$.

21. True.

22. False. The Fermi factor as a function of temperature $T$ is given by $f(E) = \left(e^{(E-E_F)/(kT)} + 1\right)^{-1}$. As shown in Figure 38-5, the Fermi factor at $T > 0$ differs from the Fermi factor at $T = 0$, which is $f(E) = 1$ for $E \le E_F$ and $f(E) = 0$ for $E > E_F$.

23. True.

24. True.

25. True.

26. False. The Madelung constant depends only on the type of geometry of the crystal and is a constant for a particular geometry such as a face-centered-cubic structure.

27. True.

28. True.

29. False. The number is eight.

30. False. The controlled addition of impurities to semiconductors is the predominant way of making semiconductor devices. But one should get rid of any other unwanted impurities.

31. True.

32. False. The current is very small when a diode is reverse biased.

33. True.

34. True.

# V.  Questions and Answers

**Questions**

1.  Why does the classical model fail in explaining the observed results of electrical conduction?

2.  Explain the difference between mean velocity and drift velocity.

3.  What is a Fermi electron gas?

4.  What is the origin of the Fermi energy at $T = 0$, and on which properties of a conductor does the Fermi energy depend?

5.  Explain the difference between a conductor and an insulator.

6.  Explain the difference between an insulator and a semiconductor.

7.  Explain how holes participate in electrical conduction.

8.  What is the critical temperature of a superconductor?

9.  What is a Cooper pair?

10. Describe the differences between the dc and ac Josephson effects.

11. What is the density of states?

12. How does the density of states vary with temperature?

13. What is the Fermi factor?

14. How does the Fermi factor vary with temperature?

15. What is the Fermi–Dirac distribution function?

16. How does the Fermi–Dirac distribution function vary with temperature?

17. What are some of the differences between an amorphous and a crystalline solid?

18. Explain how the Madelung constant originates.

19. What is the difference between intrinsic and impurity semiconductors?

20. What is the difference between the forward and the reverse biasing of a *pn*-junction diode?

21. Describe the roles of the emitter, the base, and the collector when a transistor is used as an amplifier.

22. What is an integrated circuit?

**Answers**

1. The classical model assumes that electrons behave like Newtonian point particles obeying Maxwell–Boltzmann statistics. However, electrons have wave properties and obey the Pauli exclusion principle and quantum-mechanical Fermi–Dirac statistics.

2. Mean velocity is the velocity of electrons occurring because of thermal motion. Drift velocity is the velocity in addition to the mean velocity that electrons acquire when an electric field is established in a conductor. The drift velocity is of the order of $10^{-9}$ times smaller than the mean velocity.

3. A Fermi electron gas is a collection of noninteracting spin-$\frac{1}{2}$ electrons, which are fermions that obey the Pauli exclusion principle.

4. Because of the Pauli exclusion principle, there can be at most two electrons with opposite spins in each available energy level. At $T = 0$ the electrons in a conductor fill the available energy levels sequentially. The energy of the highest occupied energy level is the Fermi energy $E_F$, which depends on the number of electrons per unit volume ($N/V$) in the conductor.

5. In a conductor, there is no energy gap between the valence and conduction bands, so the electrons in the conduction band can easily acquire enough kinetic energy to participate in electrical conduction. In an insulator, there is a large energy gap between the filled valence band and the conduction band, so almost no electrons are in the conduction band to participate in electrical conduction.

6. In an insulator, the energy gap between the filled valence band and the conduction band is so large that at normal temperatures few electrons are excited into the conduction band to participate in electrical conduction. In a semiconductor, the energy gap between the filled valence band and the conduction band is very small, so at normal temperatures even thermal energy can excite a large number of electrons into the conduction band to participate in electrical conduction, leaving holes in the valence band that also can participate in electrical conduction.

7.  Holes are vacant energy states in the valence band. If an electric field is applied, electrons in the valence band can acquire energy from the field by being excited to a hole in a vacant energy level. Although conduction really occurs when the negatively charged electrons move in a direction opposite to that of the applied field, the net effect is equivalent to positively charged holes moving in the same direction as the applied field.

8.  It is the temperature at which a superconducting material changes from its normal-conducting state to a superconducting state, in which it has zero resistance.

9.  A Cooper pair consists of two electrons of opposite spins that overcome their Coulomb repulsion by interacting with the lattice to bind together.

10. When there is no voltage across a Josephson junction (formed by two superconductors), a dc current is observed; this is the dc Josephson effect. When a dc voltage is applied across a Josephson junction, an ac current is observed; this is the ac Josephson effect.

11. The density of states $g(E)$ is a function of energy such that the product $g(E)dE$ gives the number of available states in the energy interval between $E$ and $E + dE$.

12. It does not vary with temperature.

13. The Fermi factor $f(E)$ is a function of energy that gives the probability that a state of energy $E$ is occupied.

14. The Fermi factor varies with temperature according to $f(E) = \left( e^{(E - E_{\rm F})/(kT)} + 1 \right)^{-1}$.

15. The Fermi–Dirac distribution function $n(E)$ is a function of energy such that the product $n(E)dE$ gives the number electrons having an energy between $E$ and $E + dE$.

16. The Fermi–Dirac distribution function varies with temperature in the same manner as the Fermi factor.

17. The molecules of an amorphous solid are randomly oriented; the molecules of a crystalline solid are arranged in regular arrays. An amorphous solid does not have a well-defined melting point and merely softens as its temperature increases; the melting point of a crystalline solid occurs at a well-defined temperature.

18. The potential energy of a given ion in a crystal due to another ion in the crystal is $U = \pm ke^2 / R$, where $R$ is the distance from the given ion to the other ion, and the sign depends on the charges of the two ions. Because the ions are arranged regularly, groups of ions of the same charge are located at the same distance from the given ion; so all ions in a particular group contribute the same value toward the overall potential energy. When the contributions from all the groups throughout the crystal are added to get the net potential energy of the given ion, the result is $U_{\rm net} = -\alpha ke^2 / r$, where $r$ is the separation distance between neighboring atoms, and $\alpha$ is the Madelung constant, which depends only on the geometry of the crystal structure.

19. In intrinsic semiconductors, the material is pure and has no added impurities. Electrical conduction occurs because at normal temperatures there are thermally excited electrons in the

conduction band and holes left behind in the nearly filled valence band, both of which can participate in electrical conduction. In an impurity semiconductor, controlled amounts of particular impurity atoms are deliberately mixed with the original semiconductor material to produce additional donor levels near the conduction band (to add extra electrons) or to produce additional acceptor levels near the valence band (to add extra holes).

20. To obtain a forward-biased *pn*-junction diode, the positive terminal of a voltage source is connected to the *p* side of the diode, resulting in a large current. To obtain a reverse-biased diode, the positive terminal of a voltage source is connected to the *n* side of the diode, resulting in essentially no current.

21. A small input signal is established in a circuit between the emitter and the base, with the emitter-base junction being forward biased; this results in a relatively large current into the emitter and a much smaller current from the base through the input source. Most of the current across the emitter-base junction goes across the base-collector junction, emerging from the collector as an output current. The net result is that a small input current produces a much larger output current.

22. An integrated circuit is a collection of many thousands of resistors, capacitors, diodes, and transistors that are interconnected through circuits on a semiconductor crystal, which can have an area less than $1\,cm^2$.

# VI. Problems

**Example #1.** Find the drift speed of electrons when an electric field of $9.00\times10^{-3}$ V/m is established in a copper wire at $20^\circ C = 293\,K$. Take the mean free path equal to 0.400 nm, the electron density equal to $8.47\times10^{22}$ electrons/$cm^3$, and the resistivity equal to $1.70\times10^{-8}$ $\Omega\cdot m$.

**Picture the Problem.** Use the expression that relates the resistivity to the electric field and the drift speed.

| 1. Solve for the drift speed. | $\rho = E/(n_e e v_d)$ $v_d = \dfrac{E}{n_e e \rho} = \dfrac{\left(9\times10^{-3}\,\text{V/m}\right)}{\left(8.47\times10^{28}\,\text{m}^{-3}\right)\left(1.602\times10^{-19}\,\text{C}\right)\left(1.7\times10^{-8}\,\Omega\cdot\text{m}\right)}$ $= 3.91\times10^{-5}\,\text{m/s}$ |
| --- | --- |

**Example #2—Interactive.** Find the mean speed of the electrons in the copper wire of the previous example.

**Picture the Problem.** Use the expression relating the resistivity to the mean speed and the mean free path. **Try it yourself.** Work the problem on your own, in the spaces provided, to get the final answer.

| 1. Solve for the mean speed. | |
|---|---|
| | $v_{av} = 1.6 \times 10^4$ m/s, about $10^9$ time larger than the drift velocity. |

**Example #3.** The density of potassium is $0.870 \, \text{g/cm}^3$ and its atomic mass is 38.96. Assuming that each potassium atom contributes one valence electron, find the number density of free electrons.

**Picture the Problem.** Avogadro's number can be used to find the number density of potassium atoms, and hence electrons.

| 1. Find the number of potassium atoms per unit volume. | $\dfrac{N_K}{V} = \dfrac{\left(6.022 \times 10^{23} \, \text{atoms/mol}\right)\left(0.870 \, \text{g/cm}^3\right)}{38.96 \, \text{g/mol}}$ <br> $= 1.35 \times 10^{22} \, \text{atoms/cm}^3$ |
|---|---|
| 2. If each potassium atom provides one free electron, the number density of free electrons is equal to the number density of atoms. | $N_{electrons} / V = 1.35 \times 10^{22} \, \text{electrons/cm}^3$ |

**Example #4—Interactive.** Using values provided in the previous example, calculate the Fermi energy for potassium.

**Picture the Problem.** Use the expression relating the Fermi energy to the number density of electrons. **Try it yourself.** Work the problem on your own, in the spaces provided, to get the final answer.

| 1. Solve for the Fermi energy. | |
|---|---|
| | $E_F = 2.07 \, \text{eV}$ |

**Example #5.** The critical superconducting temperature for lead is 7.19 K. Find the energy in eV required to break up a lead Cooper pair at $T = 0$.

**Picture the Problem.** The energy required to break up the pair is equal to the superconducting energy gap.

| 1. Find the superconducting energy gap. | $E_g = 3.5\,kT_c = 3.5\left(1.38\times10^{-23}\text{ J/K}\right)\left(7.19\text{ K}\right)$ |
| --- | --- |
| | $= 3.47\times10^{-22}\text{ J} = 2.17\times10^{-3}\text{ eV}$ |

**Example #6—Interactive.** The actual energy required to break up a Cooper pair in lead at $T=0$ is measured to be $2.73\times10^{-3}$ eV. Use this result to calculate the critical temperature of lead, and compare your answer with the experimental value of 7.19 K.

**Picture the Problem.** See Example #5. **Try it yourself.** Work the problem on your own, in the spaces provided, to get the final answer.

| 1. Find the critical temperature from the superconducting energy gap. | |
| --- | --- |
| | $T_c = 9.06\,\text{K}$ |

**Example #7.** In a conductor of volume $V$ containing $N$ free electrons, the density of states is $g(E) = 3NE^{1/2}/\left(2E_F^{3/2}\right)$ where $E$ is measured from the bottom of the conduction band. Find the number of states from the bottom of the conduction band to the Fermi energy.

**Picture the Problem.** Find the number of states in a differential energy interval $dE$. Integrate this expression from zero to the Fermi energy.

| 1. Determine the number of states in a small energy interval. | $dN = g(E)\,dE = \dfrac{3N}{2E_F^{3/2}}E^{1/2}\,dE$ |
| --- | --- |
| 2. Integrate the expression from step one to find the total number of states. | $N_s = \displaystyle\int dN = \int_0^{E_F}\dfrac{3N}{2E_F^{3/2}}E^{1/2}\,dE = \dfrac{3N}{2E_F^{3/2}}\int_0^{E_F}E^{1/2}\,dE$ |
| | $= N$ |

**Example #8—Interactive.** How many free electrons are below the Fermi energy in $5.00\,\text{cm}^3$ of copper at $T=0$? The number density of electrons in copper is $8.47\times10^{22}$ electrons/$\text{cm}^3$, and the Fermi energy is 7.04 eV.

**Picture the Problem.** At $T=0$ there are no electrons above the Fermi energy, so all the electrons must be below the Fermi energy. **Try it yourself.** Work the problem on your own, in the spaces provided, to get the final answer.

| 1. Determine the number of electrons in the copper from the number density. This is the number of electrons below the Fermi level. | |
|---|---|
| | $N = 4.23 \times 10^{23}$ |

**Example #9.** Find the ratio of the number of free electrons in silver that are in a small energy interval $dE$ located at 0.1 eV above and 0.1 eV below the Fermi level of 5.50 eV at a temperature of 300 K.

**Picture the Problem.** This will be the ratio of the Fermi-Dirac distribution function at each of the two energies.

| 1. Write an expression for the ratio of the two Fermi-Dirac distribution functions. | $\dfrac{n(E_U)}{n(E_L)} = \left[\dfrac{E_U^{1/2}}{e^{(E_U-E_F)/(kT)}+1}\right] \Big/ \left[\dfrac{E_L^{1/2}}{e^{(E_L-E_F)/(kT)}+1}\right]$ |
|---|---|
| 2. Evaluate each of the exponents. | $(E_U - E_F)/(kT) = \dfrac{(5.60\,\text{eV})-(5.50\,\text{eV})}{(8.62\times10^{-5}\,\text{eV/K})(300\,\text{K})} = 3.87$ <br><br> $(E_L - E_F)/(kT) = \dfrac{(5.40\,\text{eV})-(5.50\,\text{eV})}{(8.62\times10^{-5}\,\text{eV/K})(300\,\text{K})} = -3.87$ |
| 3. Now the ratio can be calculated. | $\dfrac{\left[\dfrac{(5.6\,\text{eV})^{1/2}}{e^{3.87}+1}\right]}{\left[\dfrac{(5.4\,\text{eV})^{1/2}}{e^{-3.87}+1}\right]} = 0.0212$ |

**Example #10—Interactive.** The ratio of the number of free electrons in a material that are in a small energy interval $dE$ located 0.100 eV above and 0.100 eV below the Fermi level of the material is 0.02113 at $T = 300\,\text{K}$. Find the Fermi level of the material, and from this value identify the material.

**Picture the Problem.** To find the Fermi level, proceed in a similar fashion to the previous example. Relate the Fermi level to the number density of electrons. Assuming each atom provides one free electron, this number density is the same as the number density of the material. **Try it yourself.** Work the problem on your own, in the spaces provided, to get the final answer.

| 1. From the ratio of the numbers of electrons, find the Fermi energy of the material. | |
|---|---|
| | $E_F = 7.2\,\text{eV}$ |

| | |
|---|---|
| 2. From the Fermi energy, determine the number density of the electrons, and hence the atoms. | |
| 3. From the number density and Avogadro's number, determine the atomic mass of the material, which should identify the metal. | Magnesium |

**Example #11.**  The density of a LiCl crystal is $2.07\,\text{g/cm}^3$ and the distance between ions is $r_0 = 0.257\,\text{nm}$. Determine the molecular mass of the crystal.

**Picture the Problem.**  Determine the volume of 1 mol of LiCl. Knowing the volume and density, you can calculate the molecular mass.

| | |
|---|---|
| 1. Calculate the volume of 1 mol of LiCl. | $V = 2N_A r_0^3 = 2\left(6.02 \times 10^{23}\,\text{atoms/mol}\right)\left(0.257 \times 10^{-7}\,\text{cm}\right)^3$ $= 20.4\,\text{cm}^3$ |
| 2. Find the molecular mass from the volume and the provided density. | $m = DV = \left(2.07\,\text{g/cm}^3\right)\left(20.4\,\text{cm}^3\right) = 42.3\,\text{g}$ |

# Chapter 39

# Relativity

## I. Key Ideas

*Section 39-1. Newtonian Relativity.* To measure distances and times an observer needs a set of coordinate axes to measure distance and a set of synchronized clocks to measure time. The coordinate axes and clocks make up a **reference frame.**

Reference frames are used to specify the location of events in space and time. An **event** might be the striking of a lightning bolt, the collision of two particles, or the explosion of a distant supernova. To specify an event you assign it four space–time coordinates $(x, y, z, t)$. The three position coordinates $x$, $y$, and $z$ determine the distance from the origin of your reference frame to the event. The time coordinate $t$, measured on a clock stationary in your reference frame *that is at the location of the event,* tells the instant when the event takes place.

An **inertial reference frame** is a reference frame in which objects move with a constant velocity when no net force acts on them, in accordance with Newton's first law. A spaceship coasting in outer space far away from any planets or stars is an inertial reference frame. Release an object in the spaceship and it will remain motionless in front of you. If you want to accelerate the object, you have to exert forces on it.

If the spaceship's rockets are fired, causing it to accelerate, the spaceship is no longer an inertial reference frame; it is an **accelerated reference frame.** Release an object in an accelerating spaceship and the object will appear to accelerate away from you without a net force acting on it. A rotating merry-go-round is also an accelerated reference frame. Place a marble on the floor of the merry-go-round and it will accelerate toward the perimeter, even though the net force acting on the marble appears to be zero.

According to Newtonian (and Einsteinian) relativity,

A reference frame moving at a constant velocity relative to a known inertial reference frame is also an inertial reference frame.

A consequence of this principle is that

Absolute motion cannot be detected.                Principle of Newtonian relativity

Suppose you play a game of billiards in a rocket ship moving at constant velocity, and someone else in another rocket ship moving relative to you at a different very high constant speed is also playing billiards. When you and the other person each strike a billiard ball at the same angle and with the same speed, afterwards each billiard ball will move in exactly the same manner.

Newton believed that the principle of relativity applied only to mechanical motions such as colliding billiard balls and oscillating pendula. Einstein extended the principle of relativity to *all* areas of physics, including light and other electromagnetic phenomena. Indeed, the paper in which he started the whole notion of "relativity" was entitled "On the Electrodynamics of Moving Bodies."

***Ether and the Speed of Light.*** Before Einstein put forth his theory of special relativity in 1905 it was generally believed that light and other electromagnetic phenomena did not obey a principle of relativity. This can be seen by looking at the speed of light, $c = 1/\sqrt{\varepsilon_0 \mu_0} = 3 \times 10^8$ m/s, which follows from Maxwell's equations that describe electromagnetic waves. According to Newtonian concepts this velocity can be measured only by a privileged observer who is at rest in "absolute space" in which there is a medium called the **ether** that supports the propagation of light and other electromagnetic waves. Other observers moving relative to this privileged observer will necessarily measure the speed of light to be different from the value $c$, and consequently will also find Maxwell's equations to have a form different from what the "privileged" observer obtains.

Michelson in 1881 and later Michelson and Morley in 1887 developed an interferometer sensitive enough to measure the expected change of the speed of light as the earth moved in different directions through the hypothesized ether. The result of these and many other subsequent experiments was that the speed of light did not change, and no motion of the earth through the ether was ever detected. This is one of the most famous null experimental results ever obtained.

***Section 39-2. Einstein's Postulates.*** Einstein's theory of special relativity follows from two postulates that unit all realms of physics—mechanics, electromagnetism, nuclear physics, and every other field of physics—under the common umbrella of relativity:

Postulate 1: Absolute, uniform motion cannot be detected.

Postulate 2: The speed of light is independent of the motion of the source.

The second postulate expresses a property of waves that you are probably familiar with from your previous studies. When a wave such as a sound wave leaves its source, its movement through the propagating medium is independent of how the source is moving. In particular, the speed with which the wave moves through the medium is not related to how fast or slow the source is moving.

The two postulates also imply that the speed of light is independent of the motion of the observer. Thus, the second postulate can be expressed in an alternate form:

Postulate 2 (Alternate): Every observer measures the same value $c$ for the speed of light.

This concept requires a bit of contemplation, as it runs counter to commonsense (Newtonian) concepts of how velocities add. One thing that follows from this alternate form of the second postulate is an explanation of the null result of the Michelson–Morley experiment. If all observers measure the same value for the speed of light, it follows that no experiment will ever show two observers measuring different values for the speed of light.

***Section 39-3. The Lorentz Transformation.*** An important question in both Newtonian and Einsteinian relativity is how the coordinates assigned to a particular event by two observers in different inertial frames moving relative to each other are interrelated. Figure 39-1 shows a reference frame $S'$ moving with a velocity $\vec{v}$ relative to a reference frame $S$ along the collinear $x$-$x'$ axes. Along with the rectangular coordinate axes $xyz$ in reference frame $S$ is a set of stationary synchronized clocks that measure the time $t$. A different set of synchronized clocks measuring the time $t'$ is located at rest relative to the rectangular coordinate axes $x'y'z'$ of reference frame $S'$.

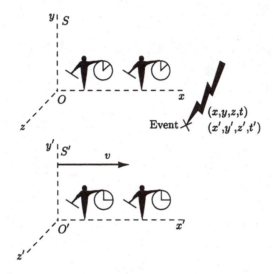

Figure 39-1

Suppose an observer in each reference frame observes a particular event, say the striking of a lightning bolt at a given instant. Each observer describes where and when the event takes place by assigning it four space–time coordinates: $(x, y, z, t)$ are assigned in $S$, and $(x', y', z', t')$ are assigned in $S'$. The interrelationship between these sets of coordinates is called a coordinate transformation.

Using his postulates, Einstein found that the space–time coordinates measured in $S$ and $S'$ are related by the Lorentz transformation

$$x = \gamma(x' + vt')$$
$$y = y'$$
$$z = z'$$
$$t = \gamma(t' + vx'/c^2)$$

Lorentz transformation

where

$$\gamma = \frac{1}{\sqrt{1 - v^2/c^2}} \qquad\qquad \text{$\gamma$-factor}$$

In these expressions $v$ is the velocity of $S'$ along the common $x$-$x'$ axis relative to $S$. $v$ is positive if $S'$ is moving in the positive $x$-$x'$ direction and negative if $S'$ is moving in the negative $x$-$x'$ direction.

The inverse of the Lorentz transformation is

$$
\begin{aligned}
x' &= \gamma(x - vt) \\
y' &= y \\
z' &= z \\
t' &= \gamma(t - vx/c^2)
\end{aligned}
\qquad\qquad \text{Inverse Lorentz transformation}
$$

If $v$ is much smaller than the speed of light ($v \ll c$), then $\gamma \rightarrow 1$ and the Lorentz transformation reduces to the classical Galilean transformation

$$
\begin{aligned}
x &= x' + vt' \\
y &= y' \\
z &= z' \\
t &= t'
\end{aligned}
\qquad\qquad \text{Galilean transformation}
$$

The inverse Galilean transformation is

$$
\begin{aligned}
x' &= x - vt \\
y' &= y \\
z' &= z \\
t' &= t
\end{aligned}
\qquad\qquad \text{Inverse Galilean transformation}
$$

*Time Dilation.* In general, two events will take place at different locations. Two separated clocks are then needed in a reference frame to measure the time interval between the two events, one clock located at one event and a second clock located at the other event.

When two events occur at the same place in a particular reference frame, the time interval between the two events can be measured with a *single* clock located at the position where the two events occur. This time interval $\Delta t_p$ measured by a single clock is called the **proper time** between the two events.

All other observers moving relative to the reference frame in which proper time is measured will find the time interval $\Delta t$ between the two events, as measured by their clocks that are necessarily separated, to be larger than $\Delta t_p$ according to the time dilation expression

$$\Delta t = \gamma\, \Delta t_p = \frac{\Delta t_p}{\sqrt{1 - v^2/c^2}} \qquad\qquad \text{Time dilation}$$

*Length Contraction.* An unaccelerated object can be at rest in only one inertial reference frame. An observer at rest in this particular inertial frame can measure the spatial coordinates of the end points of the object at her leisure at any time. The length of the object measured in the reference frame in which the object is at rest is called its **proper length** $L_p$.

Consider a rod moving along the $x$-$x'$ axis with a velocity $v$ relative to frame $S$. In order for an observer in $S$ to determine the length $L$ of the moving rod, she must measure the $x$ coordinates of the ends of the moving rod at the same time, that is, simultaneously. The observer in $S$ will find that $L$ is shorter than the proper length $L_p$ of the rod by an amount given by the **length contraction** expression.

$$L = \frac{1}{\gamma} L_p = L_p \sqrt{1 - v^2 / c^2}$$    *Length contraction*

*The Relativistic Doppler Effect.* The Doppler effect for sound results from the motion of the source and/or observer through air, the medium in which the sound propagates. With electromagnetic waves there is no medium of propagation—there is no ether. As a consequence, the Doppler equations relating the frequency of light and other electromagnetic waves are different from the Doppler-effect relations for sound.

Let $f_0$ be the frequency of a light source that is measured by an observer at rest with respect to the source. If this source moves toward another observer with a velocity $v$, the frequency $f'$ that the latter observer measures is

$$f' = \sqrt{\frac{1 + v/c}{1 - v/c}} f_0$$    Doppler effect for an approaching source

If the source is moving away from an observer with a velocity $v$, the frequency $f'$ that the observer measures is

$$f' = \sqrt{\frac{1 - v/c}{1 + v/c}} f_0$$    Doppler effect for a receding source

In the latter case, $f' < f$, which is known as a redshift. This is most commonly observed in the light we see from distant receding galaxies.

*Section 39-4. Clock Synchronization and Simultaneity.* In a reference frame you would not want some of your clocks to be running ahead of or behind others. In a "good" reference frame, all clocks should be running together, that is, all clocks should be synchronized with each other.

The Galilean transformation $t = t'$ tacitly assumes that it is obvious how to accomplish clock synchronization in all reference frames. If an observer in one reference frame adjusts his clocks to be synchronized, so that all clocks record the same time $t$ for a given event, an observer in another reference frame also will "obviously" be able to adjust his clocks to be synchronized and measure the same time $t = t'$ for the event.

Einstein realized that an important quantity such as time should not be left imprecisely defined, as it was in Newtonian physics. Einstein pointed out that operational procedures must be used to define when two clocks are synchronous with each other.

Einstein carefully defined clock synchronization using light signals. Clock synchronization is closely tied to the notion of simultaneity. If two events are simultaneous, it means that the two events occur at the same time in a given reference frame. The Galilean transformation $t = t'$ assumes that simultaneity is absolute. That is, it assumes that if one observer finds two events to be simultaneous, then all other observers moving relative to this observer will also find the two events to occur simultaneously.

Einstein defined simultaneity as follows:

> Two events in a reference frame are simultaneous if light signals from the events reach an observer halfway between the events at the same time.

Note that this definition involves an interweaving of the concepts of time, distance, and the speed of light.

A consequence of Einstein's operational definition of simultaneity is that two events that are simultaneous in one inertial reference frame will in general not be simultaneous in another inertial frame moving relative to the first. Since synchronization of clocks involves the notion of simultaneity, it also follows that two clocks that are synchronized in one reference frame will not be synchronized in any other frame moving relative to the first frame.

> If two clocks are synchronized in the frame in which they are both at rest, in a frame in which they are moving along the line through both clocks, the chasing clock leads (shows a later time) the front clock by an amount $\Delta t_s = L_p (v/c)$ where $L_p$ is the proper distance between the clocks.

*The Twin Paradox.* One of the most famous results of time dilation is the twin paradox. One twin leaves her twin brother and takes a journey at a high constant speed $v$ in a spaceship to a distant point, quickly turns around in a very small time interval, and returns to her twin brother at the same speed. When the twins compare their ages upon being reunited, they find that the brother has aged more than the sister according to the time dilation expression

$$\Delta t_{\text{brother}} = \gamma \, \Delta t_{\text{sister}} = \frac{\Delta t_{\text{sister}}}{\sqrt{1 - v^2 / c^2}} > \Delta t_{\text{sister}}$$

The "paradox" arises because, from the point of view of the twin sister, her brother moves away at the speed $v$, turns around, and then returns at the same speed. It therefore appears that the twin sister should find her brother to be younger. but you can't have it both ways. When the twins get together, one twin will definitely be younger than the other, or they will both have aged the same amount.

The paradox is resolved by noting that the motion of the two twins is not symmetrical. In order to return to her twin brother, the sister must turn around. The turning around is very real, resulting in the sister experiencing very real forces and accelerations during her turnaround period. She is equivalent to two observers in different inertial reference frames, one as she moves away and another as she moves toward her twin brother at the constant speed $v$. In contrast, the twin brother experiences no real net force nor acceleration throughout his twin's trip, and is equivalent to one single inertial observer.

***Section 39-5. The Velocity Transformation.*** Suppose observers in reference frames $S$ and $S'$ measure the velocity of a single object, such as a spaceship. The observer in $S$ measures the three components of the velocity to be $\left(u_x, u_y, u_z\right)$, whereas the observer in $S'$ measures in general three different components of velocity $\left(u'_x, u'_y, u'_z\right)$.

By differentiating the Lorentz transformation equations, the following relativistic velocity transformation equations relating the velocity components are obtained:

$$u_x = \frac{u'_x + v}{1 + vu'_x / c^2}$$

$$u_y = \frac{u'_y}{\gamma(1 + vu'_x / c^2)} \qquad \text{Relativistic velocity transformation}$$

$$u_z = \frac{u'_z}{\gamma(1 + vu'_x / c^2)}$$

In these equations $v$ is the velocity of frame $S'$ relative to $S$ in the positive $x$ direction.

The inverse velocity transformation equations are

$$u'_x = \frac{u_x - v}{1 - vu_x / c^2}$$

$$u'_y = \frac{u_y}{\gamma(1 - vu_x / c^2)} \qquad \text{Relativistic inverse velocity transformation}$$

$$u'_z = \frac{u_z}{\gamma(1 - vu_x / c^2)}$$

For low velocities ($v \ll c$) the relativistic velocity transformation equations reduce to the following classical velocity transformation equations:

$$u_x = u'_x + v$$

$$u_y = u'_y \qquad \text{Classical velocity transformation}$$

$$u_z = u'_z$$

The inverse equations are

$$u'_x = u_x - v$$

$$u'_y = u_y \qquad \text{Classical inverse velocity transformation}$$

$$u'_z = u_z$$

*Section 39-6. Relativistic Momentum.* If an object is at rest with respect to an inertial observer, the observer can measure its mass in a straightforward manner, for instance by putting it on a scale. This mass is the **rest mass** $m_0$ of the object.

If an object is moving with a velocity $\vec{u}$ relative to an observer, the observer must infer its mass indirectly from, say, momentum experiments. The result is that the **relativistic mass** $m_r$ is larger than the rest mass $m_0$, and is given by

$$m_r = \frac{m_0}{\sqrt{1 - u^2 / c^2}}$$                    Relativistic mass

The **relativistic momentum** $\vec{p}$ of an object moving with a velocity $\vec{u}$ is the product of the relativistic mass and the velocity:

$$\vec{p} = \frac{m_0 \vec{u}}{\sqrt{1 - u^2 / c^2}}$$                    Relativistic momentum

For small velocities $u \ll c$, this reduces to the usual classical momentum expression $\vec{p} = m_0 \vec{u}$. Relativistic momentum is conserved if there are no external forces acting on a system.

*Section 39-7. Relativistic Energy.* It follows from a relativistic analysis of work–energy principles that a particle with rest mass $m_0$ has a rest energy $E_0$ given by Einstein's famous mass–energy relationship:

$$E_0 = m_0 c^2$$                    Rest energy

If the particle is moving with a speed *u*, so that it has a kinetic energy *K*, its **total relativistic energy** $E = K + E_0$ is given by

$$E = K + m_0 c^2 = \frac{m_0 c^2}{\sqrt{1 - u^2 / c^2}}$$                    Total relativistic energy

If the object is not moving, $u = 0$ and $E = E_0 = m_0 c^2$.

Turning this around, you find that the **relativistic kinetic energy**—the energy due to the particle's motion—is the difference between the total energy *E* and the rest energy $E_0$ :

$$K = E - E_0$$
$$K = E - m_0 c^2 = \frac{m_0 c^2}{\sqrt{1 - u^2 / c^2}} - m_0 c^2$$                    Relativistic kinetic energy

For small velocities $u \ll c$, application of the binomial expansion shows that the relativistic kinetic energy reduces to the usual standard classical expression $K_{classical} = \frac{1}{2} m_0 u^2$.

The relativistic momentum and relativistic energy expressions are related. A useful expression for the speed *u* of a particle is

$$u/c = pc/E \qquad \text{Relativistic velocity–momentum–energy relationship}$$

When the speed $u$ is eliminated between the momentum and energy relations, the following expression relating the energy and momentum is obtained:

$$E^2 = p^2c^2 + (m_0c^2)^2 \qquad \text{Relativistic energy–momentum relationships}$$

or, since $E = K + m_0c^2$,

$$(K + m_0c^2)^2 = p^2c^2 + (m_0c^2)^2$$

If you know a particle's momentum, these expressions allow you to find its total energy $E$ or kinetic energy $K$, and vice versa.

The constituent parts of an atom or a nucleus, or any other particle, have less potential energy when together than when separated. The difference is called the **binding energy** ($E_b$) of the system. The expression $E_0 = m_0c^2$ shows that binding energy is equivalent to mass. As a result, the rest mass of a composite particle will be less than the rest masses of its constituent parts by an amount equal to the binding energy BE:

$$E_b = \left[\Sigma(\text{rest masses of constituent parts})\right]c^2$$
$$- (\text{rest mass of composite particle})c^2 \qquad \text{Binding energy}$$

***Section 39-8.  General Relativity.***  Einstein's **general theory of relativity** is a theory of gravitation that supercedes Newton's theory of gravitation. Newton's theory of gravitation states that the magnitude of the gravitational attractive force $F_g$ between two massive bodies $m_1$ and $m_2$ separated by a distance $R$ is

$$F_g = G\frac{m_1 m_2}{R^2} \qquad \text{Newton's law of gravitational attraction}$$

General relativity describes what happens in the realm of very strong gravitational fields where Newton's theory of gravitation no longer holds.

In a certain sense, special relativity and general relativity overthrow Newton's two major theories. Einstein's special theory of relativity shows that Newton's dynamical laws do not hold in the realm where velocities approach the speed of light. Einstein's general theory of relativity shows that Newton's law of universal gravitational attraction does not hold for strong gravitational fields.

Einstein was guided to his general theory of relativity by the principle of equivalence. In brief, the principle of equivalence states that in a small region of space and a short interval of time you cannot distinguish between real gravitational fields and accelerated motion:

A homogenous gravitational field is completely equivalent to a uniformly accelerated reference frame.

Principle of equivalence

If you are in a spaceship undergoing uniform acceleration, any experiment you perform will be exactly the same as if you were at rest in a uniform gravitational field. For example, if you release an object in the spaceship, it moves away from you with constant acceleration in the same manner that an object moves away from you when you release it in a gravitational field, as shown in Figure 39-2.

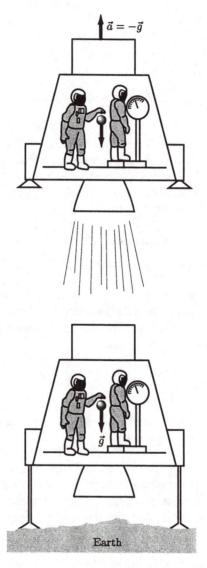

Figure 39-2

One consequence of the principle of equivalence is that light will follow a curved trajectory in a gravitational field, since light follows such a trajectory in an accelerated reference frame. The bending of a light beam as observed in an accelerated elevator is illustrated in Figure 39-3. Figure

39-3*a* shows the light beam moving in a straight line, while the elevator accelerates upward. Figure 39-3*b* shows how the light beam follows a curved trajectory when viewed in the accelerating elevator. Since the light beam is curved in the accelerated elevator, the principle of equivalence states that light should also bend in a gravitational field.

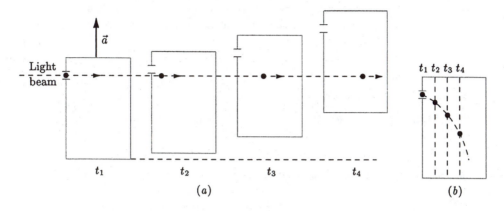

Figure 39-3

The bending of light in a gravitational field was first observed in 1919, about three years after Einstein published his theory of general relativity, by a team of scientists led by Sir Arthur Eddington who made observations of the bending of light during an eclipse of the sun that agreed exactly with Einstein's prediction. This observation immediately catapulted Einstein into worldwide fame.

Another prediction of Einstein's theory of general relativity is that clocks in a region of low gravitational potential will run slower than clocks in a region of higher gravitational potential. This slowing down of time is also often referred to as the gravitational redshift. An atomic clock measures time based on the frequency of vibrations of an atom. If an atomic clock moves to a region of low gravitational potential, the frequency of this vibration slows down, resulting in a longer wavelength of light being emitted by the atom. This slowing down of time was experimentally observed with an atomic clock in 1976. Geographical positioning system (GPS) receivers and satellites must account for this gravitational redshift in order to provide accurate position data. If the gravitational affect was not accounted for, the resulting errors in position would make the system useless in less than one day.

Yet another prediction of Einstein's theory of general relativity is the possibility of the existence of **black holes.** The concept of a black hole is built around the **Schwarzschild radius** $R_S$ associated with the mass $M$ of an astronomical object:

$$R_S = 2GM / c^2$$                            Schwarzschild radius

where $G$ is Newton's gravitational constant and $c$ is the speed of light. Given the mass $M$ of the astronomical object, its Schwarzschild radius can be calculated from this expression. For most astronomical objects, the Schwarzschild radius is much smaller than the actual radius of the body. For example, the Schwarzschild radius for our sun is about 3 km, which is much smaller than the sun's radius of about 700,000 km.

If an astronomical object with a given mass is sufficiently dense, it is possible for the Schwarzschild radius calculated from its mass to be larger than the radius of the object. For such a situation, the present standard view is that the region extending from immediately outside the surface of the object to the Schwarzschild radius lies in a black hole, and that light emitted from the surface of the object cannot escape from the black hole to the outer region. In principle, a black hole can arise from an object collapsing under the influence of gravitational forces to smaller and smaller radii, while its mass remains constant, until a black hole is formed around the collapsing object. At the center of the Milky Way is a supermassive black hole with a mass of about two million solar masses.

## II.  Physical Quantities and Key Equations

### Physical Quantities

*Rest energies*          electron $m_0c^2 = 0.511$ MeV

proton   $m_0c^2 = 938.3$ MeV

neutron $m_0c^2 = 939.6$ MeV

### Key Equations

*$\gamma$-factor*                     $\gamma = \dfrac{1}{\sqrt{1 - v^2/c^2}}$

*Lorentz transformation*

$$x = \gamma(x' + vt') \qquad\qquad x' = \gamma(x - vt)$$
$$y = y' \qquad\qquad\qquad\quad y' = y$$
$$z = z' \qquad\qquad\qquad\quad z' = z$$
$$t = \gamma(t' + vx'/c^2) \qquad t' = \gamma(t - vx/c^2)$$

Galilean transformation

$$x = x' + vt' \qquad\qquad x' = x - vt$$
$$y = y' \qquad\qquad\qquad y' = y$$
$$z = z' \qquad\qquad\qquad z' = z$$
$$t = t' \qquad\qquad\qquad t' = t$$

*Time dilation*              $\Delta t = \gamma\,\Delta t_{\mathrm{p}} = \dfrac{\Delta t_{\mathrm{p}}}{\sqrt{1 - v^2/c^2}}$

*Length contraction*              $L = \dfrac{1}{\gamma}L_{\mathrm{p}} = L_{\mathrm{p}}\sqrt{1 - v^2/c^2}$

*Relativistic Doppler effect, approaching*      $f' = \sqrt{\dfrac{1 + v/c}{1 - v/c}}\,f_0$

Relativistic Doppler effect, receding

$$f' = \sqrt{\frac{1 - v/c}{1 + v/c}} f_0$$

Relativistic velocity transformation

$$u_x = \frac{u_x' + v}{1 + vu_x'/c^2} \qquad u_x' = \frac{u_x - v}{1 - vu_x/c^2}$$

$$u_y = \frac{u_y'}{\gamma(1 + vu_x'/c^2)} \qquad u_y' = \frac{u_y}{\gamma(1 - vu_x/c^2)}$$

$$u_z = \frac{u_z'}{\gamma(1 + vu_x'/c^2)} \qquad u_z' = \frac{u_z}{\gamma(1 - vu_x/c^2)}$$

Classical velocity transformation

$$u_x = u_x' + v \qquad u_x' = u_x - v$$

$$u_y = u_y' \qquad u_y' = u_y$$

$$u_z = u_z' \qquad u_z' = u_z$$

Relativistic mass

$$m_r = \frac{m_0}{\sqrt{1 - u^2/c^2}}$$

Relativistic momentum

$$\vec{p} = \frac{m_0 \vec{u}}{\sqrt{1 - u^2/c^2}}$$

Rest energy

$$E_0 = m_0 c^2$$

Total relativistic energy

$$E = K + m_0 c^2 = \frac{m_0 c^2}{\sqrt{1 - u^2/c^2}}$$

Relativistic kinetic energy

$$K = E - E_0$$

$$K = E - m_0 c^2 = \frac{m_0 c^2}{\sqrt{1 - u^2/c^2}} - m_0 c^2$$

Relativistic velocity–momentum–energy relationship

$$u/c = pc/E$$

Relativistic energy–momentum relationships

$$E^2 = p^2 c^2 + (m_0 c^2)^2$$

$$(K + m_0 c^2)^2 = p^2 c^2 + (m_0 c^2)^2$$

Binding energy

$$E_b = \left[\Sigma(\text{rest masses of constituent parts})\right]c^2$$
$$- (\text{rest mass of composite particle})c^2$$

Schwarzschild radius

$$R_S = 2GM/c^2$$

# III. Potential Pitfalls

When dealing with problems involving the relationship between space and time measurements made by various observers, it is important to keep in mind the concept of "event." An event could be anything, like the location of a lightning bolt striking a point at a given time. The location of each end of a moving rod at the same time constitutes two different events. Each event has four coordinates $(x, y, z, t)$ assigned to it by an observer in reference frame $S$, and four other coordinates $(x', y', z', t')$ assigned to it by an observer in reference frame $S'$. The relationship between the two sets of coordinates assigned to an event is given by the Lorentz transformation.

Often you can answer questions about the space separation or time separation between two events by subtracting two appropriate Lorentz transformation equations from each other. The trick is to determine which are the "appropriate" Lorentz transformation equations to subtract. For example, if two events take place at the same location in the reference frame $S'$, a subtraction that involves $x_2' - x_1' = 0$ will probably prove useful in answering questions relating to the two events.

Do not confuse the "time separation" between two events with the "proper time interval" between the two events. If an observer in reference frame $S$ and a second observer in reference frame $S'$ measure the time interval between two events that occur at different places for both observers, their time intervals will not be related by a simple multiplication or division by the factor $\sqrt{1 - v^2/c^2}$. To get the relationship between the time interval measured by each observer, you must use the previously discussed subtraction technique.

Similarly, do not confuse the "spatial separation" between two events with "proper length." If an observer in reference frame $S$ and a second observer in reference frame $S'$ measure the spatial separation between two events that do not occur simultaneously, the spatial intervals will not be related by a simple multiplication or division by the factor $\sqrt{1 - v^2/c^2}$. To get the relationship between the spatial intervals measured by each observer, you must use the previously discussed subtraction technique.

Since two events that occur simultaneously for one observer will not in general be simultaneous for other observers, you must make sure you are clear about which observer determines that the two events are simultaneous.

Do not confuse the time at which you see an event take place with the time at which the event actually takes place. For example, suppose you see a distant supernova explode, and record the time on a clock at your location when you make your observation. To determine the actual time that the supernova explosion occurred, you must correct for the travel time that it took the light signal to reach your location.

Often, problems can be solved using the simple expression distance = velocity × time, provided that the distance, velocity, and time all refer to the same reference frame. For example, suppose you are a muon, moving at a speed $0.998c$ relative to the earth, who has to traverse a distance of 9000 m measured by an earth observer. In your frame of reference, the distance to be traversed is foreshortened and is only $\left(\sqrt{1 - 0.998^2}\right)(9000\,\text{m}) = 600\,\text{m}$, so the time $\Delta t$ you would need to traverse this distance is found from

$$\text{distance} = \text{velocity} \times \text{time}$$

$$600\,\text{m} = 0.998\left(3\times10^{8}\,\text{m/s}\right)\Delta t$$

$$\Delta t = 2\times10^{-6}\,\text{s}$$

Velocity problems usually involve two reference frames $S$ and $S'$, and a particle $P$. In the velocity transformation expressions, the quantity $v$ is the velocity of reference frame $S'$ relative to reference frame $S$. The quantities $\left(u_x, u_y, u_z\right)$ and $\left(u_x', u_y', u_z'\right)$ are the components of the particle's velocity relative to reference frames $S$ and $S'$, respectively. Don't mix up $v$ with $u$ or $u'$. Also, make sure you clearly understand which objects are to be associated with $S$, $S'$, and $P$.

In relativistic energy problems do not mistakenly use the classical expression $K_{\text{classical}} = \frac{1}{2}m_0 u^2$. Whenever $u$ is comparable to $c$, roughly $u \geq 0.1c$, you must use relativistic expressions.

Often, particles are described in terms of their kinetic energies rather than their speeds. For example, you may be told that a particle of charge $e$ has been accelerated from rest through a certain voltage, which is numerically equal to the kinetic energy in electron volts that the particle acquires. Relativistic expressions must be used when the kinetic energy is comparable to the rest energy of an object. For example, you must use relativistic expressions for a 2-MeV electron, because an electron's rest mass is about 0.5 MeV. For a 2-MeV proton, however, you can use nonrelativistic expressions to a good approximation because a proton's rest mass is about 938 MeV.

Do not mistakenly use the expression $\vec{p} = m_0 \vec{u}$ for an object's momentum in relativistic situations. The correct relativistic expression is

$$\vec{p} = \frac{m_0 \vec{u}}{\sqrt{1 - v^2/c^2}}$$

## IV. True or False Questions and Responses

**True or False**

_____ 1. All observers in inertial reference frames will measure the speed of light in any direction to be $3\times10^{8}$ m/s.

_____ 2. The medium in which light was (mistakenly) assumed to propagate was called the ether.

_____ 3. The Michelson and Morley experiment was devised primarily to make very precise measurements of the speed of light.

_____ 4. An observer finds that the time interval between two events that take place at the same location is 2 s. Another observer moving relative to this observer will measure the time interval between the two events to be less than 2 s.

_____ 5. An observer in reference frame $S$ finds that an event $A$ takes place at her origin simultaneously with a second event $B$ located at a distance along her $+x$ axis. Another observer in reference frame $S'$ moving in the positive direction along the common $x$-$x'$ axis finds that event $A$ occurs after event $B$.

_____ 6. An observer in reference frame $S$ finds that two lightning bolts strike the same place simultaneously. An observer in $S'$ moving relative to $S$ will determine that the two bolts strike at different times.

_____ 7. The reason length contraction occurs is that light from the front and back of an object take different times to reach an observer.

_____ 8. You have a meterstick located at rest along your $y$ axis. Another person moving along your $x$ axis will measure the length of the meterstick to be less than 1 m.

_____ 9. A clock moving along the $x$-$x'$ axis at constant speed is struck by a lightning bolt and is later struck by a second lightning bolt. The time interval recorded by this very sturdy clock between the two strikes is the proper time interval.

_____ 10. An object moving at a speed of $0.8c$ takes 5 years to traverse a distance of $4c \cdot$yr.

_____ 11. You measure that a rocket ship moving at $0.8c$ takes 5 yr to travel from one star to another. The time measured by the pilot of the rocket ship is 3 yr.

_____ 12. Rocket ship $A$ moves away from you to your right with a speed $0.5c$, while rocket ship $B$ moves away from you toward your left with a speed $0.4c$. The speed of $A$ as determined by $B$ is $0.9c$.

_____ 13. In the twin paradox, the twin who does not undergo real accelerations will be younger when the twins return to each other.

_____ 14. Two observers moving relative to each other measure the same value for their relative speed.

_____ 15. A consequence of the relativistic velocity transformation is that if one observer measures something to move at a speed $c$ in an arbitrary direction, then an observer in any other reference frame will also measure the speed of the object to be $c$.

_____ 16. An electron and a proton are each accelerated through a potential difference of 50,000 V and are then injected into the magnetic field of a cyclotron. You must use relativistic expressions to analyze the motion of the electron, but the motion of the proton can be treated with classical expressions to a good approximation.

_____ 17. An object's kinetic energy is equal to the difference between its total energy and its rest energy.

_____ 18. A deuterium nucleus consists of a proton (mass $938.3$ MeV/$c^2$) and a neutron (mass $939.6$ MeV/$c^2$). The mass of the deuterium nucleus will be less than $938.3$ MeV/$c^2 + 939.6$ MeV/$c^2 = 1877.9$ MeV/$c^2$.

_____ 19. The kinetic energy of a proton that has been accelerated from rest through a potential difference of $500 \times 10^6$ V is 500 MeV.

_____ 20. To a good approximation, the momentum of a 500-MeV electron is 500 MeV/$c$.

_____ 21. To a good approximation, the momentum of a 500-MeV proton is 500 MeV/$c$.

**Responses to True or False**

1. True.

2. True.

3. False. The Michelson–Morley experiment was designed to measure the speed of the earth relative to the ether, which did involve high precision.

4. False. Since both events take place at the same location in the reference frame of the first observer, 2 s is the proper time interval between them. Any observer moving relative to the first observer will find a time interval larger than 2 s because of time dilation.

5. True. The Lorentz transformation yields $t'_A - t'_B = \gamma v (x_B - x_A)/c^2$.

6. False. Since both events occur at the same place, if one observer finds the two events to be simultaneous, then all other observers will also find them to be simultaneous. Disagreements about simultaneity arise only concerning events that are spatially separated in at least one of the reference frames.

7. False. The length of a moving object is determined by finding the spatial coordinates of its end points at the same time, and need not involve light coming from its end points.

8. False. According to the Lorentz transformation, $y' = y$, so all observers will agree about lengths along the $y$ axis.

9. True.

10. True.

11. True.

12. False. The relativistic velocity transformation gives the velocity of $A$ relative to $B$ to be $0.75c$.

13. False. It's the other way around.

14. True.

15. True.

16. True.

17. True.

18. True.

19. True.

20. True.

21. False. Since the proton's kinetic energy of 500 MeV is comparable to its rest energy of 938 MeV, you must use the relativistic relation between energy and momentum to calculate the proton's momentum. The result is $p = 1090$ MeV/$c$.

# V.   Questions and Answers

**Questions**

1. An observer stationary in reference frame $S'$ measures a light signal emitted at $t' = 0$ to move from his origin directly along his $+y'$ axis at, of course, a speed $c$. Describe what an observer in $S$, moving at a constant velocity $v$ along the common $x$-$x'$ axis, measures.

2. An observer in reference frame $S'$ is moving relative to reference frame $S$ at 0.8$c$. At a certain instant a red flash is emitted at the 10-m mark on the $x'$ axis of $S'$, and 5 s later, as determined by clocks in $S'$, a blue flash is emitted at the same 10-m mark. What is the proper time interval between the red and blue flashes?

3. Is it possible for one observer to find that event $A$ happens after event $B$ and another observer to find that event $A$ happens before event $B$?

4. Suppose that an observer finds that two events $A$ and $B$ occur simultaneously at the same place. Will the two events occur at the same place for all other observers?

5. Explain how the measurement of time enters into the determination of the length of an object.

6. Triplets $A$ and $B$ take trips in high-speed spaceships while triplet $C$ stays at home. Triplet $A$ moves along the positive $x$ axis to a far galaxy and then returns home at the same speed. Triplet $B$ moves along the negative $x$ axis at the same speed as $A$ to the same distance as $A$, and then returns home at the same speed. Compare the ages of the triplets when they are all together again.

7. You find that it takes a time interval $\Delta t$ for a spaceship to move through a distance $L$. How does the time interval elapsed on the clock of the pilot of the spaceship compare to what you measure?

8. In terms of energies, when can you use $p = m_0 u$ to a good approximation?

9. In terms of energies, when can you use $p = E/c$ to a good approximation?

**Answers**

1. The coordinates of the light signal at a particular time $t'$ as determined by the observer in $S'$ are $x' = 0$ and $y = ct'$. To find the coordinates as determined by an observer in $S$, use the Lorentz transformation to get

   $$x = \gamma(x' + vt') = \gamma vt'$$
   $$y = y' = ct'$$

   This shows that the light signal moves in the $xy$ plane. To express $t'$ in terms of $t$, use the Lorentz transformation

   $$t = \gamma(t' + vx'/c^2) = \gamma t'$$
   $$t' = t/\gamma$$

   Substitute this expression for $t'$ into the expressions for $x$ and $y$ to get

   $$x = \gamma v(t/\gamma) = vt$$
   $$y = c(t/\gamma) = ct\sqrt{1 - v^2/c^2}$$

   As determined by an observer in $S$, the variation of distance with time is found from

   $$d = \sqrt{x^2 + y^2} = \sqrt{(vt)^2 + (1 - v^2/c^2)(ct)^2} = ct$$

   showing that the observer in $S$ also measures the light signal to move with speed $c$, as expected.

2. Since the red and blue flashes are both emitted at the same place as determined in reference frame $S'$, the time interval of 5 s determined by a single clock in $S'$ is the proper time interval between the two events.

3. Yes. Subtraction of the inverse Lorentz transformation expression for time gives

   $$t'_A - t'_B = \gamma\left[(t_A - t_B) - (v/c^2)(x_A - x_B)\right]$$

   Suppose in reference frame $S$ event $A$ occurs after event $B$ so that $t_A - t_B$ is positive. The time ordering $t'_A - t'_B$ between the two events as determined in reference frame $S'$ can be positive, zero, or negative depending on the spatial separation $x_A - x_B$ in reference frame $S$ and the relative velocity $v$.

4. Yes. Suppose in reference frame $S$ two events $A$ and $B$ occur simultaneously ($t_A - t_B = 0$) and at the same place $x'_A - x'_B = 0$, so the two events also take place at the same location in $S'$.

5. If the object is moving relative to you, you measure its length by finding the difference between the coordinates of its end points at the same time.

6. The time dilation effect depends on the square of the speed, so the direction of the motion makes no difference. Since $A$ and $B$ move with the same speed over the same distance, they will be the same age when the triplets get together, and they will be younger than $C$, who stayed at home.

7. The pilot measures the time interval with a single clock that records the proper time between the beginning and end points. This proper time interval will be smaller than your time interval $\Delta t$.

8. When the kinetic energy $K \ll m_0 c^2$ you can use the expression $p = m_0 u$ to a good approximation.

9. When the total energy $E \gg m_0 c^2$, or equivalently when the kinetic energy $K \gg m_0 c^2$, you can use $p = E/c$ to a good approximation.

# VI. Problems

**Example #1.** An observer in reference frame $S$ observes that a lightning bolt $A$ strikes his $x$ axis and $10^{-4}$ s later a second lightning bolt $B$ strikes his $x$ axis $1.5 \times 10^5$ m farther from the origin than $A$. What is the time separation between the two lightning bolts determined by a second observer in reference frame $S'$ moving at a speed of $0.8c$ along the collinear $x$-$x'$ axis?

**Picture the Problem.** Use the Lorentz transformation equation relating $t$, $t'$, and $x$. Subtract the two times in the prime reference frame.

| 1. Write the expression for the $t'$ transformation. | $t' = \gamma\left(t - vx/c^2\right)$ |
|---|---|
| 2. Subtract the expressions for events $A$ and $B$ to get the time difference in the prime reference frame. Because the time difference is negative, the observer in reference frame $S'$ sees that event $B$ occurs *before* event $A$. | $t'_B - t'_A = \gamma\left[\left(t_B - t_A\right) - \dfrac{v}{c^2}\left(x_B - x_A\right)\right]$ $= \dfrac{1}{\sqrt{1 - 0.8^2}}\left[10^{-4}\,\text{s} - \dfrac{0.8}{3 \times 10^8\,\text{m/s}}\left(1.5 \times 10^5\,\text{m}\right)\right]$ $= -5 \times 10^{-4}\,\text{s}$ |

**Example #2—Interactive.** How fast would the second observer in Problem 1 have to be moving to find that the two lightning bolts strike simultaneously in reference frame $S'$?

**Picture the Problem.** Set the time difference to be zero, and solve for the speed $v$. **Try it yourself.** Work the problem on your own, in the space provided, and check your answer.

| 1. Write the expression for the $t'$ transformation. | |
|---|---|
| | |

| 2. Set the time difference $t'_B - t'_A = 0$ and solve for the speed $v$. | |
|---|---|
| | $v = 6 \times 10^7 \text{ m/s} = 0.2c$ |

**Example #3.** A super rocket car traverses a straight track $2.4 \times 10^5 \text{ m}$ long in $10^{-3}$ s as measured by an observer next to the track. How much time elapses on a clock in the rocket car during this run?

**Picture the Problem.** Keep clear in your mind who is measuring each distance and time interval. Because the two clocks are moving with respect to each other, we will need to use time dilation. However, in order to determine the amount of time dilation, we need to know the relative speed of the clocks.

| 1. Determine the speed of the car as measured by the observer at the side of the track. | $v = \dfrac{\text{distance}}{\text{time}} = \dfrac{2.4 \times 10^5 \text{ m}}{10^{-3} \text{ s}}$ $= 2.4 \times 10^8 \text{ m/s} = 0.8c$ |
|---|---|
| 2. The clock in the car measures the proper time required to travel that distance. | $\Delta t = \dfrac{\Delta t_p}{\sqrt{1 - (v^2/c^2)}}$ $\Delta t_p = \Delta t \sqrt{1 - (v^2/c^2)}$ $= (10^{-3} \text{ s})\sqrt{1 - 0.8^2} = 6 \times 10^{-4} \text{ s}$ |

**Example #4—Interactive.** How far does the driver of the rocket car in the above example determine she travels in traversing the track?

**Picture the Problem.** Keep clear in your mind who is measuring each distance and time interval. The rocket car will see a contracted length of the track. **Try it yourself.** Work the problem on your own, in the space provided, and check your answer.

| 1. Determine the contracted length of the track from the velocity found in the previous example. | |
|---|---|
| | $L = 1.44 \times 10^5 \text{ m}$ |

**Example #5.** A radioactive nucleus moving at a speed of $0.8c$ in a laboratory decays and emits an electron in the same direction as the nucleus is moving. The electron moves at $0.6c$ relative to the nucleus. How fast is the electron moving according to an observer in the laboratory?

**Picture the Problem.** Associate the objects in the problem with observers in reference frames $S$ and $S'$, and particle $P$. Determine which velocities correspond to $v$, $u$, and $u'$, and apply the appropriate velocity transformation equations.

| | |
|---|---|
| 1. Make the appropriate reference frame assignments. | $S$ = laboratory observer<br>$S'$ = nucleus<br>$P$ = electron |
| 2. Make the corresponding velocity assignments. | $v = 0.8c$<br>$u'_x = 0.6c$<br>$u_x$ = speed of electron in laboratory frame |
| 3. Use the velocity transformation equation. | $u_x = \dfrac{u'_x + v}{1 + vu'_x / c^2}$<br><br>$= \dfrac{0.6c + 0.8c}{1 + (0.8c)(0.6c)/c^2} = 0.946c$ |

**Example #6—Interactive.** Rocket $A$ travels away from the earth at $0.6c$, and rocket $B$ travels away from the earth in exactly the opposite direction at $0.8c$. What is the speed of the rocket $B$ as measured by the pilot of rocket $A$?

**Picture the Problem.** Follow the same approach as in the previous example. **Try it yourself.** Work the problem on your own, in the space provided, and check your answer.

| | |
|---|---|
| 1. Make the appropriate reference frame assignments. | |
| 2. Make the corresponding velocity assignments. | |
| 3. Use the velocity transformation equation. | $v = 0.946c$ |

**Example #7.**  What is the velocity of a 2.00-MeV electron?

**Picture the Problem.**  If $K \ll m_0 c^2$, then you can use the nonrelativistic expression for kinetic energy.  Otherwise, you must use the relativistic expression for kinetic energy to determine the electron's velocity.

| 1.  Look up the rest mass of the electron. | $m_0 c^2 = 0.511 \, \text{MeV}$ |
|---|---|
| 2.  The given energy is much larger than the rest energy of the electron, so to determine its speed we must use the relativistic expression. | $K = \dfrac{m_0 c^2}{\sqrt{1 - \left(v^2 / c^2\right)}} - m_0 c^2$ <br><br> $2.00 \, \text{MeV} = \dfrac{0.511 \, \text{MeV}}{\sqrt{1 - \left(v^2 / c^2\right)}} - 0.511 \, \text{MeV}$ <br><br> $v = 0.979 c$ |

**Example #8—Interactive.**  How fast is a 1,000,000-MeV proton moving?

**Picture the Problem.**  If $K \ll m_0 c^2$, then you can use the nonrelativistic expression for kinetic energy.  Otherwise, you must use the relativistic expression for kinetic energy to determine the proton's velocity.  **Try it yourself.**  Work the problem on your own, in the spaces provided, and check your answer.

| 1.  Look up the rest mass of the proton. | |
|---|---|
| 2.  The given energy is much larger than the rest energy of the proton, so to determine its speed we must use the relativistic expression. | |
| | $v = 0.999999 c$ |

**Example #9.**  What is the kinetic energy of a proton whose momentum is 500 MeV/$c$ ?

**Picture the Problem.**  If $pc \gg m_0 c^2$, then you can use $E = pc$ or $K = pc$ to a good approximation.  Otherwise, you will need to use the relativistic expression relating energy and momentum to find the proton's energy.

| 1.  Look up the rest mass of the proton. | $m_0 c^2 = 938 \, \text{MeV}$ |
|---|---|
| 2.  The given energy is on the order of the rest energy, so the relativistic expression must be used. | $E^2 = \left(K + m_0 c^2\right)^2 = p^2 c^2 + m_0^2 c^4$ <br><br> $\left(K + 938 \, \text{MeV}\right)^2 = \left(500 \, \text{MeV}/c\right)^2 c^2 + \left(938 \, \text{MeV}\right)^2$ <br><br> $K = 125 \, \text{MeV}$ |

**Example #10—Interactive.**   What is the kinetic energy of an electron whose momentum is 300 MeV/$c$ ?

**Picture the Problem.**   If $pc \gg m_0 c^2$, then you can use $E = pc$ or $K = pc$ to a good approximation.   Otherwise, you will need to use the relativistic expression relating energy and momentum to find the electron's energy.   **Try it yourself.**   Work the problem on your own, in the spaces provided, and check your answer.

| | |
|---|---|
| 1.  Look up the rest mass of the electron. | |
| 2.  The given energy is much larger than the rest energy, so an approximation can be used. | $K = 300\,\text{MeV}$ |

# Chapter **40**

# Nuclear Physics

## I.   Key Ideas

***Section 40-1. Properties of Nuclei.*** Nuclei are composed of positively charged protons and uncharged neutrons, referred to collectively as **nucleons,** whose main properties are given in Table 40-1.

Table 40-1 **Properties of Nucleons**

| Property | Proton | Neutron |
|---|---|---|
| Charge | $+1.602 \times 10^{-19}$ C | 0 |
| Rest Mass | $1.672623 \times 10^{-27}$ kg | $1.674929 \times 10^{-27}$ kg |
| | $938.2723 \,\mathrm{MeV}/c^2$ | $939.5656 \,\mathrm{MeV}/c^2$ |
| | $1.007277 \,\mathrm{u}$ | $1.008665 \,\mathrm{u}$ |
| Spin | $\frac{1}{2}$ | $\frac{1}{2}$ |

The number of protons and neutrons in a nucleus is designated as follows:

$Z$ = number of protons, called the **atomic number**

$N$ = number of neutrons

$A = N + Z$ = number of nucleons, called the **mass number**

A particular nuclear species, called a **nuclide,** is designated by giving the atomic symbol $X$ of the nucleus, with the mass number $A$ as a presuperscript and the atomic number $Z$ as a presubscript, in the form $^{A}_{Z}X$. Often the presubscript $Z$ is not used because it is associated uniquely with the atomic symbol $X$.

**Isotopes** are nuclides with the same atomic number $Z$ but different numbers $N$ of neutrons (and correspondingly different mass numbers $A$). For example $^{12}_{6}\mathrm{C}$, $^{13}_{6}\mathrm{C}$, and $^{14}_{6}\mathrm{C}$ are all isotopes of carbon.

Nucleons in a nucleus attract each other with a **strong nuclear force** (also called a **hadronic force**). Because this attractive force is independent of the charge of the nucleons, proton–proton, proton–neutron, and neutron–neutron strong nuclear forces are all roughly equal. The nuclear force has a short range, being essentially zero between nucleons separated by more than a few femtometers ($1 \, \text{fm} = 10^{-15} \, \text{m}$).

To a very good approximation a nucleus can be treated as a sphere of radius $R$ given by

$$R = R_0 A^{1/3} \qquad\qquad\qquad \text{Nuclear radius}$$

where $R_0$ is about 1.2 fm. The value of $R_0$ depends on which nuclear property is being measured: charge distribution, mass distribution, region of influence of the strong nuclear force, and so forth. Because the mass and volume of a nucleus are both proportional to $A$, the densities of all nuclei are approximately the same.

For light stable nuclei the number of protons $Z$ and neutrons $N$ are about equal to each other. As the number of nucleons increases, $N$ becomes larger than $Z$ for stable nuclei, as shown in the plot of $N$ versus $Z$ in Figure 40-1.

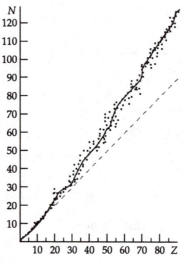

Figure 40-1

The rest mass of a stable nucleus is less than the sum of the rest masses of its constituent nucleons because of the energy required to bind the nucleons together (see Section 39-6). The **binding energy** of a nucleus of atomic mass $M_A$ can be written as

$$E_b = \left( Z M_H + N m_n - M_A \right) c^2 \qquad\qquad \text{Binding energy of a nucleus}$$

where $M_H$ is the mass of a $^1\text{H}$ atom and $m_n$ is the mass of a neutron. The reason for using atomic masses for $^1\text{H}$ and $M_A$ instead of nuclear masses is that it is the atomic masses that are directly measured and listed in tables. The mass of the extra $Z$ electrons in the $Z M_H$ term is canceled by the mass of the extra $Z$ electrons in the $M_A$ term.

A quantity of interest is the **average binding energy per nucleon**, $E_b / A$, obtained by dividing the binding energy of a nucleus by the number of nucleons in the nucleus. A plot of $E_b / A$ versus the mass number $A$ is shown in Figure 40-2. For $A > 50$, $E_b / A$ is roughly constant at about 8 MeV per nucleon, indicating that $E_b$ is approximately proportional to the number of nucleons $A$ in a nucleus.

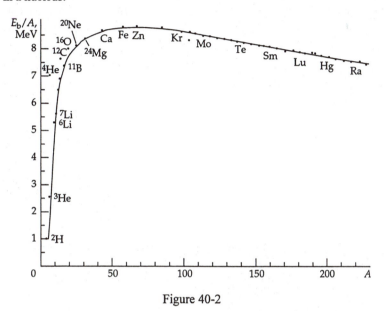

Figure 40-2

***Section 40-2. Radioactivity.*** An unstable nucleus decays into another nucleus accompanied by the emission of a **decay product**. The decay is characterized by the decay product according to the nomenclature shown in Table 40-2.

Radioactive decay is a statistical process, and as such it follows an exponential decay dependence. If there are $N_0$ radioactive nuclei at time $t = 0$, the number $N$ of radioactive nuclei remaining after a time $t$ is given by

$$N = N_0 e^{-\lambda t}$$                                                                    Radioactive decay law

where the constant $\lambda$, called the **decay constant**, depends on the particular nucleus and the decay process taking place.

Table 40-2 **Decay of Nuclei**

| Type of decay | Decay product |
| --- | --- |
| Alpha ($\alpha$) decay | $^4\text{He}$ nuclei (alpha particles) |
| Beta ($\beta$) decay | Electrons, $e^-$ or $\beta^-$, Positrons, $e^+$ or $\beta^+$, (beta particles) |
| Gamma ($\gamma$) decay | Photons (gamma rays) |

The reciprocal of the decay constant is the **average** or **mean lifetime** $\tau$.

$$\tau = 1/\lambda \qquad\qquad \text{Mean lifetime}$$

At the end of a mean lifetime, a radioactive sample has decayed to $1/e = 0.37$ or 37% of the original number of nuclei. Another time that characterizes decay rate is the half-life $t_{1/2}$, which is the time it takes for the number of radioactive nuclei in a sample to decrease to one-half the original number. As shown in Figure 40-3, if you start with $N_0$ nuclei, after one half-life $N_0/2$ nuclei will remain, after two half-lives $N_0/4$ nuclei will remain, and so forth. The half-life $t_{1/2}$ is related to the decay constant $\lambda$ by

$$t_{1/2} = \frac{\ln 2}{\lambda} = \frac{0.693}{\lambda} = 0.693\tau \qquad\qquad \text{Half-life}$$

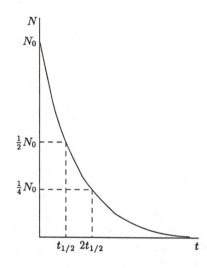

Figure 40-3

The number of decays per second, called the decay rate $R$, is found by differentiating the expression for $N$ to get

$$R = -dN/dt = \lambda N = \lambda N_0 e^{-\lambda t} = R_0 e^{-\lambda t} \qquad\qquad \text{Radioactive decay rate}$$

where

$$R_0 = \lambda N_0 \qquad\qquad \text{Initial decay rate}$$

is the decay rate at $t = 0$. The SI unit of measure of decay rate is the **becquerel** (Bq), defined as one decay per second:

$$1\,\text{Bq} = 1\,\text{decay/s} \qquad\qquad \text{Definition of SI unit of decay rate}$$

Another common unit by which decay rates are measured is the **curie** (Ci), defined as

$$1\,\text{Ci} = 3.7 \times 10^{10}\,\text{decays/s} = 3.7 \times 10^{10}\,\text{Bq} \qquad\qquad \text{Definition of the curie}$$

A curie is approximately equal to the rate of decay of 1 g of radium. After a time equal to one half-life, the decay rate decreases by a factor of $\frac{1}{2}$.

In every type of radioactive decay, a **parent nucleus** emits a decay product and decays into a **daughter nucleus.** Essentially, the birth of a daughter results from the death of the parent. Let us look at three types of decay in more detail: beta, gamma, and alpha decay. In each decay the laws of conservation of mass–energy, charge, and linear and angular momentum hold. In addition, there is a law of conservation of nucleons, according to which the number of nucleons after a decay equals the number of nucleons before the decay.

*Beta Decay.* In $\beta^-$ decay, a parent $P$ decays into a daughter $D$ by conversion of a neutron into a proton with the emission of an electron ( $\beta^-$ ) and an antineutrino:

$$^A_Z P \rightarrow \,^A_{Z+1} D + \beta^- + \bar{v}_e \qquad\qquad \beta^- \text{ decay}$$

In $\beta^+$ decay, a proton is converted into a neutron with the emission of a positron ( $\beta^+$ ) and a neutrino:

$$^A_Z P \rightarrow \,^A_{Z-1} D + \beta^+ + v_e \qquad\qquad \beta^+ \text{ decay}$$

The electron-associated **neutrino** ( $v_e$ ) or **antineutrino** ( $\bar{v}_e$ ) that appears as one of the decay products was initially thought to have zero rest mass, zero charge, spin $\frac{1}{2}$, and to move at the speed of light. It is now believed that the neutrino has some very small, but finite mass. If the neutrino or antineutrino were not emitted as a decay product in $\beta$ decay, energy, momentum, and spin would not be conserved.

It was the necessity to maintain conservation of energy and momentum in $\beta$ decay that motivated Pauli in 1930 to postulate the existence of a neutrino particle. Eventually, in 1956, a neutrino was observed experimentally by Cowan and Reines. The distinction between the various types of neutrinos will be explained in Chapter 41.

*Gamma Decay.* In **gamma ($\gamma$) decay,** a nucleus in an excited state emits a photon, called a $\gamma$ ray, as it makes a transition to a lower energy state, analogous to an atom in an excited state emitting a photon as it decays. The gamma decay reaction can be written as

$$\left(^A_Z X\right)^* \rightarrow \,^A_Z X + \gamma \qquad\qquad \text{Gamma decay}$$

where the * denotes an excited state of the nucleus. In $\gamma$ decay the parent and daughter are the same nucleus. The $\gamma$ rays are observed to be emitted with discrete wavelengths, showing that nuclei possess discrete energy levels typically separated by energies on the order MeV.

*Alpha Decay.* In **Alpha ($\alpha$) decay** a parent nucleus decays into a daughter nucleus with the emission of an $\alpha$ particle ( $^4$He nucleus):

$$^A_Z P \rightarrow \,^{A-4}_{Z-2} D + \,^4_2 \text{He} \qquad\qquad \text{Alpha decay}$$

Thus in $\alpha$ decay, $N$ and $Z$ both decrease by 2, and $A$ decreases by 4.

In many cases the daughter of a radioactive nucleus is also unstable and decays further by $\alpha$ or $\beta$ decay.

***Section 40-3. Nuclear Reactions.*** Information about a nucleus can be obtained by bombarding it with a known projectile and analyzing the resulting nuclear reaction. Many types of nuclear reactions can take place. In one typical nuclear reaction, after the projectile interacts with the target, one or more particles are detected with experimental apparatus and a residual nucleus is left unobserved:

target nucleus + projectile

$\quad\quad\quad\rightarrow$ undetected residual nucleus + detected particle(s)     Typical nuclear reaction

Sometimes the residual nucleus is radioactive and decays by the emission of other particles that can be detected. In any nuclear reaction equation, the total charge $Z$ and total number of nucleons $A$ must be the same on both sides of the equation.

The **$Q$ value** of a nuclear reaction is the difference between the total rest masses before and after the reaction:

$$Q = \left(\Sigma m_0 c^2\right)_{\text{before}} - \left(\Sigma m_0 c^2\right)_{\text{after}}$$     $Q$ value

The $Q$ value can be positive or negative. For an **exothermic reaction** the $Q$ value is positive, and energy is released in the reaction. An exothermic reaction can occur even when both initial particles are at rest. In an **endothermic reaction** the $Q$ value is negative, and energy is absorbed in the reaction. An endothermic reaction cannot occur unless the bombarding particle has a kinetic energy greater than a certain threshold value.

A **cross section** $\sigma$ is a measure of the probability that a bombarding particle will interact with the target nucleus to produce a specific nuclear reaction. The cross section for a reaction is defined as the ratio of the number of reactions per second per nucleus, $R$, to the number of projectiles incident per second per unit area, $I$ (the incident intensity):

$$\sigma = R/I$$     Cross section

The cross section has the dimensions of area and is commonly measured in a unit called the **barn**:

$$1\,\text{barn} = 10^{-28}\,\text{m}^2$$     Unit of cross section

which is of the order of the square of a nuclear radius. The larger the value of $\sigma$, the more likely a reaction will occur. A nuclear cross section of 1 barn is relatively large, so the expression became a nuclear target with a cross section of 1 barn is as easy to hit as the "broad side of a barn."

***Section 40-4. Fission and Fusion.*** Figure 40-4 shows a typical **nuclear fission reaction**: (*a*) a heavy nucleus ($A > 200$) such as $^{235}_{92}\text{U}$ absorbs a neutron, producing (*b*) an intermediate nucleus that is in an excited state; the excited intermediate nucleus (*c*) then splits into two medium-mass nuclei accompanied by the emission of several neutrons (*d*). A typical fission reaction is

$$n + {}^{235}_{92}U \rightarrow \left({}^{236}_{92}U\right)^{*} \rightarrow {}^{141}_{56}Ba + {}^{92}_{36}Kr + 3n + KE \qquad \text{Typical fission reaction}$$

Typically, about 200 MeV of energy is released in each fission.

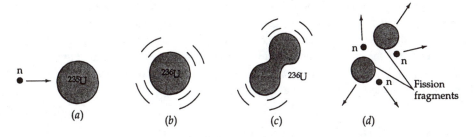

Figure 40-4

The two medium-mass nuclei have roughly the same neutron–proton ratio ($N/Z$) as the original nucleus, so they lie above the stability line of the $N$ versus $Z$ plot of Figure 40-1. Consequently, these fission fragments undergo further decay, usually beta decay, until a stable $N/Z$ ratio is reached.

The neutrons emitted in a fission reaction are available to induce more fissions in the sample. The average number of neutrons from each fission that succeed in producing a subsequent fission is called the **reproduction constant** $k$. If $k < 1$, the nuclear reactions die out. But if each of the two or three neutrons released in a fission results in a further fission—that is, $k > 1$—the number of fissions will increase exponentially. In this case, all the available energy is released in an uncontrolled manner during a very short time interval, resulting in a nuclear explosion.

*Nuclear Fission Reactors.* A **nuclear fission reactor** is designed so that, on the average, one neutron from a fission reaction produces only one new fission, that is, $k = 1$. In this way the fission reactions are controlled, the reactor is self-sustaining, and the excess energy, which appears in the form of heat, is used to heat water to make steam to drive turbines that generate electrical power.

Because the fission of ${}^{235}_{92}U$ (for example) is induced most easily by low-energy neutrons, nuclear fission reactors employ **thermal neutrons,** which have energies $kT \approx 0.025\,\text{eV}$. A thermal neutron results when the energy of a fission-produced neutron is reduced from its initial value of about 1 MeV to a thermal value by repeated collisions with light-weight nuclei in a **moderator.** The moderator, which may be water or carbon, is placed around the fissionable material in the core of the reactor.

*Fusion.* In a **fusion reaction** two light nuclei ($A < 20$) fuse together to form a heavier nucleus, with a release of energy. An example of a fusion reaction is the combination of deuterium and tritium to form a helium nucleus and a neutron:

$$^2H + {}^3H \rightarrow {}^4He + n + 17.6\,\text{MeV} \qquad \text{Deuterium–tritium fusion reaction}$$

The energy per unit mass released in a fusion reaction is larger than in a fission reaction. For example, in the deuterium–tritium fusion, $(17.6\,\text{MeV})/(5\,\text{nucleons}) = 3.52\,\text{MeV/nucleon}$ is released. As a comparison, in a typical fission reaction, about 200 MeV is released per 200 nucleons, giving roughly 1 MeV per nucleon released.

For fusion to occur, the particles must be heated to a temperature corresponding roughly to a thermal energy $kT \approx 10\,\text{keV}$, or about $10^8$ K. At this temperature, atoms are separated into a gas of positive ions and negative electrons called a plasma. Temperatures of $10^8$ K occur in stars, where fusion reactions play an important part in a star's evolution, and are the primary sources of a star's energy. On the earth, the confinement of a plasma for a period of time long enough for fusion to take place is a problem that we are only beginning to solve, so practical fusion reactors will probably not be available for several decades.

## II.  Physical Quantities and Key Equations

### Physical Quantities

| | |
|---|---|
| *Rest mass of a proton* | $1.672623 \times 10^{-27}\,\text{kg} = 938.2723\,\text{MeV}/c^2 = 1.007277\,\text{u}$ |
| *Rest mass of a neutron* | $1.674929 \times 10^{-27}\,\text{kg} = 939.5656\,\text{MeV}/c^2 = 1.008665\,\text{u}$ |
| *Definition of SI unit of decay rate* | $1\,\text{Bq} = 1\,\text{decay/s}$ |
| *Definition of the curie* | $1\,\text{Ci} = 3.7 \times 10^{10}\,\text{decays/s} = 3.7 \times 10^{10}\,\text{Bq}$ |
| *Unit of cross section* | $1\,\text{barn} = 10^{-28}\,\text{m}^2$ |

### Key Equations

| | |
|---|---|
| *Nuclear radius* | $R = R_0 A^{1/3}$ |
| *Binding energy of a nucleus* | $E_b = \left( ZM_H + Nm_n - M_A \right) c^2$ |
| *Radioactive decay law* | $N = N_0 e^{-\lambda t}$ |
| *Mean lifetime* | $\tau = 1/\lambda$ |
| *Half-life* | $t_{1/2} = \left( \ln 2 \right)/\lambda = 0.693/\lambda = 0.693\tau$ |
| *Radioactive decay rate* | $R = -dN/dt = \lambda N = \lambda N_0 e^{-\lambda t} = R_0 e^{-\lambda t}$ |
| *Initial decay rate* | $R_0 = \lambda N_0$ |
| $\beta^-$ *decay* | $^A_Z P \rightarrow\ ^A_{Z+1} D + \beta^- + \bar{\nu}_e$ |
| $\beta^+$ *decay* | $^A_Z P \rightarrow\ ^A_{Z-1} D + \beta^+ + \nu_e$ |

| | |
|---|---|
| *Gamma decay* | $\left(^A_Z X\right)^* \rightarrow ^A_Z X + \gamma$ |
| *Alpha decay* | $^A_Z P \rightarrow ^{A-4}_{Z-2} D + ^4_2 He$ |
| *Q value* | $Q = \left(\Sigma m_0 c^2\right)_{before} - \left(\Sigma m_0 c^2\right)_{after}$ |
| *Cross section* | $\sigma = R/I$ |
| *Typical fission reaction* | $n + ^{235}_{92}U \rightarrow \left(^{236}_{92}U\right)^* \rightarrow ^{141}_{56}Ba + ^{92}_{36}Kr + 3n$ |
| *Deuterium–tritium fusion reaction* | $^2H + ^3H \rightarrow ^4He + n + 17.6\,MeV$ |

## III. Potential Pitfalls

In the expression for the binding of energy of a nucleus, do not mistakenly use the mass of a proton instead of the atomic mass of hydrogen. Make sure you understand how the use of atomic masses (of which there are tables) instead of nuclear masses is justified: the masses of the electrons cancel.

Do not confuse the total binding energy of a nucleus with the binding energy per nucleon of the nucleus.

Understand that the average binding energy per nucleon is different from the energy required to remove a single nucleon from a nucleus.

Do not confuse the spin quantum number of a nucleon, which is $\frac{1}{2}$, with the nuclear spin $I$, which is the vector sum of the nucleon spins.

Do not confuse nuclear magnetic resonance with nuclear fission or nuclear fusion.

Understand that the symbols $\beta^-$ and $e^-$ are used interchangeably for electrons, as are the symbols $\beta^+$ and $e^+$ for positrons.

Do not think that the electrons or positrons emitted in $\beta$ decay are contained inside a nucleus. They are not. Beta particles are *created* in the process of decay much as photons, which do not reside in an atom, are created when an atom makes a transition from a higher to a lower energy level. The same holds for alpha decay and gamma decay.

Do not confuse an exothermic with an endothermic reaction. The former emits energy; the latter absorbs it.

# IV. True or False Questions and Responses

**True or False**

____ 1.  The mass number minus the atomic number equals the total number of nucleons in a nucleus.

____ 2.  The strong nuclear force holds electrons inside a nucleus until the electrons are emitted in beta decay or nuclear reactions.

____ 3.  A nucleus is approximately a sphere with constant density.

____ 4.  The binding energy per nucleon continuously increases as the number of nucleons in a nucleus increases.

____ 5.  After $\beta^-$ decay a nucleus has one more proton and one less neutron than before the decay.

____ 6.  The cross section for an endothermic reaction is zero below the reaction's threshold energy.

____ 7.  The cross section for a particular reaction depends only on the size of the nuclei being bombarded and not on the energy of the incident particles.

____ 8.  A thermal neutron has an energy of about 0.025 eV.

____ 9.  Fission is induced most readily with neutrons of high energy of the order of 1 MeV.

____ 10. Fission and fusion reactions release energy because the rest mass per nucleon of the resultant products is less than the rest mass per nucleon of the original nuclei.

**Responses to True or False**

1.  False. The difference equals the total number of neutrons in the nucleus.

2.  False. There are no electrons inside a nucleus. The strong nuclear force holds together the protons and neutrons inside the nucleus; it does not affect electrons.

3.  True.

4.  False. The binding energy per nucleon rises until around $A = 60$, and then gradually decreases as $A$ increases further.

5.  True.

6.  True.

7.  False. The cross section for a particular reaction is a function of the energy of the bombarding particles and can be much less than the geometrical cross section of the target nuclei.

8. True.

9. False. Fission occurs most readily with low-energy thermal neutrons (0.025 eV).

10. True.

## V.  Questions and Answers

**Questions**

1.  What is the difference between atomic number and mass number?

2.  How do the numbers of neutrons and protons compare in stable nuclides?

3.  What is an isotope?

4.  What is the difference between mean lifetime and half-life in radioactive decay?

5.  Explain how conservation of energy and momentum would be violated if a neutrino were not emitted in beta decay.

6.  What is the difference between fission and fusion?

7.  Explain how variations in the reproduction constant affect the operation of a nuclear reactor.

8.  Why are thermal neutrons used in nuclear reactors?

9.  What is the major impediment to building practical fusion reactors?

**Answers**

1.  Atomic number $Z$ equals the number of protons in a nucleus. Mass number $A$ equals the number of protons plus neutrons in a nucleus.

2.  For light nuclei, the numbers of neutrons and protons are approximately equal. As the number of nucleons increases, the number of neutrons exceeds the number of protons.

3.  Isotopes are nuclides with the same atomic number $Z$ but a different number of neutrons $N$, and so they have different mass numbers $A = Z + N$.

4.  The mean lifetime $\tau$ is the reciprocal of the decay constant $\lambda$. After a time $\tau$, both the number of radioactive nuclei and the decay rate decrease to $1/e$ of their original values. A half-life $t_{1/2}$ is the time for both the number of nuclei in a radioactive sample and the decay rate to decrease to half their original value. The two times are related by $t_{1/2} = (\ln 2)\tau$.

5.  Consider a typical $\beta$ decay, such as the decay of a neutron:  $n \rightarrow p + \beta^- + \bar{v}_e$. If the electron ($\beta^-$) were the only decay product, application of conservation of energy and momentum to the two-body decay would require that the $\beta^-$ particle be ejected with a single unique energy.

Instead, it is observed experimentally that $\beta^-$ particles are produced with energies that range from zero to a maximum value. Further, because the original neutron had spin $\frac{1}{2}$, conservation of angular momentum would be violated if the final decay products consisted of only the two particles p and $\beta^-$, each with spin $\frac{1}{2}$. To preserve conservation of energy and momentum, Pauli in 1930 postulated the existence of a third particle in the decay process, the neutrino, which was experimentally observed by Cowan and Reines in 1956.

6. In fission, a heavy nucleus splits, usually by absorbing a neutron, into two lighter nuclei, with a corresponding release of neutrons and energy. In fusion, two light nuclei fuse together to form a heavier nucleus, with a corresponding release of energy.

7. Nuclear reactors should have a reproduction constant $k$ about equal to 1, so that on the average each fission will result in one neutron that produces a subsequent fission. If $k$ is less than 1, the reaction will die out. If $k$ is greater than 1, there will be an exponential increase in fissions, and the reactor will "run away."

8. The cross section for neutron capture by $^{235}_{92}\text{U}$ is largest when the neutrons have small energies.

9. The inability to confine a high-temperature plasma for a long enough time for fusion to take place.

# VI. Problems

**Example #1.** Given that a nucleus is approximately spherical, with a radius $R = R_0 A^{1/3}$, where $R_0$ is about 1.2 fm, determine its approximate mass density. Express your answer in SI units and also in tons per cubic inch.

**Picture the Problem.** Density is mass per unit volume.

| 1. Determine the mass of the nucleus. | $\text{mass} = A(\text{mass of nucleon}) = A(1.7 \times 10^{-27}\,\text{kg})$ |
|---|---|
| 2. Determine the volume of the nucleus. | $V = \frac{4}{3}\pi R^3 = \frac{4}{3}\pi \left(R_0 A^{1/3}\right)^3 = \frac{4}{3}\pi \left(R_0^3 A\right)$ |
| 3. Determine the density of the nucleus. | $\text{density} = \dfrac{\text{mass}}{\text{volume}} = \dfrac{A(1.7 \times 10^{-27}\,\text{kg})}{\frac{4}{3}\pi (1.2 \times 10^{-15}\,\text{m})^3 A}$ $= 2.35 \times 10^{17}\,\text{kg/m}^3$ $= (2.35 \times 10^{17}\,\text{kg/m}^3)(1\,\text{m}/39.37\,\text{in})^3 (1\,\text{ton}/907\,\text{kg})$ $= 4.2 \times 10^9\,\text{ton/in}^3$ |

**Example #2—Interactive.** The following fission reaction of $^{235}_{92}\text{U}$ takes place: $\text{n} + ^{235}_{92}\text{U} \rightarrow \left(^{236}_{92}\text{U}\right)^* \rightarrow ^{141}_{56}\text{Ba} + ^{92}_{36}\text{Kr} + 3\text{n}$. Find the Coulomb energy of the two fully ionized fission fragments, assuming that they are spheres just touching each other.

**Picture the Problem.** Treat the fragments as positively charged point particles, separated by a distance equal to the sum of the radii of the two fragments. **Try it yourself.** Work the problem on your own, in the spaces provided, and check your answer.

| | |
|---|---|
| 1. Determine the radius of the Barium nucleus. | |
| 2. Determine the charge of the Barium nucleus. | |
| 3. Determine the radius of the Krypton nucleus. | |
| 4. Determine the charge of the Krypton nucleus. | |
| 5. Determine the Coulomb energy between these two charged spheres. | $U = 3.99 \times 10^{-11} \, \text{J} = 250 \, \text{MeV}$ |

**Example #3.** How much energy in MeV is released or absorbed (state which) in the reaction $^{150}_{62}\text{Sm} + \text{p} \rightarrow ^{147}_{61}\text{Pm} + \alpha$ given the atomic masses: $^{150}_{62}\text{Sm} = 149.917276 \, \text{u}$, and $^{147}_{61}\text{Pm} = 146.915108 \, \text{u}$.

**Picture the Problem.** Determine the $Q$ value. If it is positive, the reaction is exothermic, and if it is negative, the reaction is endothermic. Although this is a nuclear reaction, atomic masses are used, so we must conserve the number of electrons, as well.

| | |
|---|---|
| 1. Write an expression for the $Q$ value of the nuclear reaction in terms of atomic masses. | $Q = \left[ \left( M_{\text{Sm}} + M_{\text{H}} \right) - \left( M_{\text{Pm}} + M_{\text{He}} \right) \right] c^2$ |
| 2. Look up the atomic masses of hydrogen and helium, the atoms that correspond to the proton and alpha particle. | $^{1}_{1}\text{H} = 1.007825 \, \text{u}$ <br> $^{4}_{2}\text{He} = 4.002603 \, \text{u}$ |
| 3. Substitute values for the atomic masses into the expression for the $Q$ value and solve. Because the $Q$ value is positive, this is an exothermic reaction, and energy is released by this reaction. | $Q = \left[ \left( 149.917276 \, \text{u} + 1.007825 \, \text{u} \right) \right.$ <br> $\left. - \left( 146.915108 \, \text{u} + 4.002603 \, \text{u} \right) \right] \times \left( 931.5 \, \text{MeV/u} \right)$ <br> $= +6.88 \, \text{MeV}$ |

**Example #4—Interactive.** How much energy in MeV is released in the alpha decay of $^{242}_{94}$Pu given the atomic masses of: $^{242}_{94}$Pu $= 242.058737\,$u, and $^{238}_{92}$U $= 238.050783\,$u.

**Picture the Problem.** Write an expression for the alpha decay, and follow the same procedure as in the previous example. **Try it yourself.** Work the problem on your own, in the spaces provided, and check your answer.

| | |
|---|---|
| 1. Write an expression for the alpha decay. | |
| 2. Write an expression for the $Q$ value of the alpha decay. | |
| 3. Substitute the atomic masses given, and that of helium, to determine the $Q$. | $Q = 4.98\,$MeV |

**Example #5.** The curie unit is defined as $1\,$Ci $= 3.70 \times 10^{10}$ decays/s, which is about the rate at which radiation is emitted by $1.00\,$g of radium. Calculate the half-life of radium from this definition and compare your answer with the measured valued of $t_{1/2} = 1620\,$yr.

**Picture the Problem.** Avogadro's number can be used to find the number of atoms. Use this to find the decay constant, which can be used to determine the half-life.

| | |
|---|---|
| 1. Find the number of atoms in 1 g of radium. | $N = (1\,\text{g})(1\,\text{mol}/226\,\text{g})(6.02 \times 10^{23}\ \text{atoms/mol})$ <br> $\quad = 2.66 \times 10^{21}\ \text{atoms}$ |
| 2. Find the decay constant from the number of atoms and the decay rate. | $R = \lambda N$ <br> $\lambda = R/N = (3.7 \times 10^{10}\ \text{decays/s})/(2.66 \times 10^{21}\ \text{atoms})$ <br> $\quad = 1.39 \times 10^{-11}\ \text{atoms/s}$ |
| 3. Determine the half-life. We don't quite get 1620 yr, because the decay rate of 1 g of radium is not exactly equal to $3.7 \times 10^{10}$ decay/s. | $t_{1/2} = (\ln 2)/\lambda = 0.693/(1.39 \times 10^{-11}\ \text{s}^{-1})$ <br> $\quad = 4.98 \times 10^{10}\ \text{s} = 1580\ \text{yr}$ |

**Example #6—Interactive.** The decay rate of a sample of radioactive $^{200}_{79}$Au is measured as $15\,$mCi, and 10 min. later as $13\,$mCi. Determine the half-life of $^{200}_{79}$Au.

**Picture the Problem.** Find the decay constant from the expression for the decay rate. Use the decay constant to determine the half-life. **Try it yourself.** Work the problem on your own, in the spaces provided, and check your answer.

| 1. Determine the decay constant from the two decay rates separated by time. | |
|---|---|
| 2. Use this decay constant to find the half-life. | $t_{1/2} = 48\,\text{min}$ |

**Example #7.** Assume in a typical alpha decay that the parent with $A$ nucleons decays at rest. In the approximation that the mass of each nucleus in the reaction is proportional to its number of nucleons, determine the kinetic energies of the alpha particle and daughter nucleus in terms of the original $A$ number and the $Q$ value of the reaction. Use nonrelativistic calculations.

**Picture the Problem.** Energy and momentum must be conserved. The $Q$ value will give the sum of the kinetic energies of the alpha particle and daughter nucleus.

| 1. Write a generic expression for the alpha decay described. | $_Z^A P \rightarrow {}_{Z-2}^{A-4}D + {}_2^4\text{He}$ |
|---|---|
| 2. Assuming the masses of the particles are proportional to the number of nucleons, determine the ratio of alpha mass to the daughter mass. | $\dfrac{M_\alpha}{M_D} = \dfrac{4}{A-4}$ |
| 3. Conserve momentum | $p_\alpha = p_D$ |
| 4. Find the ratio of the kinetic energies of the particles. We can use nonrelativistic expressions. | $\dfrac{K_D}{K_\alpha} = \dfrac{p_D^2/(2M_D)}{p_\alpha^2/(2M_\alpha)} = \dfrac{M_\alpha}{M_D} = \dfrac{4}{A-4}$ |
| 5. Determine the $Q$ value of the reaction, which is the total kinetic energy of the particles after the reaction. | $Q = K_D + K_\alpha = \left[\dfrac{4}{A-4}K_\alpha + K_\alpha\right] = \dfrac{AK_\alpha}{A-4}$ |
| 6. Solve the above expression for the kinetic energy of the alpha particle in terms of $Q$ and $A$. | $K_\alpha = \dfrac{A-4}{A}Q$ |

| 7. Use the expressions in steps 5 and 6 to solve for the kinetic energy of the daughter particle. Since for a given alpha decay reaction the $Q$ value is fixed, the alpha particle comes off with a precise kinetic energy, that is, it is monoenergetic. The larger the value of $A$, the smaller is the daughter's kinetic energy and the kinetic energy of the alpha particle is more nearly equal to the $Q$ value of the reaction. | $K_D = Q - K_\alpha = Q - \dfrac{A-4}{A}Q = 4Q/A$ |
|---|---|

**Example #8—Interactive.** Assuming that the daughter nucleus has negligible kinetic energy, determine the maximum energy of an electron emitted in the beta decay of $^{14}_6C$, given the atomic masses of $^{14}_6C = 14.003242\,u$, and $^{14}_7N = 14.003074\,u$.

**Picture the Problem.** Because the daughter has essentially no kinetic energy, the energy released by this reaction will be split between the beta particle and the neutrino. To find the maximum energy of the beta particle, assume the energy of the neutrino is zero. **Try it yourself.** Work the problem on your own, in the spaces provided, and check your answer.

| 1. Write the reaction for the beta decay. Add equal numbers of electrons to both sides of the reaction so that the reaction is expressed in terms of atomic masses. | |
|---|---|
| 2. Find the $Q$ value of the reaction, which is the maximum value of the energy of the beta particle. | $Q = 0.156\,\text{MeV}$ |

**Example #9.** Assuming that in a fission reactor a neutron loses half its energy in each collision with an atom of the moderator, determine how many collisions are required to slow a 200-MeV neutron to an energy of 0.04 eV.

**Picture the Problem.** After each collision, $K_f = 0.5K_i$, and $K_f$ becomes the initial kinetic energy for the next collision.

| 1. Write an expression for the kinetic energy after $n$ collision. | $K_n = 0.5^n K_i$ |
|---|---|

| 2. Find the ratio of the final and initial kinetic energies. | $\dfrac{K_n}{K_i} = 0.5^n = \dfrac{0.04\,\text{eV}}{2\times10^8\,\text{eV}} = 2\times10^{-10}$ |
|---|---|
| 3. Take the log of both sides of the expression in step 2 and solve for n. | $(-.301)n = -9.70$ <br> $n = 32.2$ <br> $n \approx 33$ |

**Example #10—Interactive.** How many days will it take for 1.00 kg of $^{235}_{92}$U to be used up in a nuclear reactor that generates 50.0 MW of power, if each fission produces 200 MeV of energy?

**Picture the Problem.** Use dimensional analysis to determine the number of fissions per second required to generate 50.0 MW. Use Avogadro's number to determine the number of uranium nuclei. **Try it yourself.** Work the problem on your own, in the spaces provided, and check your answer.

| 1. Determine the number of fissions per second required to produce 50.0 MW of power. | |
|---|---|
| 2. Determine the number of uranium particles in the 1.00 kg of fuel. | |
| 3. Use dimensional analysis with the previous two results to determine the time it takes to use up the fuel. | $t = 19\,\text{days}$ |

**Example #11.** You want to produce fusion with deuterium nuclei. Estimate the temperature required to bring two deuterium nuclei together so that they can fuse.

**Picture the Problem.** Assume that fusion will take place when the two deuterium nuclei are separated by a distance of two radii, and that the only force you need to overcome is their Coulomb repulsion.

| 1. Determine the radius of deuterium. | $R = R_0 A^{1/3} = \left(1.2\times10^{-15}\,\text{m}\right)\left(2\right)^{1/3} = 1.5\times10^{-15}\,\text{m}$ |
|---|---|
| 2. Find the Coulomb energy at this distance. | $U = \dfrac{kq_1 q_2}{r} = \dfrac{ke^2}{2R} = \dfrac{1.44\,\text{MeV}\cdot\text{fm}}{3.0\,\text{fm}} = 0.48\,\text{MeV}$ |
| 3. Equate this potential energy to the thermal energy, and solve for $T$. | $U = k_B T$ <br> $T = \dfrac{U}{k_B} = \dfrac{0.48\,\text{MeV}}{8.62\times10^{-11}\,\text{MeV/K}} = 5.7\times10^9\,\text{K}$ |

**Example #12—Interactive.** In Example #11, how much energy is released when two deuterium nuclei fuse? Is this larger or smaller than the energy required to start the fusion process, as determined in Example #11?

**Picture the Problem.** The energy released is equal to the $Q$ value of the reaction. **Try it yourself.** Work the problem on your own, in the spaces provided, and check your answer.

| | |
|---|---|
| 1. Write out the fusion reaction. | |
| 2. Determine the $Q$ value.<br><br>This value is much larger than the energy need to initiate the reaction as calculated in Example #11. | $Q = 23\,\text{MeV}$ |

# Chapter **41**

# Elementary Particles and the Beginning of the Universe

## I.   Key Ideas

Our picture of elementary particles was quite simple before the 1930s. An atom was composed of a small nucleus that contained protons and neutrons and a set of electrons that swarmed around the nucleus. A fourth particle, the photon, carried electromagnetic energy. Later, research teams found other elementary particles: the positron (the antiparticle of the electron) predicted by Dirac, the neutrino predicted by Pauli, and more and more elementary particles, until at present they number in the hundreds.

While the elementary particle zoo was burgeoning, physicists also searched for theories to explain what was being observed experimentally and predicted the existence of even more new particles. Theories rose and fell with new discoveries. At present, the so-called standard model of elementary particles has emerged from the efforts of many minds—and the expenditure of enormous quantities of taxpayers' dollars—to construct particle accelerators with higher and higher energies. The superconducting supercollider (SSC) that was proposed for the mid-1990s to produce particles with energies of the order of 100 TeV ($1\,\text{TeV} = 10^{12}\,\text{eV}$) was canceled by Congress because its cost would have been more than 15 billion dollars.

The story of elementary particles is by no means finished. Many predictions made by the standard theory have not been confirmed, and as experiments continue, there is a good chance of obtaining results that the standard model cannot explain. For example, at this writing there is some experimental evidence that the "massless" neutrino has a small mass, of the order of a few $\text{eV}/c^2$. So keep in mind that the particle picture presented here could change with new discoveries and new ways of thinking.

*Section 41-1. Hadrons and Leptons.* One way of classifying elementary particles is according to the way they participate in **interactions** between particles. The word "interaction" is used instead of "force" because the quantum picture of how elementary particles interact is quite different from the traditional action-at-a-distance forces of classical physics. Four basic interactions exist in nature:

1.   The strong nuclear interaction
2.   The electromagnetic interaction

3.     The weak (nuclear) interaction
4.     The gravitational interaction

Let's look at each of these interactions and the types of particles that participate in them.

The **strong nuclear interaction** is responsible for the attractive force that holds nucleons together in the nucleus. The interaction has a short range, its strength dropping to zero when participating particles are more than a few femtometers apart. The interaction is independent of the charge of the nucleons—that is, the proton–proton, neutron–proton, and neutron–neutron forces are all about equal. Particles that decay via the strong interaction have very short decay times of the order of $10^{-23}$ s, which is roughly the time it takes for light to travel across a nucleus.

Particles that interact via strong interactions are called **hadrons.** A hadron with a half-integral spin is called a **baryon;** neutrons and protons are common examples. A hadron with an integral or zero spin is called a **meson,** with examples being the pion ($\pi$ meson) and kaon ($K$ meson). Baryons are the most massive of elementary particles, whereas mesons have masses intermediate between electrons and protons. The properties of some hadrons and their antiparticles are given in Table 41-1. This table includes only particles that are stable against decay via the strong interaction. It does not include the particles with extremely short lifetimes, of the order of $10^{-23}$ s, that decay through the strong interaction. An example of a strong interaction between hadrons is

$$p + \pi^- \rightarrow n + \pi^0 \qquad\qquad \text{Strong interaction}$$

As we shall discuss later in more detail, hadrons are thought to be constructed of entities called quarks.

The **weak interaction** occurs in beta decay (discussed in Chapter 40), which results in the production of an electron or positron, and a neutrino. The interaction is termed weak because its strength is about $10^{-13}$ times that of the strong interaction. The range of the weak interaction is around 100 times smaller than the short range of the strong interaction. Particles that decay via the weak interaction have lifetimes on the order of $10^{-10}$ s, which is much longer than the $10^{-23}$ s lifetimes of particles that decay via the strong interaction.

Particles that interact with each other via weak interactions are called leptons. Leptons can be classified according to their mass, which is tabulated in 41-2.

Actually, there are three different neutrinos: $v_e$ associated with an electron, $v_\mu$ associated with a muon, and $v_\tau$ associated with a tau. Each lepton has a corresponding antilepton (for example, the positron is the antiparticle of the electron). Some properties of leptons are given in Table 41-2. An example of a weak interaction is the beta decay of a neutron (hadron) into a proton (hadron) plus two leptons:

$$n \rightarrow p + e^- + \overline{v}_e \qquad\qquad \text{Weak interaction decay}$$

Unlike hadrons (which are composed of quarks), leptons are not composed of more basic particles and so can be regarded as truly elementary.

Table 41-1 **Properties of Some Hadrons and their Antiparticles ($L=0$)**

| Family | Hadron particle (antiparticle) | Rest mass, MeV/$c^2$ | Particle mean lifetime, s | Charge number | Spin, $\hbar$ | Baryon number | Strangeness |
|---|---|---|---|---|---|---|---|
| Mesons | $\pi^+$ ($\pi^-$) | 139.6 | $2.6\times10^{-8}$ | +1(−1) | 0 | | 0(0) |
| | $\pi^0$ ($\pi^0$) | 135.0 | $0.8\times10^{-16}$ | 0 | 0 | | 0(0) |
| | $\pi^-$ ($\pi^+$) | 139.6 | $2.6\times10^{-8}$ | −1(+1) | 0 | | 0(0) |
| | $K^+$ ($K^-$) | 493.7 | $1.2\times10^{-8}$ | +1(−1) | 0 | | +1(−1) |
| | $K^0$ ($\bar{K}^0$) | 497.7 | $8.8\times10^{-11}$ | 0(0) | 0 | | +1(−1) |
| | $K^0$ ($K^0$) | 497.7 | $5.2\times10^{-8}$ | 0(0) | 0 | | −1(+1) |
| | $K^-$ ($K^+$) | 493.7 | $1.2\times10^{-8}$ | −1(+1) | 0 | | −1(+1) |
| | $\eta^0$ ($\eta^0$) | 549 | $2\times10^{-19}$ | 0 | 0 | | 0(0) |
| Baryons | $p$ ($\bar{p}$) | 938.3 | Stable | +1(−1) | $\frac{1}{2}$ | +1(−1) | 0(0) |
| | $n$ ($\bar{n}$) | 939.6 | 930 | 0(0) | $\frac{1}{2}$ | +1(−1) | 0(0) |
| | $\Lambda^0$ ($\bar{\Lambda}^0$) | 1116 | $2.5\times10^{-10}$ | 0(0) | $\frac{1}{2}$ | +1(−1) | −1(+1) |
| | $\Sigma^+$ ($\bar{\Sigma}^-$) | 1189 | $8.0\times10^{-11}$ | +1(−1) | $\frac{1}{2}$ | +1(−1) | −1(+1) |
| | $\Sigma^0$ ($\Sigma^0$) | 1193 | $10^{-20}$ | 0(0) | $\frac{1}{2}$ | +1(−1) | −1(+1) |
| | $\Sigma^-$ ($\bar{\Sigma}^+$) | 1197 | $1.7\times10^{-10}$ | −1(+1) | $\frac{1}{2}$ | +1(−1) | −1(+1) |
| | $\Xi^0$ ($\bar{\Xi}^0$) | 1315 | $3.0\times10^{-10}$ | 0(0) | $\frac{1}{2}$ | +1(−1) | −2(+2) |
| | $\Xi^-$ ($\Xi^+$) | 1321 | $1.7\times10^{-10}$ | −1(+1) | $\frac{1}{2}$ | +1(−1) | −2(+2) |
| | $\Omega^-$ ($\Omega^+$) | 1672 | $1.3\times10^{-10}$ | −1(+1) | $\frac{1}{2}$ | +1(−1) | −3(+3) |

Table 41-2 **Some Properties of Leptons**

| Lepton particle (antiparticle) | Rest mass, MeV/$c^2$ | Particle mean lifetime, s | Charge number | Spin, $\hbar$ | Lepton number |
|---|---|---|---|---|---|
| $e^-$ ($e^+$) | 0.511 | Stable | −1(+1) | $\frac{1}{2}$ | +1(−1) |
| $\mu^-$ ($\mu^+$) | 105.7 | $2.2\times10^{-6}$ | −1(+1) | $\frac{1}{2}$ | +1(−1) |
| $\tau^-$ ($\tau^+$) | 1784 | $3\times10^{-13}$ | −1(+1) | $\frac{1}{2}$ | +1(−1) |
| $v_e$ ($\bar{v}_e$) | 0 | Stable | 0(0) | $\frac{1}{2}$ | +1(−1) |
| $v_\mu$ ($\bar{v}_\mu$) | 0 | Stable | 0(0) | $\frac{1}{2}$ | +1(−1) |
| $v_\tau$ ($\bar{v}_\tau$) | 0 | Stable | 0(0) | $\frac{1}{2}$ | +1(−1) |

*Section 41-2. Spin and Antiparticles.* Elementary particles are classified according to their spin. Spin-$\frac{1}{2}$ particles (or $\frac{3}{2}$, $\frac{5}{2}$, ...) are called **fermions**. They obey the Pauli exclusion principle— only one fermion can be in a given quantum state. Examples of fermions are electrons, protons,

neutrons, and neutrinos. Particles with integer spin (0, 1, 2, . . .) are called **bosons.** They do not obey the Pauli exclusion principle—any number of bosons can be in a given quantum state. Examples of bosons are mesons and photons.

Every particle has an **antiparticle** with the same mass and spin but opposite charge. A particle is usually an entity, such as an electron or proton, that is found naturally in our part of the universe. The antiparticle of a particle, such as a positron or an antiproton, is an entity that is not found in nature, and when formed quickly annihilates itself with its corresponding particle.

*Section 41-3. The Conservation Laws.* In an elementary particle reaction or decay, the familiar conservation laws hold:

**Conservation of mass–energy.** The sum of the rest masses and kinetic energies of the particles before a reaction must be equal to the sum of the rest masses and kinetic energies of the particles after the reaction.

$$\sum\left(m_0c^2 + K\right)_{\text{before}} = \sum\left(m_0c^2 + K\right)_{\text{after}} \qquad \text{Conservation of mass–energy}$$

**Conservation of linear momentum.** The linear momentum $\vec{p}$ of the particles before and after a reaction must be the same.

$$\sum\left(\vec{p}\right)_{\text{before}} = \sum\left(\vec{p}\right)_{\text{after}} \qquad \text{Conservation of linear momentum}$$

**Conservation of angular momentum ($\vec{spin}$).** Spin is a form of angular momentum. In any reaction, the vector sum of the spins of the products after the reaction must equal the vector sum of the spins of the particles before the reaction.

$$\sum\left(\vec{spin}\right)_{\text{before}} = \sum\left(\vec{spin}\right)_{\text{after}} \qquad \text{Conservation of spin}$$

**Conservation of charge.** The net charge $Q$ before and after a reaction must be the same.

$$\sum Q_{\text{before}} = \sum Q_{\text{after}} \qquad \text{Conservation of charge}$$

In addition, two conservation laws hold especially for baryons and leptons. Baryon and lepton numbers are assigned to particles as follows:

$$\begin{aligned} B &= +1 \text{ for all baryons} \\ B &= -1 \text{ for all antibaryons} \qquad\qquad \text{Baryon number B} \\ B &= 0 \text{ for all other particles} \end{aligned}$$

$$\begin{aligned} L &= +1 \text{ for all leptons} \\ L &= -1 \text{ for all antileptons} \qquad\qquad \text{Lepton number L} \\ L &= 0 \text{ for all other particles} \end{aligned}$$

The following conservation laws then hold:

**Conservation of baryon number.** The net baryon number $B$ before and after a reaction or decay must be the same.

$$\sum B_{\text{before}} = \sum B_{\text{after}}$$    Conservation of baryon number

**Conservation of lepton number.** The net lepton number $L$ before and after a reaction or decay must be the same.

$$\sum L_{\text{before}} = \sum L_{\text{after}}$$    Conservation of lepton number

An example of both of these conservation laws is seen in neutron decay:

$$\text{n} \rightarrow \text{p} + \text{e}^- + \bar{\nu}_e$$
$$B: \ +1 = +1 + 0 \ +0 \qquad \text{Conservation of baryon and lepton numbers in neutron decay}$$
$$L: \ \ \ 0 = \ \ 0 +1 \ -1$$

Another number that obeys a conservation law is **strangeness.** The notion of strangeness was introduced to account for unexpected reactions that involved baryons and mesons. For example, kaons and pions interact via strong nuclear reactions, so physicists expected that the decay $K^0 \rightarrow \pi^+ + \pi^-$ would proceed via the strong interaction with a decay time of the order of $10^{-23}$ s. Instead, the decay time is of the order of $10^{-10}$ s, a time that is characteristic of the weak interaction. Because of such experiments, a **strangeness number** $S$ was assigned to the **strange particles,** as shown in Table 41-1. For strong interactions, the following conservation law holds:

**Conservation of strangeness.** In a strong interaction the net strangeness number $S$ before and after the reaction is the same.

$$\sum S_{\text{before}} = \sum S_{\text{after}}$$    Conservation of strangeness (strong interactions)

An example of this is the reaction

$$\text{p} + \pi^- \rightarrow K^0 + \Lambda^0$$
$$S: \ 0 + 0 \ = \ +1 -1$$

For weak interactions, a **strangeness selection rule** holds: in a weak interaction the strangeness number changes by zero or one, that is:

$$\Delta S = 0, \pm 1$$    Strangeness selection rule (weak interactions)

An example of this is the weak decay of the $\Lambda^0$ particle formed in the previous reaction:

$$\Lambda^0 \rightarrow \text{p} + \pi^-$$
$$S: \ -1 \neq 0 \ +0 \ \ ; \ \ \Delta S = +1 \qquad \text{Example of strangeness selection rule}$$

***Section 41-4. Quarks.*** Hadrons (baryons and mesons) are thought to be made up of fundamental particles called **quarks.** It is convenient to divide our discussion of quarks into two parts: "old" quarks (the picture that developed before the mid-1960s) and "new" quarks (the understanding that has emerged since then).

The old quark picture contains three quarks with **flavors** labeled *u, d,* and *s* (for *u*p, *d*own, and *s*trange) and their three antiquarks; each has a spin, charge (which is a fraction of the electron charge), baryon number, and strangeness number, as listed in Table 41-3. Hadrons are composed of quark combinations according to the following rules:

- Mesons consist of a quark–antiquark pair.
- Baryons consist of three quarks.
- Antibaryons consist of three antiquarks.

Table 41-3 **Properties of Quarks and Antiquarks**

| Flavor | Spin | Charge | Baryon number | Strangeness | Charm | Topness | Bottomness |
|---|---|---|---|---|---|---|---|
| **Quarks** | | | | | | | |
| *u* (up) | $\frac{1}{2}\hbar$ | $+\frac{2}{3}e$ | $+\frac{1}{3}$ | 0 | 0 | 0 | 0 |
| *d* (down) | $\frac{1}{2}\hbar$ | $-\frac{1}{3}e$ | $+\frac{1}{3}$ | 0 | 0 | 0 | 0 |
| *s* (strange) | $\frac{1}{2}\hbar$ | $-\frac{1}{3}e$ | $+\frac{1}{3}$ | $-1$ | 0 | 0 | 0 |
| *c* (charmed) | $\frac{1}{2}\hbar$ | $+\frac{2}{3}e$ | $+\frac{1}{3}$ | 0 | $+1$ | 0 | 0 |
| *t* (top) | $\frac{1}{2}\hbar$ | $+\frac{2}{3}e$ | $+\frac{1}{3}$ | 0 | 0 | $+1$ | 0 |
| *b* (bottom) | $\frac{1}{2}\hbar$ | $-\frac{1}{3}e$ | $+\frac{1}{3}$ | 0 | 0 | 0 | $+1$ |
| **Antiquarks** | | | | | | | |
| $\overline{u}$ | $\frac{1}{2}\hbar$ | $-\frac{2}{3}e$ | $-\frac{1}{3}$ | 0 | 0 | 0 | 0 |
| $\overline{d}$ | $\frac{1}{2}\hbar$ | $+\frac{1}{3}e$ | $-\frac{1}{3}$ | 0 | 0 | 0 | 0 |
| $\overline{s}$ | $\frac{1}{2}\hbar$ | $+\frac{1}{3}e$ | $-\frac{1}{3}$ | $+1$ | 0 | 0 | 0 |
| $\overline{c}$ | $\frac{1}{2}\hbar$ | $-\frac{2}{3}e$ | $-\frac{1}{3}$ | 0 | $-1$ | 0 | 0 |
| $\overline{t}$ | $\frac{1}{2}\hbar$ | $-\frac{2}{3}e$ | $-\frac{1}{3}$ | 0 | 0 | $-1$ | 0 |
| $\overline{b}$ | $\frac{1}{2}\hbar$ | $+\frac{1}{3}e$ | $-\frac{1}{3}$ | 0 | 0 | 0 | $-1$ |

These simple quark-combination rules accounted for the structure of all the hadrons that were known before the mid-1960s.  Table 41-4 shows examples of the quark structure for the hadrons $\pi^-$ , p, and $\Xi^-$ .

Table 41-4 **Quark Structure for Some Hadrons**

| Hadron | Quark structure |
|---|---|
| $\pi^+$ | $u\overline{d}$ |
| p | $uud$ |
| $\Xi^-$ | $dss$ |

Let's look at the $\Xi^-$ structure to show that this scheme really works. In Table 41-4, you can see that the quark structure *dss* results in

charge $Q = (-\frac{1}{3} - \frac{1}{3} - \frac{1}{3})e = -e$

baryon number $B = (+\frac{1}{3} + \frac{1}{3} + \frac{1}{3}) = 1$

strangeness number $S = (0 - 1 - 1) = -2$

Also, the spin-$\frac{1}{2}$ quantum numbers can be combined as vectors to give an overall spin of $\frac{1}{2}$. The values

$(\text{spin}, B, S) = (\frac{1}{2}, 1, -2)$

are exactly those of a $\Xi^-$ particle. You should verify for yourself that the quark combination scheme also works for the hadrons $\pi^-$ and p.

By the late 1960s physicists realized that more than three quarks (and their antiquarks) were needed to explain new developments in the burgeoning elementary particle picture. So they added three new quarks, labeled *c, t,* and *b* (for *charm, top,* and *bottom,* or, what some people like better, *charm, truth,* and *beauty*). Each carries a new quantum number. The charmed quark has a *charm* number +1, the top quark has a *topness* number +1, and the bottom quark has a *bottomness* number +1, as listed in Table 41-4. Many more hadrons were discovered whose structure involved the new quarks, such as the charmed mesons $D^0\,(\bar{u}c)$, $D^+\,(\bar{d}c)$, $D^-\,(d\bar{c})$, and the charmed baryon $\Lambda_c^+\,(udc)$.

It is probably no accident that the picture of hadrons composed of six fundamental quarks meshes symmetrically with the picture of six fundamental leptons: the electron, muon, tau, and the three associated neutrinos. Deep-seated theoretical reasons, based on arguments of symmetry, suggest that there should be exactly six quarks and six leptons, though, of course, new developments could change this picture.

***Quark Confinement.*** Do quarks really exist? That depends on what you mean by *exist*. The six-quark picture derives from elaborate theoretical arguments and explains the structure of the hundreds of hadrons that have been detected. But an isolated quark has never been observed, even though many experiments have looked for them. To account for this fact, a theoretical argument called quark confinement was developed. Quark confinement states that we will never observe isolated quarks because the force between quarks gets stronger as they separate, much like the force between balls at the ends of a rubber band. The enormous energy of this restoring force confines the quarks within the bounds of their hadron.

Thus, the presently espoused quark picture follows from a theoretical underpinning which states that the particle on which the whole structure of the physical universe is based—the quark—can never be observed.

***Section 41-5. Field Particles.*** The theoretical picture of the mechanism that produces the interaction between elementary particles—leptons or quarks—is quite different from the classical picture of forces. The elementary particle theory states that interactions between particles take place via the exchange of virtual **field particles** or **field quanta.** The word *virtual* is used because the field quanta are not observed directly but are inferred from theoretical considerations. One particle emits a field quantum that is absorbed by another particle. The second particle then sends the field quantum back to the first particle. The force between the two particles results from this continuous game of catch with the field quantum, which is said to **mediate** the interaction.

An analogy can be seen in a game of catch between two persons on frictionless ice. When the two persons throw a ball back and forth, each experiences a repulsive force, as in Figure 41-1*a*. When the two persons skillfully throw a boomerang back and forth, each experiences an attractive force, as in Figure 41-1*b*.

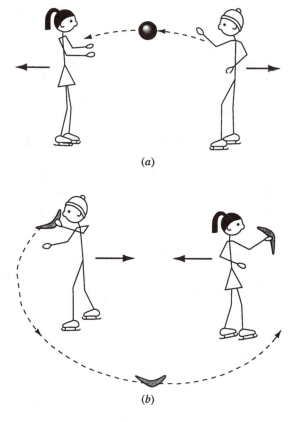

(a)

(b)

Figure 41-1

The four fundamental interactions are mediated by four different field quanta, as follows:

1.  Electromagnetic interaction—The field quantum is the photon, which has rest mass 0, charge 0, and spin 1.
2.  Gravitational interaction—The field quantum is the **graviton,** which has rest mass 0, charge 0, and spin 2. The graviton has not been observed experimentally.

3.  Weak interaction—There are three field quanta called **vector bosons:** $W^+$, $W^-$, and $Z^0$. The $W^+$ and its antiparticle $W^-$ have identical masses of 80.22 GeV/$c^2$, charges of $+e$ and $-e$ respectively, and spin 1. The $Z^0$ has a mass of 91.19 GeV/$c^2$, charge 0, and spin 1.
4.  Strong interaction—The field quantum is the **gluon,** which has rest mass 0, charge 0, and spin 1. Gluons are responsible for the strong force between quarks. A gluon has not been observed experimentally.

Note that all the field quanta have integer spins, so all are bosons.

The development of the theory of strong interactions has resulted in the introduction of another new quantum number called **color charge.** Quarks are said to come in three colors—red, blue, and green—which are the "charges" responsible for quark interactions. The field theory that describes these colors is called **quantum chromodynamics (QCD).**

***Section 41-6. The Electroweak Theory.*** The **electroweak theory** is an attempt to unite the electromagnetic and weak interactions into a single more fundamental interaction. At very high particle energies, of the order of 100 GeV, the single electroweak interaction is mediated by four field quanta: the $W^+$ and $W^-$ and new quanta called $W^0$ and $B^0$ that cannot be observed directly. The observed $Z^0$ and the photon are formed from combinations of the $W^0$ and $B^0$. The united electromagnetic and weak interaction have equal strength and a range of less than $10^{-19}$ m. As the particle energy decreases, the equal strength symmetry between the interactions is broken, and the single electroweak interaction becomes two separate interactions, electromagnetic and weak.

***Section 41-7. The Standard Model.*** The particle picture shows that the fundamental building blocks of all matter are leptons and quarks. Forces between these fundamental particles result from the exchange of field quanta in one of four basic interactions—strong, electromagnetic, weak, or gravitational—with the weak and electromagnetic interactions being combined at very high energies into the single electroweak interaction. Besides the familiar electric charge of electromagnetic interactions and the mass (charge) of gravitational interactions, there are weak flavor charges carried by leptons and color charges carried by quarks and gluons. All of this—and much more that we have not mentioned—is the **standard model.**

***Grand Unification Theories.*** The electroweak theory joins the electromagnetic and weak interactions into a single electroweak interaction. Pushing this idea still further, many people have been trying to unite the electroweak and the strong interactions under a single **grand unification theory (GUT).** Thus far, no one has been able to make a GUT work.

***Section 41-8. Evolution of the Universe.*** One of the properties of a galaxy that can be readily measured is its **redshift.** If the redshift is associated with the Doppler effect alone, the recessional velocity of the galaxy from earth can then be calculated. In addition, the distance of the galaxy from the earth can be determined from its brightness. These distance measurements are complicated. They rely on astronomical "yardsticks" which are distances to galaxies of known brightness. In many cases these distance yardsticks are not accurately known, resulting in uncertainties in the values of distances ascribed to the other galaxies.

Hubble found that the velocity $v$ of a receding galaxy (as determined from its redshift) is linearly related to the distance $r$ of the galaxy from earth (as determined from its brightness and galactic yardsticks) by **Hubble's law:**

$$v = Hr \qquad \text{Hubble's law}$$

where $H$ is the **Hubble constant,** which has units of reciprocal time. Hubble's law is illustrated in Figure 41-2 for a group of spiral galaxies.

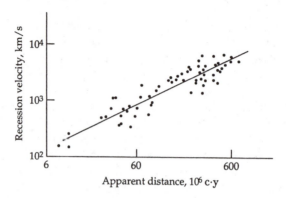

Figure 41-2

Because it is very difficult to measure astronomical distances, distances to only a small fraction of the galaxies in the observable universe have been determined. Moreover, the astronomical yardsticks are constantly being recalibrated as more data becomes accessible to astronomers. Currently, the value of the Hubble constant has been estimated to be

$$H = \frac{23 \text{ km/s}}{10^6 \text{ c·y}} \qquad \text{Present value of the Hubble constant}$$

We have talked about Hubble's law as relating velocities and distances of galaxies measured from earth, or equivalently from our Galaxy (the Milky Way). However, our Galaxy does not occupy a special place in the universe. Measurements made from any other galaxy should yield the same Hubble law with the same Hubble constant.

Hubble's law has an enormously profound consequence. If you are located on any galaxy in the universe, all other galaxies are receding from you with a speed proportional to the distance from your galaxy. The numerical value of recessional speed of any galaxy from your galaxy is given by the present value of the Hubble constant. In short, the entire universe is expanding.

***The 2.7-K Background Radiation.*** The presently observed abundance of helium in stars cannot be explained by nucleosynthesis in the stars. The helium abundance can be accounted for from fusion occurring in the extremely high temperatures in the early stages of the **Big Bang.**

If the universe were expanding as a blackbody after the Big Bang, the **blackbody radiation** still remaining should have a temperature of around 3 K. Blackbody radiation of this temperature was experimentally detected in 1965 by Penzias and Wilson at the Bell Telephone Laboratories in

New Jersey (for which they shared a Nobel Prize). Since then, more careful measurements have fixed the background blackbody radiation to have a temperature of $2.7 \pm 0.1$ K.

***The Big Bang.*** Out of many cosmological models that have been proposed to describe the structure and evolution of the universe, the Big Bang is the model that at present has obtained widespread acceptance. According to this model, our universe was born around 10–20 billion years ago in a giant explosion—the Big Bang—that occurred throughout all space, and has been expanding ever since.

The Big Bang picture is supported by several experimental observations. You have already seen one of these in Hubble's law, indicating an expanding universe.

There are more radio galaxies at far distances than close by. Since observations of far distances correspond to earlier times, this means that there were more radio galaxies at earlier times than there are now, showing that the universe is indeed evolving and is not static.

**Cosmologists,** scientists who study the universe, have developed a picture called the **standard model** that purports to describe the evolution of the universe in its early history shortly after the Big Bang. The standard model makes much use of the theory of elementary particles discussed at the beginning of this chapter. The closer the time is to the Big Bang at $t = 0$, the higher is the temperature of the universe, and the more energetic are the elementary particles.

The times discussed in the "very early" expanding universe are amazingly short, and the corresponding temperatures and energies are amazingly high. Before $10^{-43}$ s after the Big Bang, the four forces presently known today—gravitational, strong, electromagnetic, and weak—were unified in a single interaction described by a single (as yet unknown) theory. After $10^{-43}$ s, when the expanding universe had "cooled" to a temperature of about $10^{32}$ K, gravity broke free from this unification, and the remaining strong, electromagnetic, and weak forces remained unified in a single interaction described by grand unification theories (GUTs). At this stage, elementary particles had energies of the order of $10^{19}$ GeV. As the still-expanding universe cooled further to $10^{27}$ K at around $10^{-35}$ s, the strong force split away from the GUTs groups, leaving the electromagnetic and weak interactions still united as an electroweak force. Still later, around $10^{-10}$ s, the electromagnetic and weak forces parted company from their electroweak union, forming separate electromagnetic and weak forces, resulting in the four forces as we know them today.

As the universe kept expanding and cooling, the predominance of different elementary particles changed from one epoch to another. Quarks and leptons that were initially indistinguishable from each other in a "quark soup" changed into separate entities. "Ordinary" particles such as protons, neutrons, and electrons eventually emerged and combined into atoms and molecules. Eventually, as stars and galaxies formed, the density of matter grew larger than the density of radiation, resulting in the matter-dominated universe as we know it now, about 10–20 billion years after the Big Bang.

## II.  Physical Quantities and Key Equations

### Physical Quantities

Currently accepted value of the Hubble constant

$$H = \frac{23 \text{ km/s}}{10^6 \text{ c•y}}$$

Temperature of the present background blackbody radiation $\quad T = 2.7 \pm 0.1\, K$

### Key Equations

Conservation of mass–energy

$$\sum \left( m_0 c^2 + K \right)_{\text{before}} = \sum \left( m_0 c^2 + K \right)_{\text{after}}$$

Conservation of linear momentum

$$\sum \left( \vec{p} \right)_{\text{before}} = \sum \left( \vec{p} \right)_{\text{after}}$$

Conservation of spin

$$\sum \left( \vec{spin} \right)_{\text{before}} = \sum \left( \vec{spin} \right)_{\text{after}}$$

Conservation of charge

$$\sum Q_{\text{before}} = \sum Q_{\text{after}}$$

Conservation of baryon number

$$\sum B_{\text{before}} = \sum B_{\text{after}}$$

Conservation of lepton number

$$\sum L_{\text{before}} = \sum L_{\text{after}}$$

Conservation of strangeness (strong interactions)

$$\sum S_{\text{before}} = \sum S_{\text{after}}$$

Strangeness selection rule (weak interactions)  $\quad \Delta S = 0, \pm 1$

Hubble's law  $\quad v = Hr$

## III.  Potential Pitfalls

Do not confuse hadrons with leptons. Understand which particles are hadrons and which are leptons.

Do not confuse particles that decay via the strong interaction with those that decay via the weak interaction. The decay time for a strong interaction is of the order of $10^{-23}$ s, whereas the decay time for a weak interaction is of the order of $10^{-10}$ s.

Do not confuse fermions with bosons. Fermions are spin $\frac{1}{2}$, $\frac{3}{2}$, ... particles and obey the Pauli exclusion principle. Bosons are spin 0, 1, 2, ... particles and do not obey the Pauli exclusion principle.

Watch out for the change of strangeness number in reactions involving strange particles. In strong interactions, strangeness is conserved ($\Delta S = 0$). In weak interactions, strangeness may be conserved ($\Delta S = 0$) or may change by one unit ($\Delta S = \pm 1$).

Do not confuse field quanta with the fundamental lepton or quark particles. Field quanta are particles that are exchanged between leptons or quarks, thereby producing the force exerted between two fundamental particles.

Do not confuse the electroweak theory with grand unification theories (GUTs). The electroweak theory unifies the electromagnetic and weak interactions and is generally thought to be established. Grand unification theories are attempts to produce still further unification by combining electromagnetic, weak, and strong interactions into a single theory. Thus far no one has developed a successful GUT.

## IV. True or False Questions and Responses

**True or False**

_____ 1. Mesons and leptons are baryons.

_____ 2. A particle that decays via the strong interaction has a much longer lifetime than a particle that decays via the weak interaction.

_____ 3. Virtual photons violate conservation of energy.

_____ 4. Leptons interact via the weak interaction.

_____ 5. Particles and antiparticles have the same mass and charge.

_____ 6. Bosons are spin-$\frac{1}{2}$ particles.

_____ 7. A 1-MeV photon cannot create an electron–positron pair.

_____ 8. Lepton, baryon, and strangeness numbers are conserved in all reactions or decays.

_____ 9. Hadrons are composed of three quarks.

_____ 10. Quarks have fractional multiples of the electronic charge.

_____ 11. All quarks are fermions.

_____ 12. A quark never has more than one nonzero strangeness, charm, topness, or bottomness number.

_____ 13. Quarks come in six flavors, and each flavor comes in three colors (not counting antiflavors and anticolors).

_____ 14. A quark has never been observed experimentally.

_____ 15. Quark confinement explains why no one will ever observe a quark.

_____ 16. A force exists between two elementary particles because the particles "play catch" with virtual field quanta.

_____ 17. All field quanta are massless.

_____ 18. All field quanta are fermions.

_____ 19. Gluons are field quanta that "glue" quarks together, that is, gluons are responsible for the strong force between quarks.

_____ 20. A gluon has never been observed experimentally.

_____ 21. According to the electroweak theory, at very high energies symmetry considerations show that the electromagnetic and weak interactions are part of a single electroweak interaction.

_____ 22. A bumper sticker reading "Particle physicists have GUTs!" means that particle physicists have an enormous intestinal fortitude for trying to understand the complex world of elementary particles.

_____ 23. The velocity of a distant galaxy is determined from its redshift.

_____ 24. Hubble's law states that all galaxies are receding from us with velocities that are proportional to their distances from us.

_____ 25. The Big Bang refers to the enormous explosion of a supernova.

_____ 26. The observed 2.7-K background radiation is supportive evidence for the Big Bang picture.

**Responses to True or False**

1. False. Mesons and baryons are hadrons. Leptons are in a class by themselves.

2. False. It's the other way around.

3. True. But only for such a short time that the violation cannot be observed in accordance with the uncertainty principle.

4. True.

5. False. They have the same mass but opposite charge.

6. False. Bosons have integer spin.

7. True. To create an electron–positron pair, a photon has to have energy of at least $2m_ec^2 = 1.022$ MeV.

8. False. Strangeness can change by $\pm 1$ in weak interactions.

9. False. Baryons are composed of three quarks. Mesons are composed of two quarks.

10. True.

11. True.

12. True.

13. True.

14. True.

15. True.

16. True.

17. False. The $W^+$, $W^-$, and $Z^0$ field quanta that mediate the weak interaction have mass.

18. False. All field quanta are bosons.

19. True.

20. True.

21. True.

22. False. It might be true if the word were *guts*. But the word GUTs means that particle physicists are pursuing grand unification theories that unite the strong, electromagnetic, and weak interactions.

23. True.

24. True.

25. False. The Big Bang refers to the enormous explosion that resulted in the birth of the entire universe.

26. True.

## V.  Questions and Answers

**Questions**

1. Baryons and mesons are both hadrons. What is the main characteristic that distinguishes baryons from mesons?

2. What distinguishes a hadron from a lepton?

3. What are the similarities and differences between a particle and its antiparticle?

4. A positron is stable, that is, it does not decay. Why, then, does a positron have only a short existence?

5. When a positron and electron annihilate at rest, why must more than one photon be created?

6.  How many leptons are there? Name them.

7.  How many quarks are there? Name them.

8.  Hadrons are built from quarks. An isolated quark has never been observed. How does particle theory explain this?

9.  Describe how field quanta produce forces between particles.

10. What does the electroweak theory do?

11. What is the hope for GUTs?

12. How is the presently observed abundance of helium in stars related to the theory of the Big Bang?

13. How is the presently observed background blackbody radiation related to the theory of the Big Bang?


**Answers**

1.  Their spin. Baryons are fermions with spin $\frac{1}{2}$, $\frac{3}{2}$, . . . whereas mesons are bosons with spin 0, 1, . . . In addition, the masses of baryons are significantly larger than the masses of mesons.

2.  Hadrons interact with each other via the strong interaction. Leptons interact with each other via the weak interaction.

3.  A particle and its antiparticle have the same mass and spin, but opposite charge, baryon number, and lepton number.

4.  Once formed, a positron quickly meets an electron to annihilate with from the abundant supply of electrons in the matter in our universe.

5.  If only one photon were created, linear momentum could not be conserved.

6.  Six. Electron, muon, tau, and three distinct neutrinos, one associated with the electron, one with the muon, and one with the tau.

7.  Six. Up, down, strange, charmed, top, bottom.

8.  Particle theory states that quarks must remain within the bounds of a hadron. As the distance between quarks increases, they experience larger and larger attractive forces. To separate quarks completely from each other would require an infinite amount of energy.

9.  A force between two particles arises from a back-and-forth exchange of a field quantum between the particles.

10. It unifies the electromagnetic and weak interactions into a single electroweak interaction.

11. Theorists hope that grand unification theories will unify the strong, electromagnetic, and weak interactions into a single interaction.

12. Nucleosynthesis in stars cannot explain the observed abundance of helium, whereas the initially high temperatures of the Big Bang provide the reaction rates necessary to account for the present abundance.

13. Theoretical calculations show that the temperature of the blackbody radiation in the universe that has been expanding for about 10 billion years after the Big Bang will have cooled to the presently observed value of 2.7 K.

## VI. Problems

**Example #1.** Determine the unknown particle $X$ in the strong reaction: $p + \pi^- \rightarrow K^0 + X$.

**Picture the Problem.** Apply the conservation laws to determine $Q$, $s$, $B$, $L$, and $S$ for the unknown particle.

| | |
|---|---|
| 1. Apply conservation of charge. | $+1 - 1 = 0 + Q$     $Q = 0$ |
| 2. Apply conservation of spin. | $\frac{1}{2} + 0 = 0 + s$     $s = \frac{1}{2}$ |
| 3. Apply conservation of baryon number. | $1 + 0 = 0 + B$     $B = 1$ |
| 4. Apply conservation of lepton number. | $0 + 0 = 0 + L$     $L = 0$ |
| 5. Apply conservation of strangeness | $0 + 0 = 1 + S$     $S = -1$ |
| 6. Compare these quantum numbers with the available particles. | $X = \Lambda^0$   or   $X = \Sigma^0$ |

**Example #2—Interactive.** Determine the unknown particle $X$ in the strong reaction: $p + p \rightarrow n + \pi^+ + \Lambda^0 + X$.

**Picture the Problem.** Apply the conservation laws, and compare the resulting quantum numbers with the known particles. **Try it yourself.** Work the problem on your own, in the spaces provided, and check your answer.

| | |
|---|---|
| 1. Apply conservation of charge. | |
| 2. Apply conservation of spin. | |
| 3. Apply conservation of baryon number. | |
| 4. Apply conservation of lepton number. | |
| 5. Apply conservation of strangeness | |

| 6. Compare these quantum numbers with the available particles. | |
|---|---|
| | $X = K^+$ meson |

**Example #3.**  Which of the following two possibilities for the weak decay of a $\Sigma^-$ particle are possible?

$(a)\ \ \Sigma^- \rightarrow \pi^- + p$

$(b)\ \ \Sigma^- \rightarrow \pi^- + n$

**Picture the Problem.**  Apply the conservation laws and determine which of the equations satisfies all conservation laws for weak decay.

| 1. Apply conservation of charge for reaction (a). Because this expression is not true, reaction (a) is not possible. | $-1 = -1 + 1$ |
|---|---|
| 2. Apply conservation of charge for reaction (b). | $-1 = -1 + 0$ |
| 3. Apply conservation of spin for (b). | $\frac{1}{2} = 0 + \frac{1}{2}$ |
| 4. Apply conservation of baryon number for (b). | $+1 = 0 + 1$ |
| 5. Apply conservation of lepton number for (b). | $0 = 0 + 0$ |
| 6. Apply conservation of strangeness for (b). $\Delta S$ can equal $\pm 1$ for weak reactions. All quantities are conserved appropriately, so reaction (b) is allowed. | $-1 = 0 + 0$ |

**Example #4—Interactive.**  Here are two possibilities for the weak decay of a neutron. Which one is possible?

$(a)\ n \rightarrow p + e^- + v_e$

$(b)\ n \rightarrow p + e^- + \bar{v}_e$

**Picture the Problem.**  Apply the conservation laws to see which reaction is allowed. **Try it yourself.** Work the problem on your own, in the spaces provided, and check your answer.

| | |
|---|---|
| 1. Apply conservation of charge to both decay expressions. | |
| 2. Apply conservation of spin to both decay expressions. | |
| 3. Apply conservation of baryon number to both decay expressions. | |
| 4. Apply conservation of lepton number to both decay expressions. | |
| 5. Apply conservation of strangeness to both decay expressions. | |
| 6. Determine which expression is allowed | |
| | (b) |

**Example #5.** Which conservation law is violated in the strong reaction: $p + \pi^- \rightarrow \Sigma^0 + \eta^0$ ?

**Picture the Problem.** Apply the laws and see which one doesn't work.

| | |
|---|---|
| 1. Apply conservation of charge. | $1 - 1 = 0 + 0$ |
| 2. Apply conservation of spin. | $\frac{1}{2} + 0 = \frac{1}{2} + 0$ |
| 3. Apply conservation of baryon number. | $1 + 0 = 1 + 0$ |
| 4. Apply conservation of lepton number. | $0 + 0 = 0 + 0$ |
| 5. Apply conservation of strangeness. Strangeness is not conserved. | $0 + 0 \neq -1 + 0$ |

**Example #6—Interactive.** Which conservation law is violated in the strong reaction: $p + p \rightarrow \Sigma^- + K^+$ ?

**Picture the Problem.** Apply the laws and see which one doesn't work. **Try it yourself.** Work the problem on your own, in the spaces provided, and check your answer.

| | |
|---|---|
| 1. Apply conservation of charge. | |
| 2. Apply conservation of spin. | |
| 3. Apply conservation of baryon number. | |

| | |
|---|---|
| 4. Apply conservation of lepton number. | |
| 5. Apply conservation of strangeness. | |
| 6. Spin and baryon number are not conserved. | |

**Example #7.** The solid lines in Figure 41-3 show the tracks of two charged particles in a photograph of a reaction that occurred in a bubble chamber, where a magnetic field was directed into the paper. Determine the identity of the unknown neutral particle $X$, shown by the dashed line, that was one of the reaction products but was not observed in the bubble chamber.

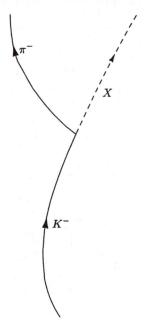

Figure 41-3

**Picture the Problem.** Write an expression for the decay shown, and apply the conservation laws to determine the quantum numbers of the unknown particle.

| | |
|---|---|
| 1. Write an expression for the decay shown. | $K^- \rightarrow \pi^- + X$ |
| 2. Apply conservation of charge. No charge agrees with the fact the particle leaves no trail. | $-1 = -1 + Q_X \qquad Q_X = 0$ |
| 3. Apply conservation of spin. | $0 = 0 + s \qquad s = 0$ |
| 4. Apply conservation of baryon number. | $0 = 0 + B \qquad B = 0$ |
| 5. Apply conservation of lepton number. | $0 = 0 + L \qquad L = 0$ |

| 6. Apply conservation of strangeness. | $-1 = 0 + S$    $S = -2, -1, 0$ |
|---|---|
| 7. Match the quantum numbers to the known particles. | $\pi^0, \bar{K}^0, \eta^0$ |
| 8. Use mass-energy conservation to narrow down the identity of the particle. Find the mass of the given particles, and the mass of the possible particles in step 6.<br><br>The only particle that does not violate mass-energy conservation is the $\pi^0$ particle. | $M_{K^-} = 494 \, \text{MeV/c}^2$<br>$M_{\pi^-} = 140 \, \text{MeV/c}^2$<br>$M_X \le M_{K^-} - M_{\pi^-} = 354 \, \text{MeV/c}^2$<br>$M_{\bar{K}^0} = 498 \, \text{MeV/c}^2$<br>$M_{\eta^0} = 549 \, \text{MeV/c}^2$<br>$M_{\pi^0} = 135 \, \text{MeV/c}^2$ |

**Example #8—Interactive.** The solid lines in Figure 41-4 show the tracks of two charged particles in a photograph of a reaction that occurred in a bubble chamber, where a magnetic field was directed into the paper. Determine the identity of the unknown neutral particle $X$, shown by the dashed line, that precipitated the reaction but was not observed in the bubble chamber.

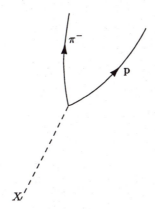

Figure 41-4

**Picture the Problem.** Write an expression for the decay shown, and apply the conservation laws to determine the quantum numbers of the unknown particle. **Try it yourself.** Work the problem on your own, in the spaces provided, and check your answer.

| 1. Write an expression for the decay shown. | |
|---|---|
| 2. Apply conservation of charge | |
| 3. Apply conservation of spin. | |
| 4. Apply conservation of baryon number. | |

| | |
|---|---|
| 5. Apply conservation of lepton number. | |
| 6. Apply conservation of strangeness. | |
| 7. Match the quantum numbers to the known particles. | |
| 8. If more than one particle is possible, use mass-energy conservation to narrow down the identity of the particle. Find the mass of the given particles, and the mass of the possible particles in step 6. | $\Lambda^0$, undergoing a weak decay |

**Example #9.** What is the quark structure of a $\pi^-$ meson?

**Picture the Problem.** Recall that a meson is composed of a quark-antiquark pair. Look up the quantum numbers of the meson, and determine the quark-antiquark pair than can produce that combination.

| | |
|---|---|
| 1. Look up the quantum numbers for the $\pi^-$ meson. | $Q = -e$ <br> $S = 0$ <br> $s = 0$ |
| 2. Use Table 41-3 to find a quark-antiquark combination that results in these quantum numbers. | $\bar{u}d$ |

**Example #10—Interactive.** What is the quark structure of a $K^0$ meson?

**Picture the Problem.** Follow the steps in Example #9. **Try it yourself.** Work the problem on your own, in the spaces provided, and check your answer.

| | |
|---|---|
| 1. Look up the quantum numbers for the $K^0$ meson. | |
| 2. Use Table 41-3 to find a quark-antiquark combination that results in these quantum numbers. | $d\bar{s}$ |

**Example #11.** Determine the particle that is composed of the quark structure *uds*.

**Picture the Problem.** Look up the resulting quantum numbers from these quarks, and compare that to the quantum numbers of known particles. Remember that a combination of three quarks means the resulting particle is a baryon or antibaryon.

| 1. Determine the quantum numbers that result from combining the three particles. | $Q = \left(\frac{2}{3} - \frac{1}{3} - \frac{1}{3}\right)e = 0$ <br> $S = 0 + 0 - 1 = -1$ <br> $s = \frac{1}{2}$ or $\frac{3}{2}$ |
|---|---|
| 2. Compare the resulting quantum numbers from step 1 with the known baryon and antibaryon particles. | $\Lambda^0$ or $\Sigma^0$ |

**Example #12—Interactive.** Determine the particle that is composed of the quark structure *uus*.

**Picture the Problem.** Follow the steps used in Example #11. **Try it yourself.** Work the problem on your own, in the spaces provided, and check your answer.

| 1. Determine the quantum numbers that result from combining the three particles. | |
|---|---|
| 2. Compare the resulting quantum numbers from step 1 with the known baryon and antibaryon particles. | $\Sigma^+$ |